STAMP 2 COMMUNICATIONS AND CONTROL PROJECTS

Tom Petruzzellis

McGraw-Hill

New York Chicago San Francisco Lisbon London Madrid
Mexico City Milan New Delhi San Juan Seoul
Singapore Sydney Toronto

The McGraw-Hill Companies

Library of Congress Cataloging-in-Publication Data

Petruzzellis, Thomas.
Stamp 2 communications and control projects / Tom Petruzzellis.
p. cm.
Includes bibliographical references.
ISBN 0-07-141197-6 (set)—ISBN 0-07-141198-4 (book)—ISBN 0-07-141199-2 (CD-ROM)
1. Programmable controllers. 2. BASIC Stamp computers. I. Title.

TJ223.P76P48 2003
629.8'95—dc21

2002045207

1 2 3 4 5 6 7 8 9 0 DOC/DOC 0 9 8 7 6 5 4 3

P/N 141198-4
Part of
ISBN 0-07-141197-6

The sponsoring editor of this book was Judy Bass. The editing supervisor was Caroline Levine, and the production supervisor was Pamela Pelton. It was set in Times New Roman per the TAB4 Design by Paul Scozzari of McGraw-Hill Professional's Hightstown, N. J. composition unit.

Printed and bound by RR Donnelley.

This book is printed on recycled, acid-free paper containing a minimum of 50% recycled, de-inked fiber.

STAMP 2 COMMUNICATIONS AND CONTROL PROJECTS

This book is dedicated to my parents,
Joseph and Antoinette,
whose enduring love and everlasting guidance
provide constant direction through the storms of life.

CONTENTS

ACKNOWLEDGMENTS

In the process of creating any book, there are usually a number people to thank. I would like to thank the following people for contributions to the projects detailed in this book: Mike Shellim, Professor Richard Clemens, J. Gary Sparks, Dr. Tracy Allen, Gerald Crenshaw, Tim Bitson, Mark Hammond, Gene Bridgman, and Roger Cameron. I would also like to thank Parallax, manufacturers of the BASIC STAMP series microprocessors as well as other companies who allowed me to use their data sheets. And, of course, all the folks at McGraw-Hill, especially Judy Bass.

INTRODUCTION

The microprocessor has entered the consciousness of almost everyone in our society and has revolutionized the design of electronic products and appliances in a way that no other device has ever done. Electronic engineers and technicians and hobbyists have seen electronic projects transition from tubes to transistors to integrated circuits and on to microprocessors. In the mid-1990s, the BASIC STAMP microprocessor came into being. The BASIC STAMP 1 microprocessor is a powerful tool for creating many different types of complex electronic circuits. Following the BASIC STAMP 1, a more powerful BASIC STAMP 2 (BS2) emerged, permitting even more complex projects to be built in a fraction of the time, and with far fewer components and less cost than previous circuitry using discrete transistors and integrated circuits. In short, the STAMP 2 has revolutionized electronic project building. Project building has seen a recent resurgence in the last few years, and the STAMP 2 controller is a very powerful and often-used building block. STAMP 2 has been discovered by engineers, technicians, and hobbyists alike and has made project building a dream. The beauty of the STAMP 2 is its reduced instruction set and its ability to perform tasks quickly and efficiently with a decent amount of memory. The STAMP 2 can run off a standard 9-V transistor radio battery and consumes little power. Utilizing the STAMP 2 microprocessor allows project designs to be powerful, flexible, and have a much lower parts count than a discrete IC design. The STAMP 2 allows project designs to be upgradable and to evolve as more features and complexity are needed—try that with discrete logic.

The STAMP 2 has had a long and varied evolution and has developed a cult status as it has become popular and widely utilized. Since their inception, the STAMP 1 and 2 have allowed many electronics-minded persons to construct all types of electronic projects and systems from simple light-emitting diode (LED) followers to more complex Internet interfaces. The STAMP 2 has had a large following over its life to date and consequently has created an extensive library of expertise and software, so that many newcomers can benefit from all the combined experience. There are very helpful newsgroups, web ring, and e-mail groups dedicated to helping STAMP 2 newcomers. Many STAMPERs are readily available and willing to help newbies. The STAMP 2, since it has had a long evolution, has many support chips, interfaces, and sensors available for those wishing to build modern complex circuits.

STAMP 2 Communications and Control Projects features 25 projects, ranging from basic serial communication to a 12-channel Internet-based alarm reporting system. This book presents many novel, interesting, and useful projects for technically minded people who are interested in discovering the joys of building hands-on projects using the powerful BASIC STAMP 2 microcontroller. There are many projects designed for electronics hobbyists, radio hams, and Internet junkies.

Hardware as well as software for each project is presented in an easy-to-understand format, which will allow anyone with a basic interest in electronics and microprocessors to assemble, as well as program, any of the projects described in this book. The CD-ROM accompanying the book also includes three appendixes that are an invaluable aid to the reader and hobbyist alike. Appendix 1 contains many full data sheets of the integrated circuits used for projects covered in the book. Appendix 2 includes many useful tables and charts for help in understanding and programming the STAMP 2. Appendix 3 presents useful contacts and names, STAMP 2 news/mail groups, and manufacturers. The CD-ROM for *STAMP 2 Communications and Control Projects* also includes all the programs for the projects in this book. You can easily load any of the programs in just a few minutes. The CD-ROM also includes a STAMP 2 manual, application notes, and many additional free programs for other interesting projects.

One of the reasons the STAMP 2 has been so successful is that it can be interfaced with many different types of sensors and external devices quite readily. The BASIC STAMP 2 can often be interfaced with the outside world with one- or two-wire interfaces. Many applications such as analog-to-digital conversion or temperature sensing, as well as external memory devices can be interfaced with two wires. One-wire devices or I2C devices are very easy to interface to the STAMP 2. The Dallas Semiconductor 1-Wire devices are available as sensors, external memory, and they even have a one-wire weather station that can be used with the STAMP 2.

Many of the projects in this book are complete and self-contained in their own right, but also could be used together with other projects to form more complex projects and systems. Some projects could be used as idea generators for more complex systems. We hope this book will act to spur your imagination for building and modifying the circuit and programs contained within, to gain a better appreciation of the power and flexibility of the STAMP 2 microprocessor. Hobbyists, engineers, technicians, and amateur radio enthusiasts alike will appreciate this fascinating new book.

Chapter 1 presents an alternative STAMP 2—a low-cost substitute for BASIC STAMP 2. This chapter will show you how to construct your own less expensive alternative for about half the price of the original! You can substitute the alternative STAMP 2 for any BS2 project in this book.

Chapter 2 covers the basics of interfacing input and output devices such as switches, sensors, LEDs, and motors. This chapter also includes an introduction to serial interfacing of the STAMP 2 microprocessor to a personal computer and other devices. A QBASIC program for uploading and downloading between the STAMP 2 and a PC is included, and a full terminal program is presented.

In Chapter 3 we explore new and interesting speciality ICs, or serial building blocks. While not a specific STAMP 2 project, this chapter presents special ICs that can be used with the BASIC STAMP 2 processor to create new and interesting serial infrared (IR) and radio remote control projects. This chapter is a project idea generator designed to stimulate your imagination in finding ways of using some new dedicated microprocessors that, combined with STAMP 2, create powerful new projects.

Chapter 4 features an RJ-45 Ethernet network cable tester. This project is a real time saver and prevents headaches. Wiring 10Base-T Ethernet networks can be a long and frustrating task, fraught with problems. The RJ-45 cable tester uses STAMP 2 to test all eight wires in an Ethernet cable using a "walking LED" pattern. The tester is a two-piece unit. One end is placed at the originating end of the cable and the second part of the tester is placed at the end user PC and will test opens and shorts.

Chapter 5 presents an infrared communicator utilizing the STAMP 2. This IrDA transceiver allows the STAMP 2 to communicate with other IrDA-enabled computers, laptops, and personal digital assistants (PDAs), for data collection and transfer.

Chapter 6 illustrates a multichannel wireless alarm system. The wireless alarm system is a two-part system that combines a 12-channel alarm system with a transmitter and a complementary receiver and liquid crystal display (LCD) unit. With this alarm system you can monitor sensors, doors, windows, floor mats, etc. The resultant alarm warning is sent via radio to a remote receiver and display. The LCD unit beeps and indicates the particular alarm sensor that was activated.

Chapter 7 describes the construction of a unique and inexpensive way to measure lightning activity up to 50 or more miles away. This monitor provides real-time data on lightning activity by counting the number of lightning strikes in the area. This lightning detector project is a two-piece system: a remote outdoor sensor and an indoor microprocessor and display unit. This system was designed to provide advanced warning for weather hobbyists or amateur radio operators who want advance notice of an approaching electrical storm. The lightning detector/display is based on the BASIC STAMP 2, which is used to display the number of lightning strikes per minute via a row of eight colored LEDs.

Chapter 8 illustrates a 12-channel remote control system using an IBM PS/2–type keyboard, interfaced to an Abacom model AM-RT4/RT5 series wireless transmitter. An Abacom receiver coupled to a BASIC STAMP 2 controller provides a means to control 12 different devices via radio. This system can be expanded to control many more devices if desired.

Chapter 9 interfaces the STAMP 2 to the Motorola MC145447 caller-ID demodulator, to provide a very useful and low-cost caller-ID project. The STAMP 2 program reads in the caller-ID data, stores it in electrically erasable and programmable read-only memory (EEPROM), and outputs the information on the serial LCD backpack. The program also allows selection of up to 10 telephone numbers, which are called "blocked numbers." If a blocked number calls, the program can choose to inhibit that call from reaching the telephone, answering machine, or other device.

Chapter 10 presents both a touch-tone encoder and a touch-tone decoder based on the new CM-8880 touch-tone encoder/decoder chip. The touch-tone decoder project features an LCD display that will display the decoded digits, which can be utilized in many applications including ham radio, robotics, and remote control projects. The touch-tone decoder with display can be used to monitor touch-tone sequences over radio or the telephone line for monitoring purposes.

Chapter 11 depicts a touch-tone controller, which can be utilized in many applications including ham radio, robotics, or remote control projects. The touch-tone controller project can be used for selective calling (sel-cal) in radio applications. The project allows a radio receiver or transceiver to remain silent, until the correct tone sequence is sent. Once the tone sequence is sent, the radio speaker comes alive. This project can also be used as a remote control device to control appliances via a radio receiver and the touch-tone controller project.

Chapter 12 presents the radio mailbox, which can be used by Ham Radio or CB enthusiasts to retrieve voice messages left for them by their radio buddies. The radio mailbox connects to any radio receiver or transceiver and it "looks" for the proper touch-tone sequence and records voice messages while you are away from your radio transceiver.

The Morse code keyer project in Chapter 13 provides a four-message keyer for amateur radio. The user can input four different messages, their character counts, and display the output and/or key a transmitter by use of four buttons. Keyer messages are limited to the upper- and lowercase letters, digits 0 to 9, space, period, comma, question mark, and slash. The Morse keyer can be expanded into a voice keyer if desired, by using the ISD 2500 voice chip.

Chapter 14 illustrates an automated "fox" for amateur radio direction finding. Fox hunting is a popular radio hobby activity where a hidden transmitter is located by using direction finding techniques. Often prizes are awarded for finding the fox quickly. This project can also be used as a Morse code ID and keyer for a propagation beacon. VHF/UHF radio propagation beacon stations are used by radio amateur operators to predict the best conditions in which to communicate. The beacon project sends Morse ID, then a tone for 1 minute, then pauses 1 minute in a loop.

In Chapter 15 we create the OrbTracker, a smart computer controller for an antenna rotator. The Orbtracker utilizes the BASIC STAMP 2 to control azimuth/elevation (AZ-EL) rotators for amateur radio satellite communication. The STAMP 2 controller can control the inexpensive Orbit 360 or the Yaesu 5400/5500 series antenna rotators. The STAMP 2 is interfaced with your PC running NOVA or WISP tracking programs.

Ham radio repeaters increase effective communication range between two stations. The amateur radio repeater controller in Chapter 16 will allow you to build a very capable system with many features, including a touch-tone decoder for remote control functions, timeout timer, and ID timers, anti-kerchunk, CTCSS encode/decode and muting, voice and CW-ID, using digital voice recorder, and a burglar alarm/tamper alarm. This is a fun project with lots of possibilities.

The remote balloon data telemetry system in Chapter 17 is a unique telemetry project that allows a balloon enthusiast to collect high-altitude atmospheric pressure and temperature and immediately send the information in real-time to a receiver on the ground. The project was designed for balloon and amateur radio enthusiasts who like to combine both hobbies.

The aerophotography project in Chapter 18 interfaces a BASIC STAMP 2 microcontroller, a radio link, and an electronic camera, which will allow you to take fantastic pictures from a model plane, a glider, or a kite. High-level commands are used for reading/writing square waves, using serial communication. The STAMP 2 interface monitors pulses on one receiver channel. When a state change occurs (corresponding to a switch being moved on the transmitter), the interface generates a command string which it outputs to the camera via the built-in RS 232 port. Alternatively the interface can trigger the camera at regular intervals without an RC link.

The Cell-Alert system in Chapter 19 monitors four different channels and will call a cell phone and alert a person at the other end of the call for each of the four channels that may have been activated with a series of different musical notes. Each channel plays a different sequence of musical notes. Simply plug Cell-Alert into a phone line and a power supply and it's ready to serve.

The Page-Alert system featured in Chapter 20 monitors up to eight different input channels and will call a pager and inform the party carrying the pager that one of eight devices has been activated, by indicating the particular channel with a one or zero in the string of eight places.

The Data-Alert system featured in Chapter 21 monitors up to four alarm sensors or channels and will call a phone number of your choice and leave a data message with the data terminal or a remote personal computer, connected via a modem. The Data-Alert utilizes a mini modem chip and is connected to your telephone line.

Chapter 22 presents a number of sensor modules that can be used to enhance the Cell-Alert, Page-Alert, and Data-Alert projects. A normally open/closed sensor input circuit, an infrared body heat detector, a comparator/threshold detector, and a listen-in module are all covered in this chapter. The listen-in module was designed for the Cell-Alert project only.

The Data-Term project in Chapter 23 utilizes a STAMP 2 controller and a modem. The Data-Term is connected to a telephone line and waits for calls from the Data-Alert unit. The Data-Term can accept data, information, and commands from the Data-Alert, and can control remote X-10 and local relay controls.

The Weblink project illustrated in Chapter 24 was designed to monitor up to 12 different channels or sensors, and will notify you via a remote web page browser. A STAMP 2 controller is coupled to a Siteplayer mini web server, connected via a broadband Internet connection, and will remotely notify the user if any of the channels has been activated. With the Weblink system you can monitor windows, doors, temperatures, voltage changes, etc.

Chapter 25 presents the Xlink project. The Weblink is a web-enabled remote control system that marries a STAMP 2 microcontroller with a Siteplayer mini web server module. With the Xlink and a laptop, you are can remotely control a matrix of 25 different X-10 control devices plus an additional six local relay-controlled devices via the Internet. The Xlink is placed at your home or office, and you can control devices by using any remote browser. This is the ultimate control project.

Appendix 1 includes many data sheets for integrated circuits used in projects in this book. A number of programming charts and tables are provided in Appendix 2. Appendix 3 features weblinks, addresses, BASIC STAMP 2 user groups, etc. The CD-ROM provided with this book includes the appendixes and all the programs for the projects. The BASIC STAMP 2 manual, applications notes, and additional programs for other interesting projects are all included.

1

THE BASIC STAMP 2

CONTENTS AT A GLANCE

The BASIC STAMP 1 (BS1) was the first of the STAMP series, using a miniversion of BASIC or BASIC tokens in order to program the microcontroller. The BASIC STAMP 1C features a 256-byte EEPROM, or about 75 instructions, and runs at 4 Mhz, performing 2000 instructions per second. A revision of the STAMP 1 later became the STAMP 1 rev. D. The latest version of the BASIC STAMP 1 is the STAMP 1 module shown in Fig. 1-1. Figure 1-2 depicts the BASIC STAMP 1 schematic. The STAMP 1 is programmed via the parallel port of a personal computer. All of the projects in this book will use the BASIC STAMP 2 (BS2) microprocessor, which sports a 2048-byte EEPROM for about 500 instructions, and runs at 20 MHz, performing 4000 instructions per second. Table 1-1 shows some comparisons between the STAMP 1 and STAMP 2 input/output instructions. The STAMP 2 runs much faster, has a number of extra commands, such as FREQOUT, SHIFTIN, DTMFOUT, and XOUT, and will hold larger programs than the original STAMP 1. The BASIC STAMP 2 has become the de facto microprocessor for sensing and control applications, a model that a number of other companies have decided to follow.

The STAMP 2 is a very versatile and popular microprocessor building block that can be used to develop many different electronic sensing and control systems (see Fig. 1-3).

Figure 1-1 BASIC STAMP I module.

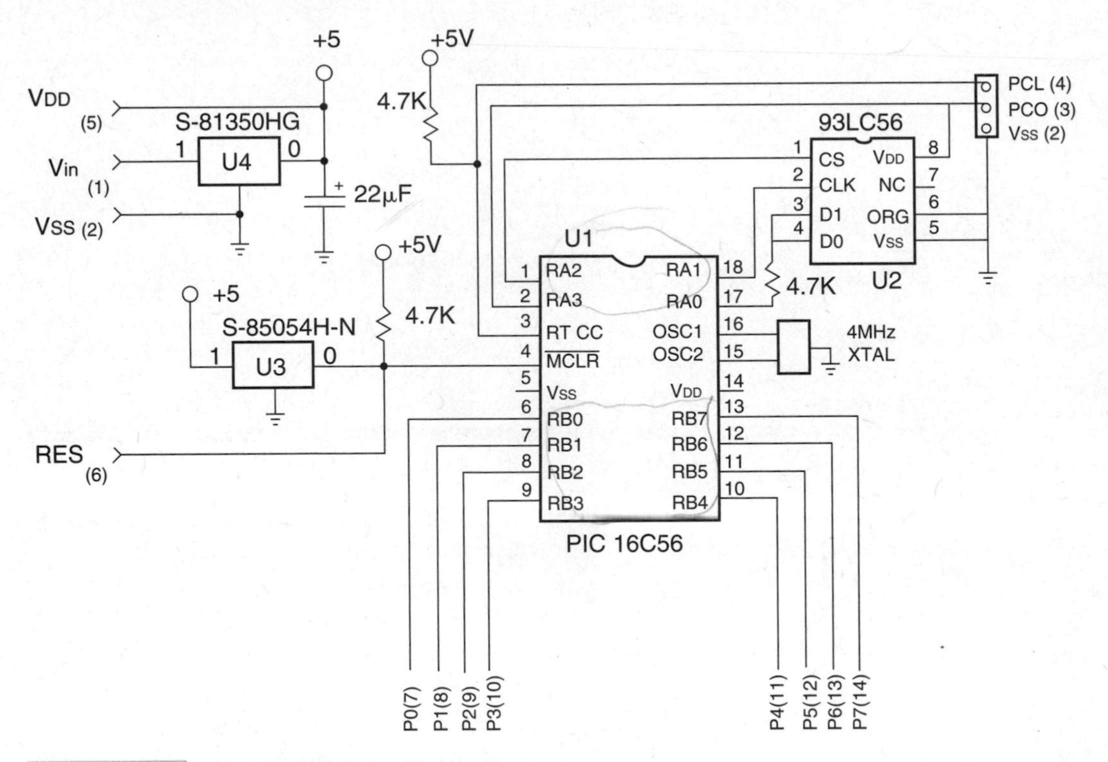

Figure 1-2 BASIC STAMP 1 schematic.

The BASIC STAMP 2 is available in two different types: dual inline package (DIP) or a small outline integrated circuit (SOIC) package. The BASIC STAMP 2 is a PIC16C57 chip with a PBASIC2 interpreter running at 20 MHz. The STAMP 2 has 16 pins for input/output applications and can talk directly to a serial port from a personal computer, for programming. Parallax, which sells the STAMP 2 products, offers the STAMP 2 in a 24-pin module, which is a self-contained STAMP 2 with onboard regulator and brownout protection. This STAMP 2 module is offered for $49. There are now many versions of the

TABLE 1-1 BASIC STAMP I/O INSTRUCTIONS

BASIC STAMP 1 (BS1)	BASIC STAMP 2 (BS2)
BUTTON	BUTTON
HIGH	COUNT
LOW	DTMFOUT
OUTPUT	FREQOUT
POT	HIGH
PULSIN	LOW
PULSOUT	OUTPUT
PWM	PULSIN
REVERSE	PULSOUT
SERIN	PWM
SEROUT	RCTIME
SOUND	REVERSE
TOGGLE	SERIN
	SEROUT
	SHIFTIN
	TOGGLE
	XOUT

STAMP 2: STAMP 2P, 2E, 2I, 2SX, and a STAMP 2 original equipment manufacturer (OEM) version, shown in Fig. 1-4—a true testament to a truly successful product.

The BASIC STAMP 2 (BS2-IC) is the single most popular BASIC STAMP module and is widely used in educational, hobby, and industrial applications. This module normally has 2 kbytes of program space and 16 I/O pins. A serial PC interface provides enhanced debugging features.

Advantages of STAMP 2

The BASIC STAMP 2P (BS2P24, BS2P40) has several advantages over all previous BASIC STAMPs, including commands for interfacing with parallel LCDs, I2C devices, Dallas Semiconductor 1-Wire parts, and a polled interrupt capability. The BS2P is 3 times faster than the BS2-IC and 20 percent faster than the BS2SX-IC and is available in a pin-compatible format to other BS2 variants, or as a 40-pin module (with 16 extra I/O pins!).

The BASIC STAMP 2SX (BS2SX-IC), using the Ubicom SX microcontroller, is 2.5 times faster than the BS2-IC. It has the same program memory as a BS2P but does not have

Figure 1-3 Electronic sensing and control systems that can be built with BASIC STAMP 2.

Figure 1-4 STAMP 2 OEM version.

commands that interface with the parts listed above. It is a great solution for those needing more power than a BS2-IC offers.

The BASIC STAMP 2E (BS2E-IC), using the Ubicom SX microcontroller, runs at the same speed as BASIC STAMP 2, but has the RAM and EEPROM size benefits of the BS2SX, without the speed and power consumption. It is ideal for those who use the BS2 and would like more variable or program space.

Most of the projects in this book will center around the BS2-IC, which was the first and most popular version of the STAMP 2. A schematic diagram of the original BASIC STAMP 2 (BS2-IC) is shown in Fig. 1-5. The original BASIC STAMP 2 contains the SOIC version of the PIC16C57 chip at U1, the external memory chip at U2, a reset chip at U3, and a voltage regulator at U4. Finally, Q1 through Q3 are used as serial port input conditioners. Figure 1-6 illustrates the serial port programming pinouts on the original BS2. The first four pins on the left of the BS2 carrier are the serial port connections. Pin 1 on the BS2 is the S_{out} or RX pin, which connects to pin 2 of an RS-232 DB-9 serial female connector. Pin 2 on the BS2 is the S_{in} or TX pin, connecting to pin 3 of the DB-9. Pin 3 on the BS2 is the ATN pin which connects to pin 4 of the DB-9, and finally pin 4 of the BS2 is the ground connection on pin 5 of the DB-9 connector. Note, that pins 6 and 7, or DSR and RTS pins, are tied together on the serial port connector.

Table 1-2 illustrates a memory map of the BS2-IC. The memory map shows where program information is stored; it also clarifies how to access inputs and outputs of the STAMP 2. Table 1-3 depicts the STAMP 2 pinouts and this chart describes each of the pins and their respective functions.

Approaches to Construction

When constructing circuits using the BASIC STAMP 2, there are a number of different approaches you can take. One often-used approach to building and testing new microprocessor and electronic circuits is using the protoboard. The protoboard allows the beginner as well as the seasoned engineer or technician to test new circuits before he or she commits them to a special circuit board. The Parallax NX1000 protoboard is shown in Fig. 1-7. This advanced protoboard is a dream to use. It contains extra components to make new circuit testing go smoothly. The NX1000 protoboard sports serial ports, voltage regulator, LEDs, switches, extra input and output connections, and the like. Using this type of protoboard you can easily change wires and add new components quite simply on the fly. Another approach is to use a simple, no-frills protoboard. The simple protoboard consists of the inner white portion of the NX1000. Using this approach, you will need to supply all the extras yourself. This approach is somewhat cheaper but takes up more real estate on your bench top.

Many hobbyists choose the STAMP 2 and a simple carrier board, depicted in Fig. 1-8, which allows building a circuit on the carrier board with the BS2 onboard. The simple BS2 carrier board incorporates a STAMP 2 socket, a 9-V battery terminal, and a serial connector, as well as single row of header jacks, to facilitate connecting between the STAMP 2 and other points on the board. Using this approach, you would solder other components to the board and run jumper wires between components. Many hobbyists add additional IC

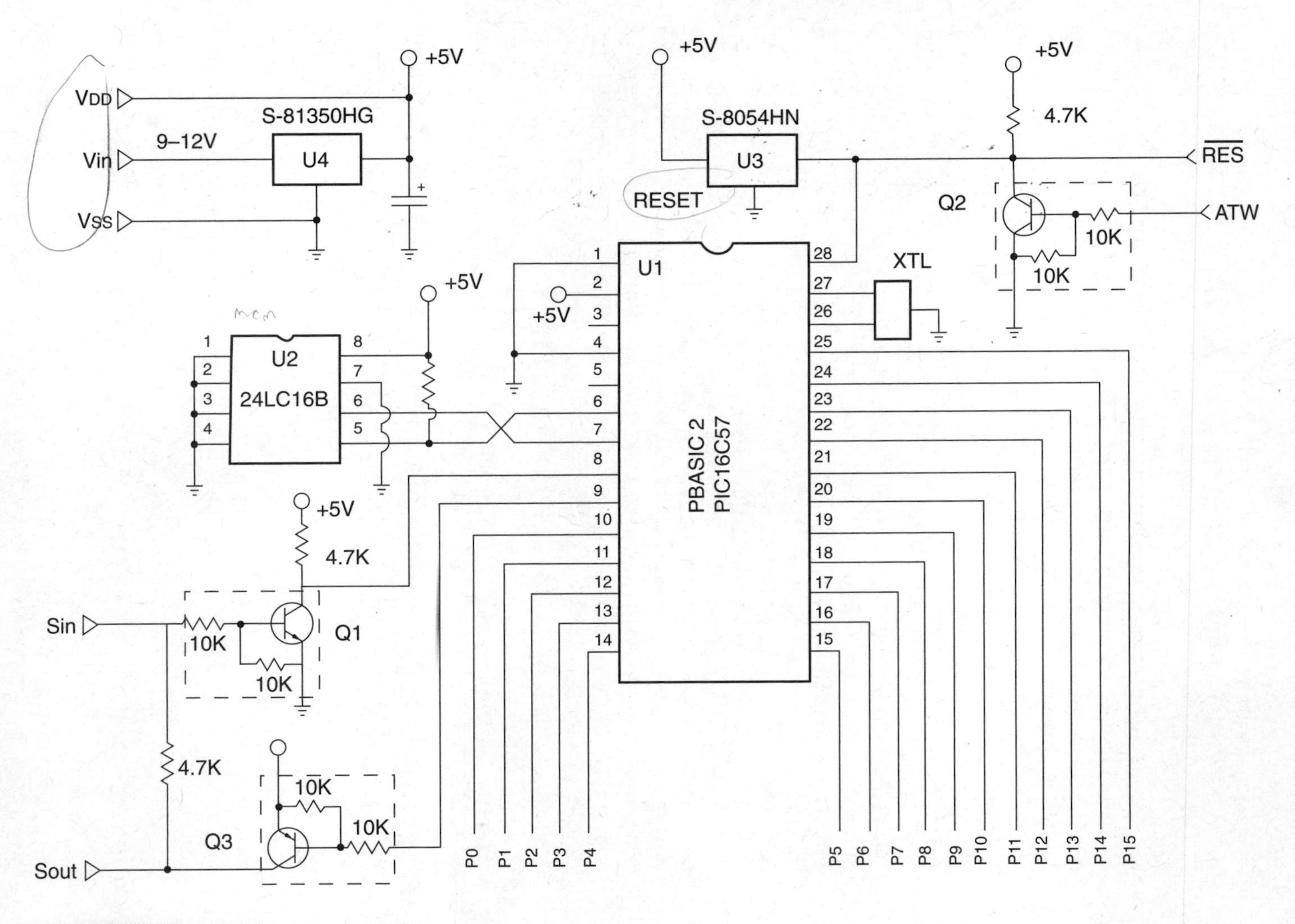

Figure 1-5 BASIC STAMP 2 schematic.

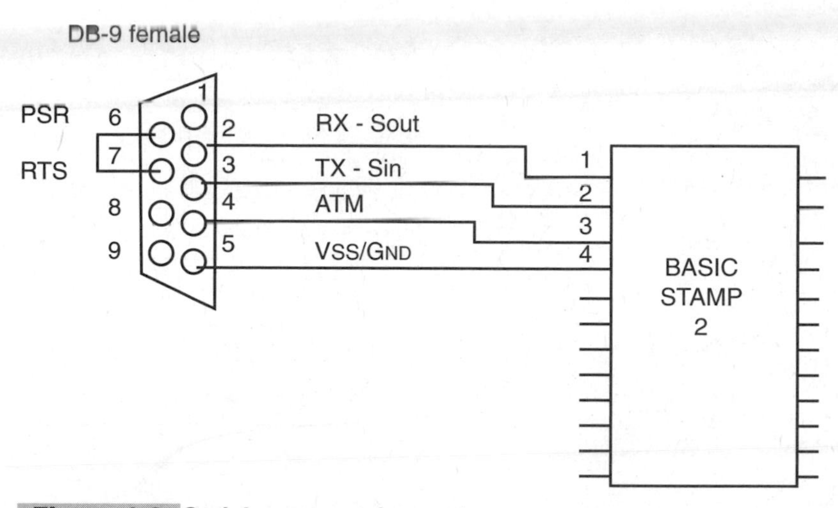

Figure 1-6 **Serial programming port.**

sockets to the board. Another approach to using the carrier board concept is to use the supercarrier board, as illustrated in Fig. 1-9. The supercarrier board contains a number of more useful aides, such as power regulators, more header pins, serial connections, and reset switch. Additional IC sockets can be placed on the board for expansion, with jumper wires added between sockets. This approach is a bit more permanent than using the protoboard. Many people use the carrier board as their protoboard for simple circuits. Other hobbyists use the carrier board as a replacement for a printed circuit board.

Alternative STAMP 2

Another approach to obtaining the STAMP 2 is to build your own "alternative" BASIC STAMP 2. In this configuration you obtain the 28-pin DIP interpreter chip and a few extra support or "glue" components, and you can save about half the cost of the original STAMP 2 by assembling or "rolling" your own. The alternative STAMP 2 project revolves around the lower-cost BASIC STAMP 2 interpreter chip, a 16C57 PIC, a 28-pin DIP-style STAMP 2, rather than the stock BASIC STAMP 2 module. The lower-cost 28-pin PIC BASIC interpreter chip requires a few external support chips and occupies a bit more real estate, but the cost is significantly less. The BASIC STAMP 2 costs $49, while rolling you own will cost about half, or $21. The 28-pin BASIC STAMP 2 interpreter chip requires a few external support components, a 24LC6B memory chip, a MAX232 serial interface chip, and a ceramic resonator. The MAX232 requires four 1-μF capacitors, a diode, and a resistor and provides a conventional serial I/O, which is used to program the microprocessor.

The alternative STAMP 2 processor circuit is shown in Fig. 1-10. Both the STAMP 2 and the alternative STAMP 2 perform equally well for all of the projects in this book. Please note that the pinout arrangement is a bit different between the original BASIC STAMP 2 and the alternative STAMP 2 (see Table 1-4). Another alternative is to purchase the original equipment manufacturer (OEM) module kit from Parallax, which is quite similar to the alternative STAMP 2.

TABLE 1-2 BASIC STAMP 2 MEMORY MAP

WORD NAME	BYTE NAME	NIBBLE NAME	BIT NAME	SPECIAL NOTES
INS	INL INH	NA,INB	IN0-IN7 IN8-IN15	Input pins, word, byte
OUTS	OUTL OUTH	OUTA,OUTB	OUT0-OUT7 OUT8-OUT15	Output pins, word, byte nibble and bit addressable
DIRS	DIRL DIRH	DIRA,DIRB DIRC,DIRD	DIR0-DIR7 DIR8-DIR15	I/O pin direction, control word, byte, nibble and bit addressable
W0	B0			General purpose, word, byte
	B1			nibble and bit addressable
W1	B2			General purpose, word, byte
	B3			nibble and bit addressable
W2	B4			General purpose, word, byte
	B5			nibble and bit addressable
W3	B6			General purpose, word, byte
	B7			nibble and bit addressable
W4	B8			General purpose, word, byte
	B9			nibble and bit addressable
W5	B10			General purpose, word, byte
	B11			nibble and bit addressable
W6	B12			General purpose, word, byte
	B13			nibble and bit addressable
W7	B14			General purpose, word, byte
	B15			nibble and bit addressable
W8	B16			General purpose, word, byte
	B17			nibble and bit addressable
W9	B18			General purpose, word, byte
	B19			nibble and bit addressable
W10	B20			General purpose, word, byte
	B21			nibble and bit addressable
W11	B22			General purpose, word, byte
	B23			nibble and bit addressable
W12	B24			General purpose, word, byte
	B25			nibble and bit addressable

TABLE 1-3 BASIC STAMP 2 PINOUTS

PIN	NAME	FUNCTION	DESCRIPTION
1	SOUT	Serial out	Temporarily connects to PCs (RX)
2	SIN	Serial out	Temporarily connects to PCs (TX)
3	ATN	Attention	Temporarily connects to PCs (DTR)
4	VSS	Ground	Temporarily connects to PCs ground
5	P0	User I/O 0	User ports which can be used for inputs or outputs.
6	P1	User I/O 1	
7	P2	User I/O 2	
8	P3	User I/O 3	*Output mode:* Pins will source from V_{dd}. Pins should not be allowed to source more than 20 mA, or sink more than 25 mA. As groups, P0–P7 and P8–P15 should not be allowed to source more than 40 mA or sink more than 50 mA.
9	P4	User I/O 4	
10	P5	User I/O 5	
11	P6	User I/O 6	
12	P7	User I/O 7	
13	P8	User I/O 8	*Input mode:* Pins are floating (less lan 1 μA leakage).The 0/1ogic threshold is approx. 1.4 V.
14	P9	User I/O 9	
15	P10	User I/O 10	
16	P11	User I/O 11	
17	P12	User I/O 12	*Note:* To realize low power during sleep, make sure that no pins are floating, causing erratic power drain. Either drive then to V_{ss} or V_{dd}, or program them as outputs that don't have to source current.
18	P13	User I/O 13	
19	P14	User I/O 14	
20	P15	User I/O 15	
21	VDD	Regulator out	Output from 5-V regulator (V_{in} powered). Should not be allowed to source more than 50 mA, including P0–P15 loads.
		Power in	Power input (V_{in} not powered). Accepts 4.5 to 5.5 V. Current consumption is dependent on run/sleep mode and I/O.
22	RES	Reset I/O	When low, all I/Os are inputs and program execution is suspended. When high, program executes from start. Goes low when V_{dd} is less than 4 V or ATN is greater than 1.4 V. Pulled to V_{dd} by a 4.7-kΩ resistor. May be monitored as a brownout/reset indicator. Can be pulled low externally (i.e., button to V_{ss}) to force a reset. Do not drive high.
23	VSS	Ground	Ground. Located next to V_{in} for easy battery backup.
24	VIN	Regulator in	Input to 5-V regulator. Accepts 5.5 to 15 V. If power is applied directly to V_{dd}, pin may be left unconnected.

Figure 1-7 Parallax NX1000 protoboard.

Figure 1-8 STAMP 2 and a simple carrier board.

Figure 1-9 STAMP 2 supercarrier board.

The cost of the OEM version is between those of the original STAMP 2 and the alternative STAMP 2. Note that the designation OEM is an indication that Parallax's intent is that you might use this board as a drop-in board in an overall new system design. In this book we chose to utilize the STAMP 2 (BS2-IC) instead of the cheaper PIC chips, since most of the projects in this book are works in progress and features can be added at any time in the future.

You can build the alternative STAMP 2 on a protoboard, perfboard, carrier board, or your own custom printed circuit board. By utilizing protoboards, you ensure that your designs are fluid and can be easily changed on the fly, until you complete and debug your project. The perfboard method of construction uses point-to-point wiring between components. Components are soldered to wires that interconnect. This method of construction is more permanent than protoboard construction, but can be modified by resoldering and repositioning. Carrier board construction is quite similar to perfboard construction and can be modified more easily than a printed circuit board. Printed circuit boards are generally used when design is complete, or finalized and debugged, and when neatness or mass production is important.

The design of the alternative STAMP 2 requires you to supply your own 5-V regulator in order to power the BASIC interpreter, memory, and serial communications chip. You would need to supply a 5-V regulator, such as LM7805, as shown in Fig. 1-13 at U4. The 5-V regulator can be supplied with power from a 9- to 12-Vdc "wall wart" power supply, or you could build your own power supply. A heat sink on the regulator is advisable, in order to dissipate heat properly. When you are constructing the alternative STAMP 2, be sure to observe the correct polarity when you install the four 1-μF electrolytic capacitors and the diode at D1. Note that the switch S1 is used to reset the microprocessor if a program stalls. The alternative

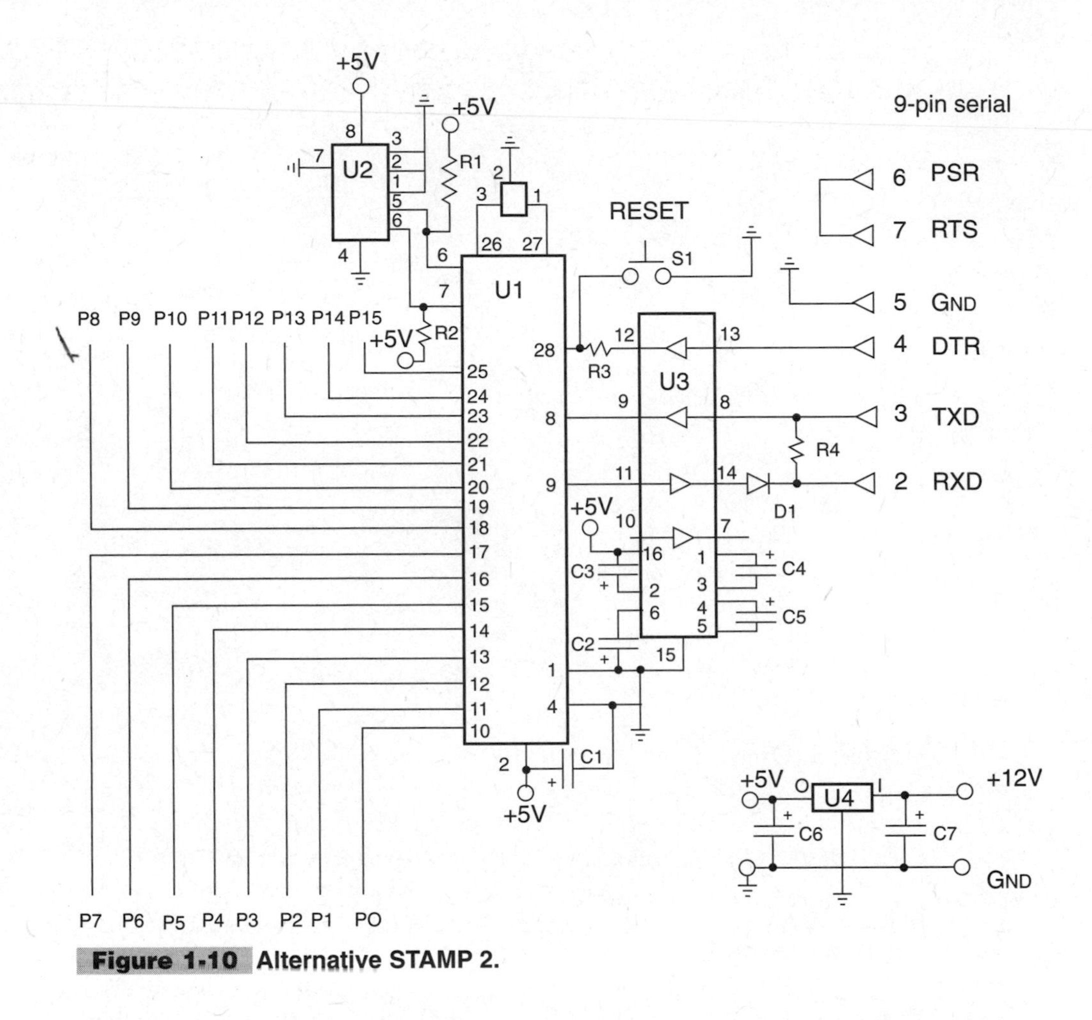

Figure 1-10 Alternative STAMP 2.

STAMP 2 has 16 I/Os, but the pinouts are different from the original STAMP 2. The alternative STAMP 2 provides P0 through P15 at the interpreter chip pins 10 through 25. A simple trick is to add 10 to the functional designation; see the chart in Table 1-4. In order to connect input switches to the STAMP 2's I/O, you will need to install 10-kΩ pull-up resistors from the 5-V supply to the STAMP 2 pins that will be used for inputs (see Fig. 1-11). The MAX232 chip at U3 is a serial interface chip that allows the 16C57 PIC processor to communicate to your personal computer, so that you can download your programs via the serial port of your PC (see Fig. 1-12).

An Exercise

Now let's have some fun and exercise the alternative STAMP 2 (refer to the diagram in Fig. 1-13). For this first project, you will need two 470-Ω resistors and two LEDs as well as

TABLE 1-4 ORIGINAL STAMP 2 VERSUS ALTERNATIVE STAMP 2 PINOUTS

P DESIGNATION	ORIGINAL STAMP 2 PIN	ALTERNATIVE STAMP 2 PIN
P0	5	10
P1	6	11
P2	7	12
P3	8	13
P4	9	14
P5	10	15
P6	11	16
P7	12	17
P8	13	18
P9	14	19
P10	15	20
P11	16	21
P12	17	22
P13	18	23
P14	19	24
P15	20	25

two 10-kΩ resistors and two momentary pushbutton switches as shown. The LEDs are connected to pins P0 and P1 and the two input switches are connected to pins P2 and P3. The program will begin blinking one of the LEDs when switch S1 is pressed. When S2 is pressed, the second LED will begin blinking with a double blink. The blink program demonstrates the use of inputs and outputs as well as whether the alternative STAMP 2 is functioning. Once the test circuit has been breadboarded or built on a perfboard or PC board, you are now ready to load your first program into the microprocessor. Once you have assembled your alternative STAMP 2, you will need to connect it to your PC's serial port, using a serial programming cable. Once your serial cable has been built and connected between your PC and the alternative STAMP 2, apply power through the 5-V regulator and you are ready to roll. Locate the Windows desktop editor software STAMP-W.EXE and call it up on your PC. Once the program has been installed and called up, you can apply power to your alternative STAMP 2 circuit and support chips. Remember, if you are using the alternative STAMP 2, you will need to provide 5 V to pin 2. You are now ready to begin programming your new STAMP 2 microcontroller. Type the BLINK.BS2 program (Listing 1-1) into the Windows STAMP 2 editor and then run the blink program to test your alternative STAMP 2. Your alternative STAMP 2 should now come to life as soon as you press switch S1 or S2. You are now on your way to programming and enjoying your STAMP 2 microprocessor.

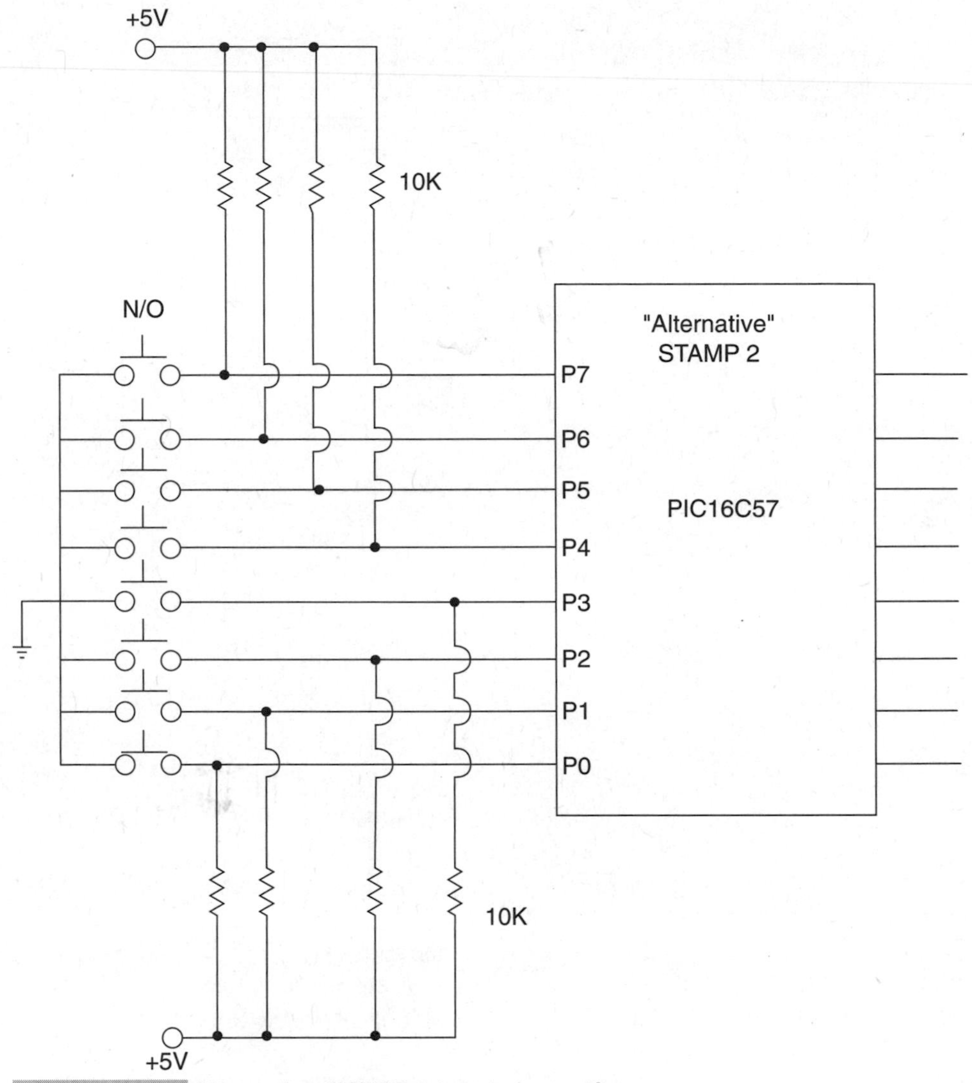

Figure 1-11 **Alternative STAMP 2 input connections.**

Parallax, the supplier of the STAMP products, has very good online documentation. The Parallax website also has many application notes and links to many other STAMP providers and hobbyists. The resources on the enclosed CD-ROM will aid the STAMP 2 project builder with additional tools. Appendix 1 contains complete data sheets that are used for projects in this book. Appendix 2 contains a number of important tables and charts and commands, which should prove be very useful. Appendix 3 lists a number of important websites and support chip manufacturers that complement the BASIC STAMP 2. The STAMP 2 manual can be found on the companion CD-ROM as well as all the program listings for the projects contained in this book. An additional section on the CD-ROM contains many additional STAMP 2 projects not covered in this book.

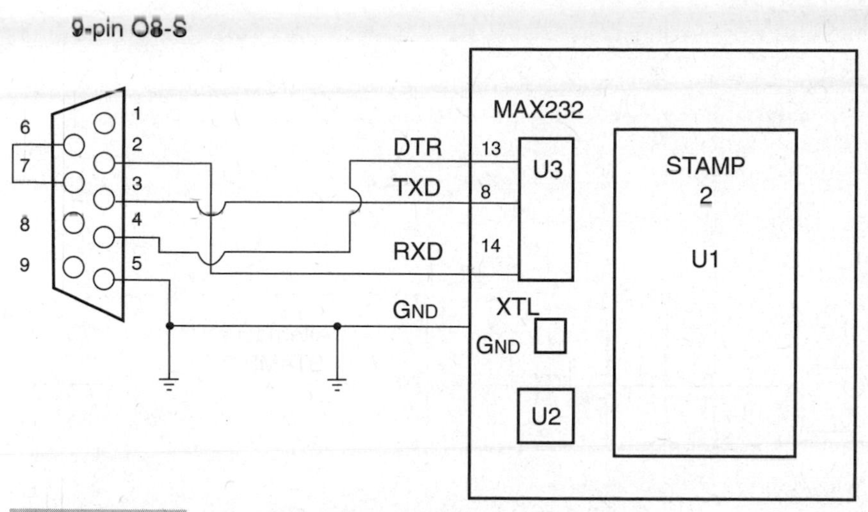

Figure 1-12 Alternative BASIC STAMP 2 serial cable.

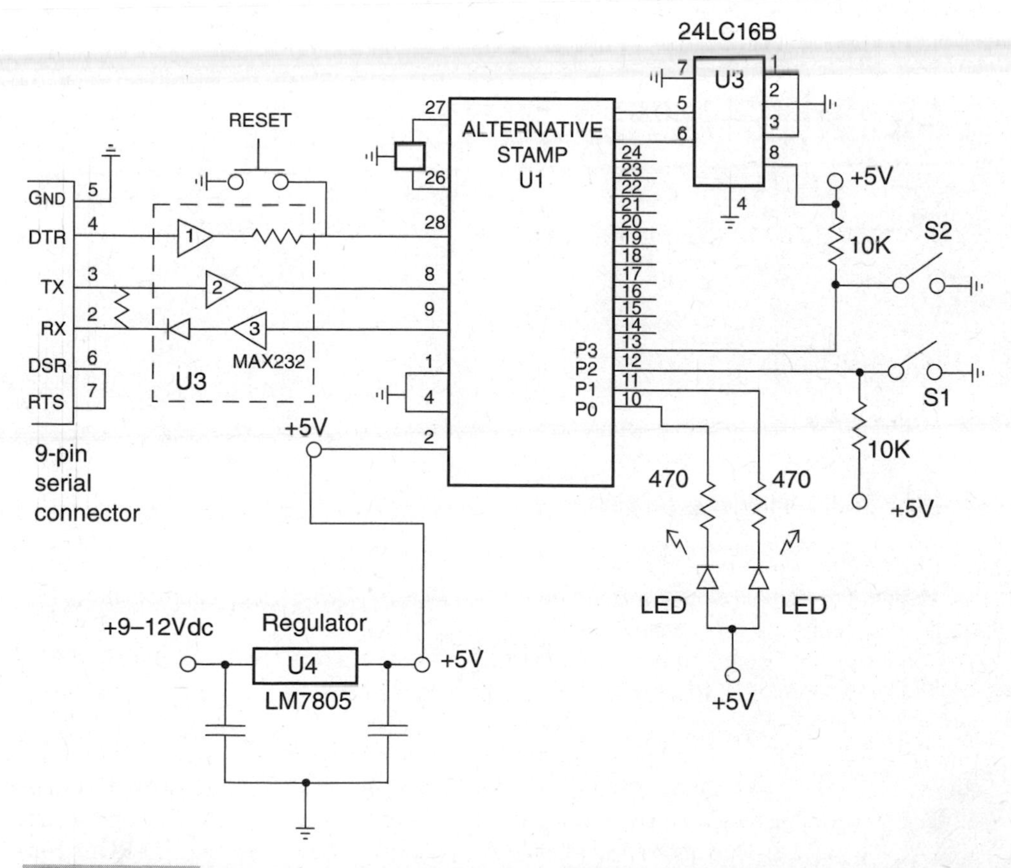

Figure 1-13 Blink test program setup.

```
'Blink.bs2
output 0
output 1
input 2
input 3
out0=1
out1=1
retest:
if in2=0 then blink
if in3=0 then
two_blink
goto retest
blink:
low 0
pause 200
high 0
pause 200
goto retest
two_blink:
low 0
low 1
pause 200
high 0
high 1
pause 200
goto retest
```

Listing 1-1 Alternative STAMP 2 blink program.

2

STAMP 2 INPUT/OUTPUT INTERFACING

CONTENTS AT A GLANCE

The BASIC STAMP 2 microprocessor has revolutionized the electronics world in many ways. The STAMP 2 will permit you to construct many types of powerful and complex projects and keep parts counts low, with the ability to redesign and/or modify your project or system on the fly.

This chapter will present STAMP 2 interfacing topics, from basic input and output devices and drivers to RS-232 serial interfacing between the STAMP 2 and other STAMP 2's, as well interfacing to personal computers. In this chapter we will also cover RS-485 serial interfacing between STAMP 2 microprocessors.

In order for the STAMP 2 to control the world, it must be able to connect or to interface with the outside world. In order for the STAMP 2 to accept data from switches or sensors at its I/O pins, the STAMP 2 must be set initially by simply addressing the direction registers. These direction registers must be addressed to act as either inputs or outputs, by placing a 0 or a 1 in the direction register. The STAMP 2 has 16 I/O pins, and you need to tell the STAMP 2 if a particular pin will be used as an input or output. The DIRS variable describes the data direction, where a 1 in a given bit makes that pin an output and a 0 makes that pin an input. The OUTS variable latches the output storing the output states 1 or 0 of the pins. The INS variable is a read-only variable reflecting the states 1 or 0 on the pin. Most programs will place these variables at the beginning of the program, along with other defining variables and reset or starting conditions. When programming the STAMP 2, you must also specify constants and variables at the beginning of the program; you can observe this by looking at some of the example programs in the STAMP 2 manual provided on the enclosed CD-ROM. A number of useful charts and tables are also offered in App. 2 on the CD-ROM.

Configuring Switches

Input switches and sensors can be configured in two different ways, in which there are two variations. The diagram in Fig. 2-1 illustrates an active low input, using a normally open type of sensor or switch. In this diagram the STAMP 2 pin P0 acts as an input pin with a 10-kΩ resistor to the +5-V supply and the switch goes low or to ground upon activation. In this configuration a normally open pushbutton switch or sensor is used to trigger the input of the STAMP 2. An example of a typical code is:

```
ACTIVE LOW—normally open input switch or sensor
     sense:
     if in0=0 then activate
     goto sense
     activate:
     debug "switch depressed"
     goto sense
```

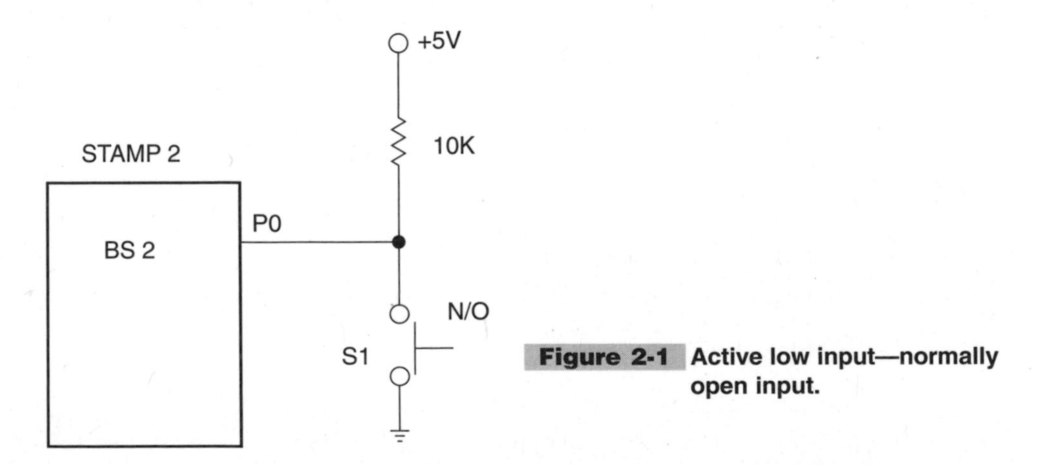

Figure 2-1 Active low input—normally open input.

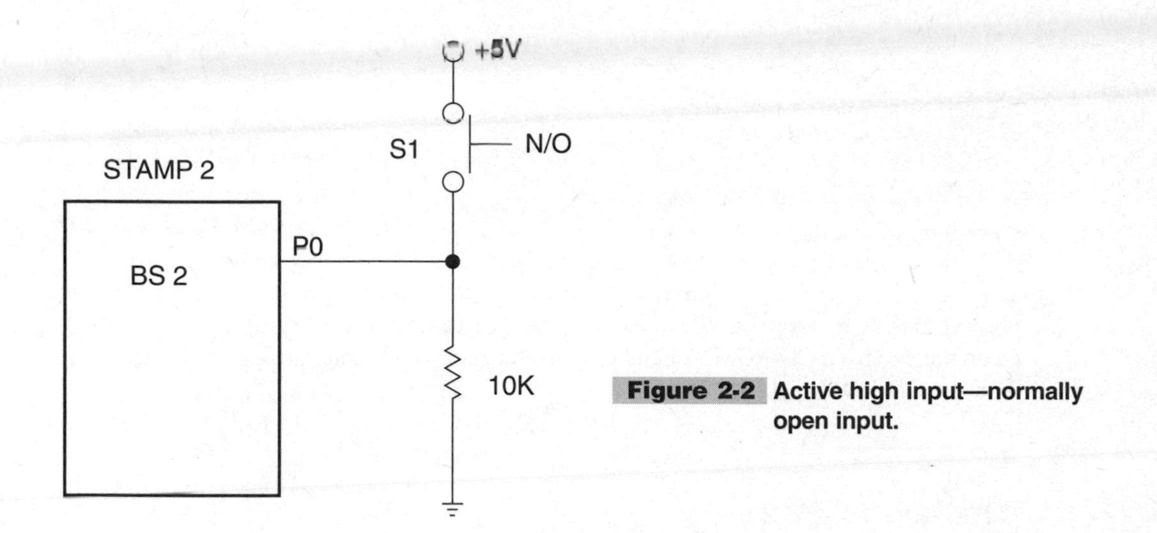

Figure 2-2 Active high input—normally open input.

The diagram in Fig. 2-2 depicts an active high input. In this diagram pin P0 is connected to a normally open switch or sensor that is directly fed from a 5-V supply. The input pin P0 is also connected to ground through a 10-kΩ resistor. The sensor or switch connects directly to the power supply pin and is thus going high upon activation. The code snippet for this condition is a bit different from the example above.

```
ACTIVE HIGH—normally open input switch or sensor
   sense:
   if in0=1 then activate
   goto sense
   activate:
   debug "switch depressed"
   goto sense
```

The input diagram in Fig. 2-3 shows the STAMP 2 pin P0 in an active high configuration but using a normally closed sensor or switch, shown at S1. Pin P0 is brought high through a 10-kΩ resistor. Activation takes place when the normally closed switch connection to ground is broken at pin P0. This configuration is ideal for normally closed alarm switches where you want a protected and supervised loop condition at all times, except when there is a true alarm condition.

```
ACTIVE HIGH—normally closed input switch or sensor
    sense:
    if in0=1 then activate
    goto sense
    activate:
    debug "switch depressed"
    goto sense
```

The input configuration in Fig. 2-4 depicts an active low-going input. A resistor is tied from pin P0 to ground, while +5 V is fed to pin P0 through a normally closed switch.

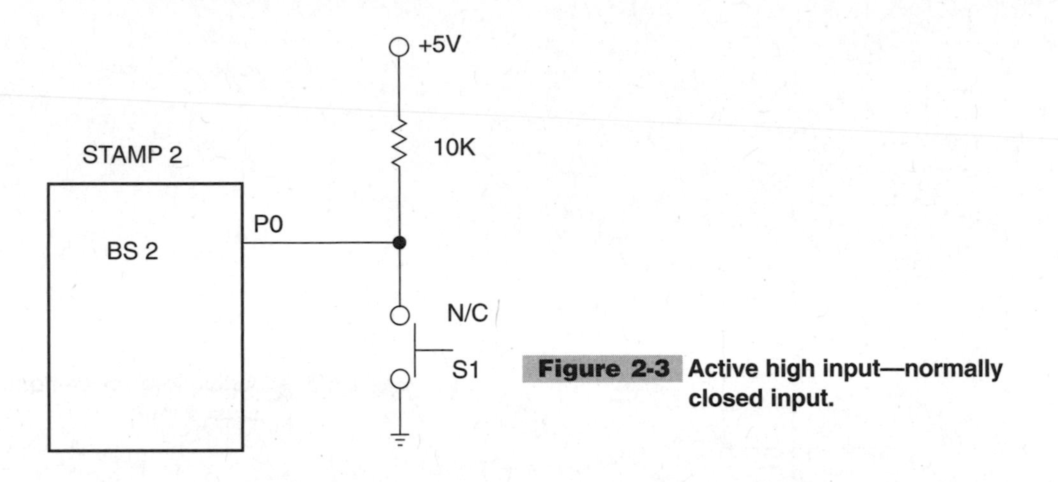

Figure 2-3 Active high input—normally closed input.

When the normally closed switch is pressed, an open condition causes the input to go low, i.e., to ground. This configuration would also be good for alarm monitoring systems, where a supervisory condition exists. The code for this configuration is as follows:

```
ACTIVE LOW—normally closed input switch or sensor
    sense:
    if in0 =0 then activate
    goto sense
    activate:
    debug "switch depressed"
    goto sense
```

In order to utilize the STAMP 2 to sense or control the outside world, you must change the direction of the I/O pins to tell the STAMP 2 that you would like to use a particular I/O pin as an output pin (see the manual on the CD-ROM). The STAMP 2 can control almost any type of device; however, it will supply only about 20 mA of current to an output device. The STAMP 2 will drive an LED or optocoupler with no problem, but almost any other output device would need to be driven by a transistor. The transistor acts a current switch that can control larger-current devices.

In Fig. 2-5, the STAMP 2 turns on the LED when P0 goes low. The cathode of the LED is sunk to ground through the output pin. Figure 2-6 illustrates the STAMP 2 sourcing current to the LED. The microprocessor in this example is sourcing current from the positive supply voltage through the LED to ground. Some microcontrollers may not have enough sourcing current capacity, so a transistor may have to be used to drive an output.

The diagram in Fig. 2-7 depicts an LED being driven by an npn transistor. When pin P0 goes high, a current is caused to flow to the base of the 2N2222 npn transistor, thus effectively connecting the cathode of the LED to ground; this is known as sinking current to ground. This example would be effective if you wished to turn on a high-current LED, or infrared LED, or other device that draws more current. Figure 2-8 shows a similar application but uses a pnp transistor to drive a current device. Note that the emitter of the transistor is connected to the plus supply voltage. The current in this example is being sourced by the transistor. Once again this application is for driving higher-current LEDs or other devices.

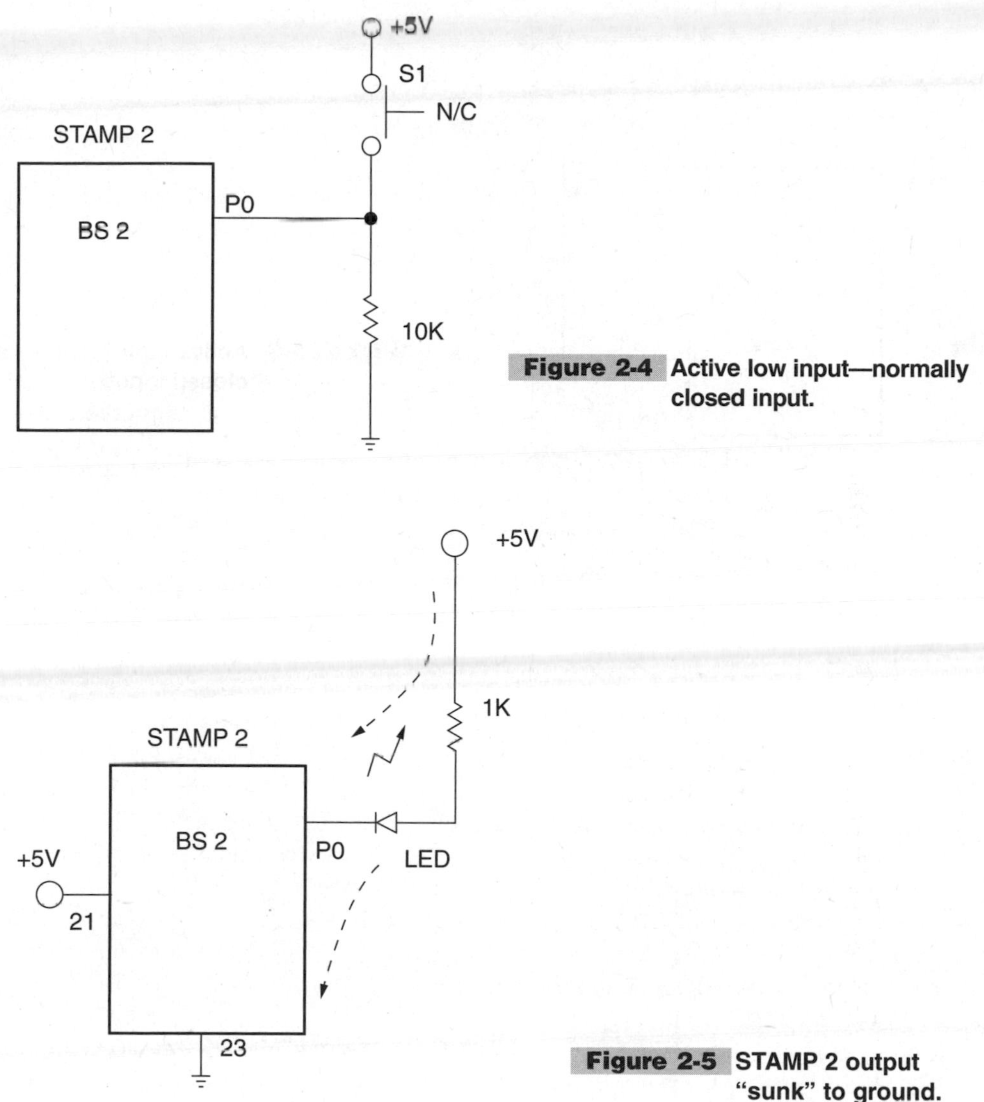

Figure 2-4 Active low input—normally closed input.

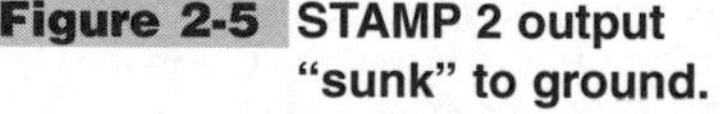

Figure 2-5 STAMP 2 output "sunk" to ground.

Interfaces for the Microprocessor

Sometimes it's necessary to interface a microprocessor to other ICs or microprocessors. The circuit in Fig. 2-9 depicts a bipolar transistor turned on by the STAMP 2 on pin P0. Transistors are ideal interfacing devices; they require a small current to turn on and can easily isolate the next circuit stage. This circuit can be thought of as an inverting buffer, since the signal output from the STAMP 2 is a positive pulse. The output after the transistor becomes a "low," thus creating an inverted and buffered signal. Many circuits require

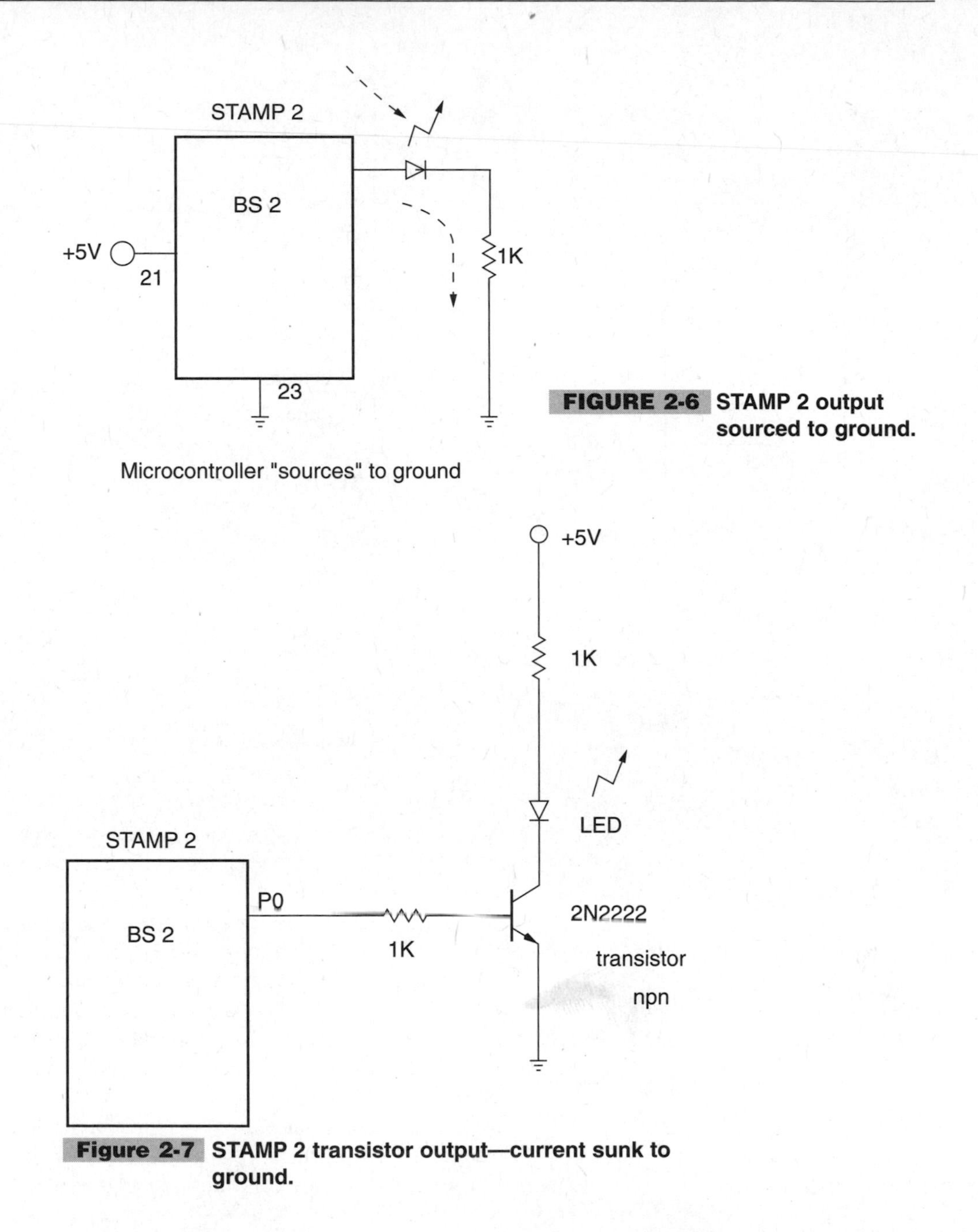

FIGURE 2-6 STAMP 2 output sourced to ground.

Figure 2-7 STAMP 2 transistor output—current sunk to ground.

a negative- or low-going input in order to be activated, and this type of circuit is perfect in this type of application.

The diagram in Fig. 2-10 shows an optoisolator inverting buffer. This type of inverting buffer is used when absolute isolation is needed between circuits. The STAMP 2 output at pin P0 is once again a positive-going pulse, and the output of the optoisolator becomes low. A PS2501 optoisolator is shown, but any general-purpose device could be substituted.

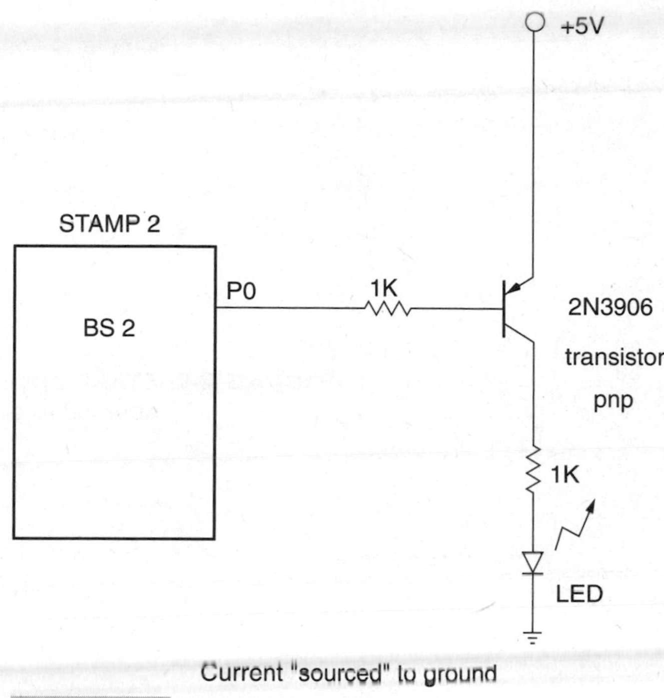

Figure 2-8 **STAMP 2 transistor output—current "sourced" to ground.**

Often it is necessary to use a microprocessor to control a relay. In order to have the STAMP 2 control a relay, a transistor driver must be used. In the next example, shown in Fig. 2-11, pin P0 is used to drive a 2N2222 npn transistor, which in turn is used to control the relay. Although some mechanical relays may draw a low enough current to control a relay directly, this is not recommended. When the electromagnetic field of a relay collapses, a voltage spike is generated. This spike can damage MOS-type circuits found in microprocessors. In this circuit a high signal on P0 causes the transistor to conduct and sink the relay to ground. The magnetic field energizes, and relay contacts close. Upon deactivation, the relay coil generates a reverse voltage spike. This voltage spike is shorted out or snubbed by the diode across the relay coil. Also note that if you want to control very large currents, then a low-current relay could be used to drive a high-current relay.

The STAMP 2 can be used to control a dc motor as shown in Fig. 2-12. In this circuit, a high at pin P0 drives the IRF511 MOSFET device, which in turn operates the dc motor. A diode is placed across the MOSFET when the motor field changes, thus protecting the MOSFET and the STAMP 2. Note that you will have to scale your motor and its current requirement with the MOSFET used in your particular application.

Serial Port Interfacing

One of the reasons the STAMP 2 is so powerful is that it can communicate with the outside world through a serial port. There are many peripherals that will communicate with the

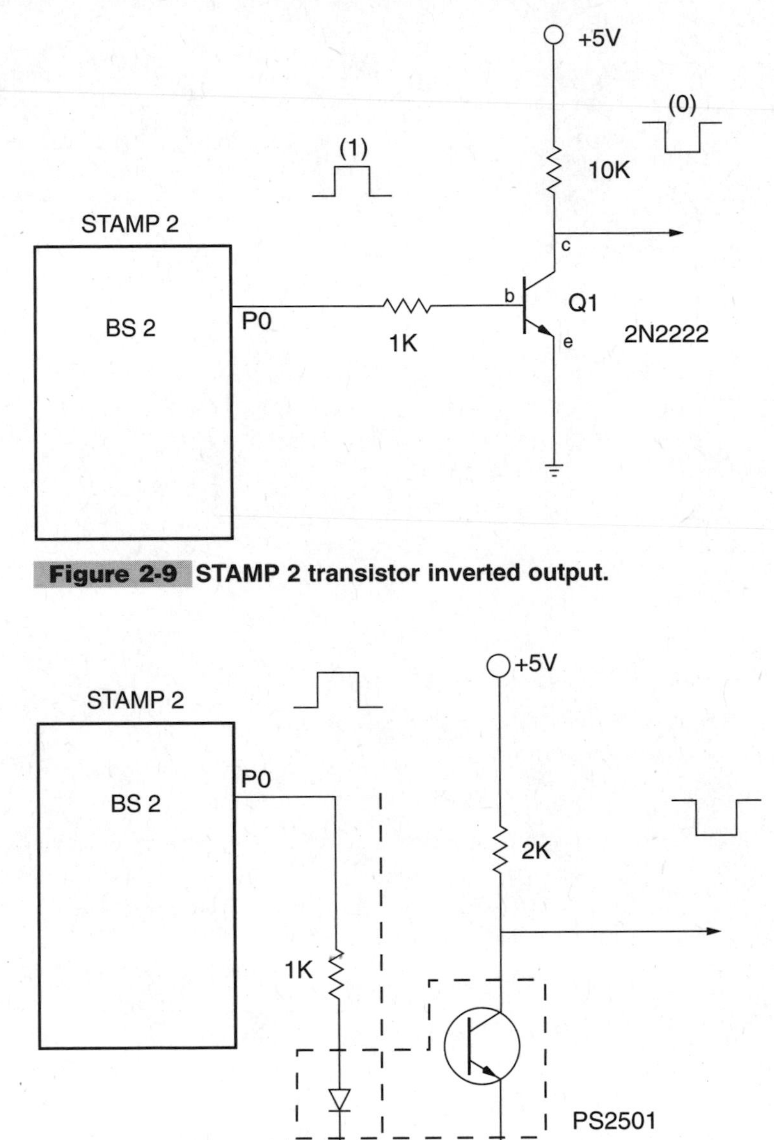

Figure 2-9 STAMP 2 transistor inverted output.

Figure 2-10 STAMP 2 optoisolator inverted output.

STAMP 2 through serial communication. This serial capability allows the STAMP 2 to communicate with personal computers, PDAs (personal data assistants), and other microprocessors as well as one-wire sensors, memory devices, keyboards, LCD displays, and display controllers. The STAMP can be set up to communicate through any of its 16 I/O pins by the serial protocol. Serial communication is the process of transmitting and receiving one data bit at a time. Serial communication is formally known as asynchronous serial commu-

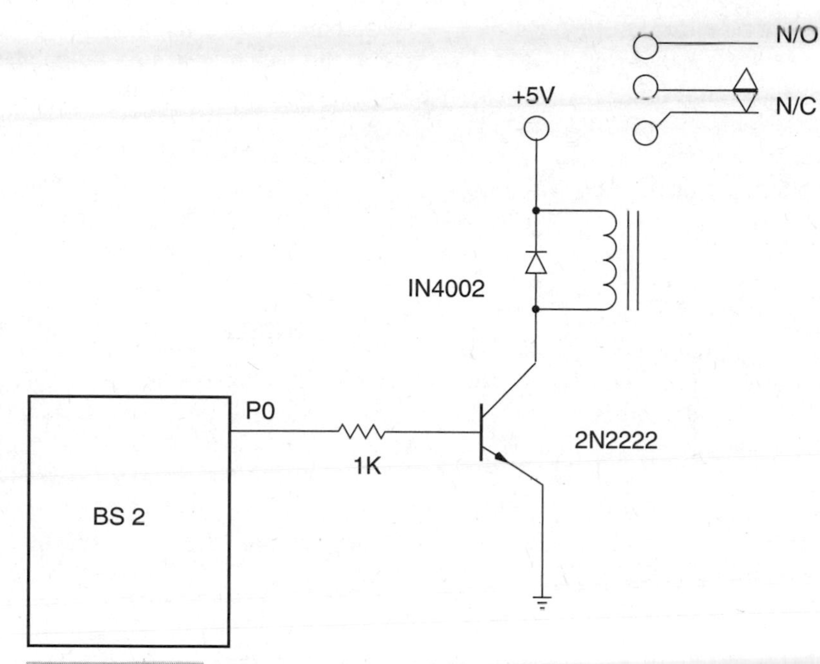

Figure 2-11 STAMP 2 transistor relay driver.

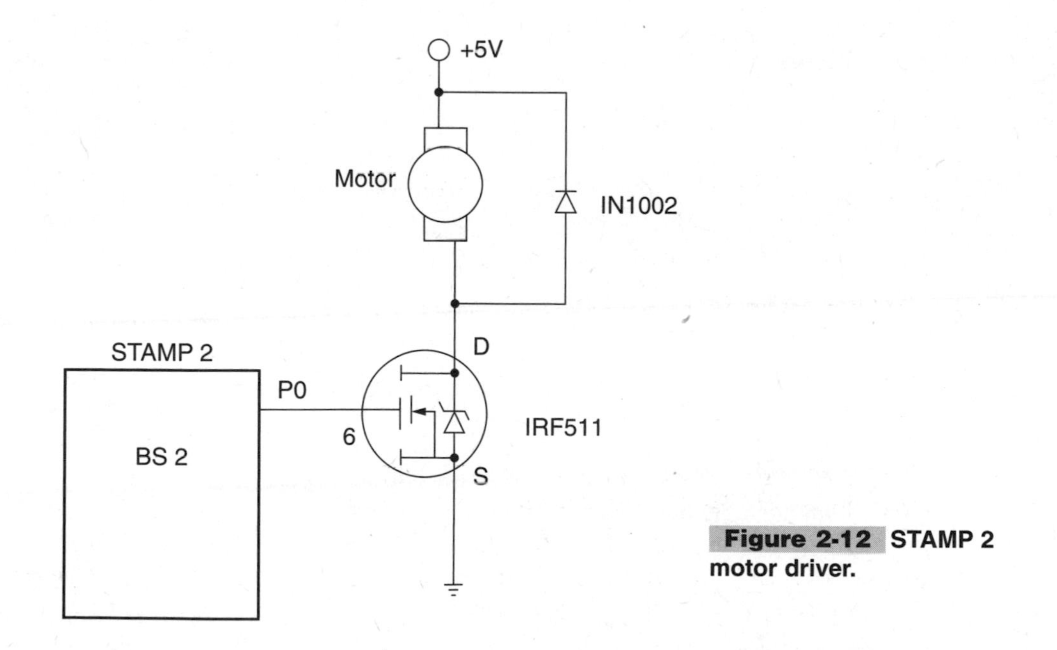

Figure 2-12 STAMP 2 motor driver.

nication, and it sends data without a separate synchronizing signal to help the receiver distinguish one bit from the next. In order to overcome this lack of synchronization, the asynchronous serial protocol applies strict rules for timing and organizing bits.

The basic principle of serial communication is to send multiple bits over a single wire. Each bit is placed on the wire for a specific amount of time. The serial protocol has a number of parameters that must be followed in order for it to work correctly. A typical serial setting might read "2400 N-8-1." The 2400 is the baud rate. Data bits are sent at a precise speed and expressed in bits per second (bps) or baud rate. It interesting to note that the bit rate is the reciprocal of the bit time or bit period or the amount of time allotted to each bit. A 2400-bps rate allows each bit 1/2400 second, or 416.67 microseconds (μs). This timing rate has to very precise for the serial link to work! In the serial setting, you will also notice an N or no-parity bit. A parity bit is an additional bit added that is compared or checked to count either an odd or an even number of data bits so that the system can ensure that the transmitted number of bits will be equal to the received number of bits. *No parity* means no extra bit is used for checking. Next you see the number 8. The number 8 is used to note the number of data bits that will be sent. Finally, you will see the number 1 or 0, and this last parameter is the number of stop bits used. A stop bit is the pause between the last data bit and next start bit. Generally stop bits are either 1 or 2 bits. The STAMP 2 supports 1 stop bit only.

Serial signals can be in either of two states, which traditionally are called *mark* and *space.* In conventional RS-232 systems a *mark* is a negative voltage, i.e., –10 V, and a *space* is a positive voltage at +10 V. When no data is being sent, the signals line idles in the *mark* state, also known as the *stop bit* condition. In starting a transmission, the sender switches to the *start bit* condition, a *space,* which holds the signal line in that state for 1 bit time. After the start bit come the data bits. A *mark* is represented by a 1 or minus (−) and a *space* is represented by a 0 or plus (+). The use of opposite start and stop bits is the key to asynchronous communication. The receiver must identify each incoming bit by its precise time of arrival. It bases its reckoning of time on the exact instant of the signal's transition from the stop to start bit state. This allows it to reset its timing with each new arrival of data, so that small timing errors can't accumulate over many frames to become big timing errors.

RECEIVING SERIAL DATA

Not every computer-type device can receive serial data and process it as fast as it might arrive. Some devices, like the STAMP, must devote all of their attention to the serial line in order to receive data. Often handshaking bits are used to control the flow of data.

The STAMP 2 uses 5-V logic. It outputs 0 V for a 0 and 5 V for a 1. When it accepts inputs from other circuits, it regards a voltage less than 1.5 V as a 0 and voltages greater than 1.5 V as a 1. In the RS-232 specification, a 0 is represented as a higher voltage, i.e., +10 V, and a 1 by a negative voltage, typically −10 V. To convert a 5-V logic level from the STAMP 2's outputs and back generally requires RS-232 line drivers and line receivers.

An RS-232 transmission is supposed to output +5 to +15 V for a 0. An RS-232 receiver is required to recognize +3 to +15 V as a 0. The 2-V difference between send and receive is accounted for by voltage drop over long wire. Note that a 5-V logic 1 is equivalent to an RS-232 logic 0. For a logic 1, an RS-232 transmission is supposed to output −5 to −15 V. There is no way to get −3 V from the STAMP 2 5-V logic without additional components, but most RS-232 devices also regard a signal close to 0 V as a logic 1. So if our 5-V logic's 0 is close to 0 V, it will suffice as an RS-232 logic 1.

The STAMP 2 will support this type of "fooling"; you simply program the STAMP 2's serial output instruction to send inverted serial data. So now when the STAMP 2 receives data from conventional RS-232 sources, it is receiving a ±15-V signal. In order to protect the STAMP 2 from destruction, a 22-kΩ resistor is placed in series with the STAMP 2 I/O line for incoming RS-232 data.

Many projects presented in this book use the alternative STAMP 2, which utilizes MAX232 serial line drivers to get serial data to and from the STAMP 2. The baudmode command is a 16-bit number that specifies all the important characteristics of the serial transmission, i.e., bit time, data and parity bits, polarity, etc. Table 2-1 illustrates more detail on decimal baud rate number information and direct versus line driver interfacing. Also check App. 2 and the STAMP 2 programming manual included on the enclosed CD-ROM.

The Serin and Serout STAMP 2 serial command lines are typically quite similar in their appearance. The Serin command to send/receive asynchronous data is represented as Serin *rpin{\fpin},baudmode,{plabel,}{timeout,label,}[inputData]*, and often the command can be as simple as Serin *rpin,baudmode,[inputData]*. The Serout command structure for transmitting asynchronous data is represented as Serout *tpin,baudmode,{pace,}[outputData]* or Serout *tpin\fpin,baudmode,{timeout,tlabel,}[outputData]*. Often the Serout command can be as simple as Serout *tpin,baudmode,[outputData]*.

Some typical STAMP 2 SERIN commands are shown below.

```
Serin rpin,baudmode, [inputdata]
```

In the above example the rpin is the STAMP 2 I/O pin that will be receiving serial data. Baudmode is the setting for the bit rate, data bits, parity, and polarity. The inputdata specifies what do to with the incoming data—to compare it, ignore it, store it, or convert it from text to a numeric value, for example.

TABLE 2-1 SINGLE FRAME OF SERIAL DATA—BAUDMODE (Decimal Numbers)

BAUD RATE	8 BITS NO PARITY	7 BITS EVEN PARITY	8 BITS NO PARITY	7 BITS EVEN PARITY
300	19697	27889	3313	11505
600	18030	26222	1646	9838
1200	17197	25389	813	9005
2400	16780	24972	396	8588
4800	16572	24764	188	8380
9600	16468	24660	84	8276
19200	16416	24608	32	8224
38400	16390	24582	6	8198

```
SerData var byte
Serin 1, 16468, [serData]
```

The baudmode in this example is set to 9600 baud, 8-N-1, with an inverted polarity. In this example, Serin waits for and receives a single byte of data through pin P1 and stores that byte in the variable serData.

```
SerData var byte
Serin 1,16468, [DEC serData]
```

In the above example, the DEC modifier tells Serin to wait for numeric text and to convert that text into a 1-byte value (range 0 to 256). Note, there are a number of other modifiers for decimal, hex, and binary numbers; see the STAMP 2 manual in App. 2.

```
Serin 1, 16468, [WAIT ("DECODE")]
Debug "password accepted"
```

In this example, Serin waits for that word, and the program will not continue until it is received. Serin is looking for an exact match; i.e., the word DECODE in capital letters has to be sent from a keyboard in order for the "password accepted" message to be displayed.

Some typical STAMP 2 Serout commands are quite similar to the Serin commands and are shown below.

```
Serout  tpin, baudmode, {pace,} [outputdata]
```

In the above example, tpin is the STAMP 2 I/O pin that will be used to send serial data. Baudmode is the setting for the bit rate, data bits, polarity, and drive, and pace is an optional delay between bytes in milliseconds.

```
Serout 1, 16468, [Greetings!"]
```

In the example above the message Greetings would be sent on pin 1 at 9600 baud, 8N inverted.

```
Serout 1, 16468, [81]
```

In the example above Serout sends a single byte whose bits would total the decimal number 81. Looking up 81 in the ASC2 chart, you will note that the letter Q would be sent.

```
Serout 1, 16468, [DEC81]
```

In this example the DEC (decimal) modifier tells the STAMP 2's BASIC to convert the single-byte value of 81 into text representing the decimal number 81.

Networking STAMP 2's

A simple STAMP 2 networking scheme can be implemented by using the open baudmodes, which can be used to connect multiple STAMP 2's to a single pair of wires to create a simple

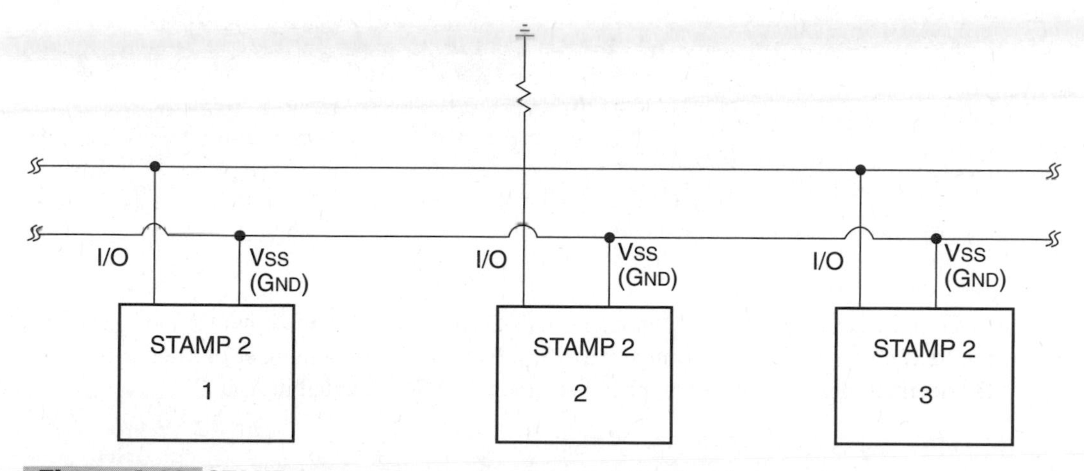

Figure 2-13 STAMP 2 network open/noninverted baudmode.

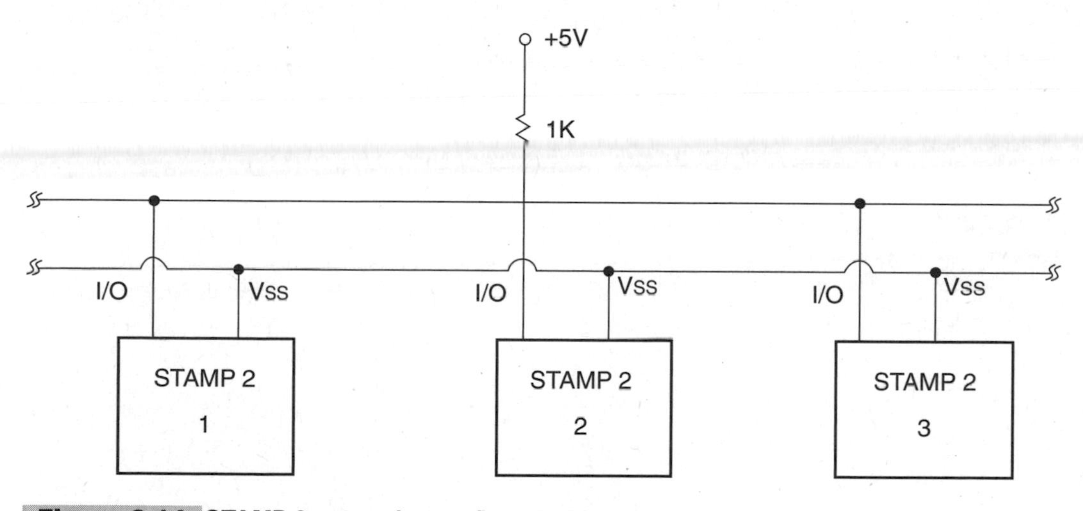

Figure 2-14 STAMP 2 network open/inverted baudmode.

party line network. Open baudmodes can actively drive the Serout pins only in one state; in the other state they disconnect the I/O pin. Since the open baudmodes drive only in one state and float in the other, there's no chance of shorting the I/O pins. Note the polarity selected for Serout determines which state is driven and which is open; see Table 2-2. Since the open baudmodes drive to only one state, they need a resistor to pull the network into the other state, as shown in Figs. 2-13 and 2-14.

Open baudmodes allow the STAMP 2 to share a party line, but it is up to the software to resolve other network problems, such as who talks when and how to prevent data errors. In the example in Fig. 2-15, and in the programs illustrated in BSNET1.BS2 and BSNET2.BS2 (Listings 2-1 and 2-2) the two STAMP 2's share a party line. They monitor the serial line for a specific word ("Yin" or "Yang") and then transmit data. A personal computer can be used to monitor the net activity via a line driver or CMOS inverter, as shown in Fig. 2-15. The result of the two programs BSNET1.BS2 and BSNET2.BS2 is that a monitoring PC would

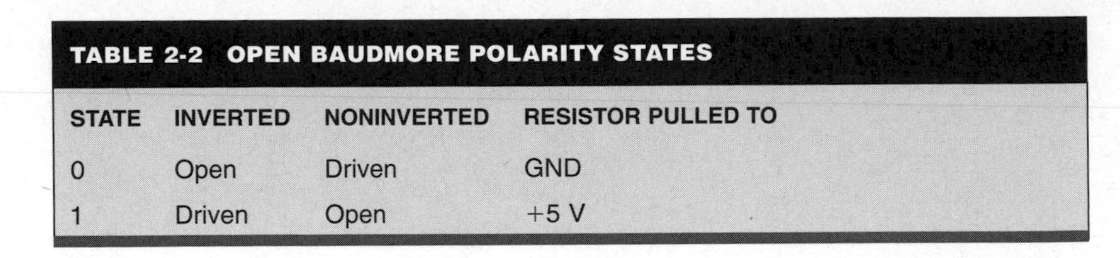

TABLE 2-2 OPEN BAUDMORE POLARITY STATES

STATE	INVERTED	NONINVERTED	RESISTOR PULLED TO
0	Open	Driven	GND
1	Driven	Open	+5 V

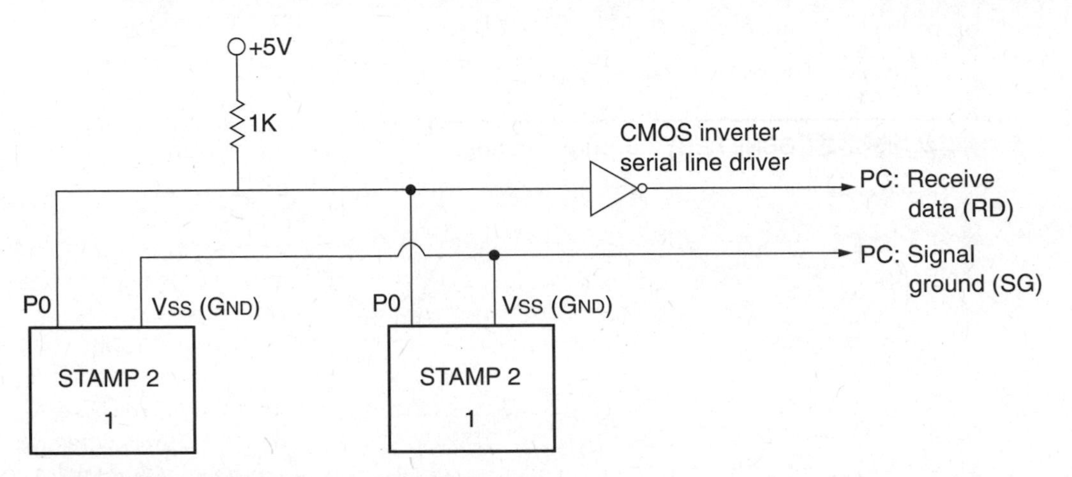

Figure 2-15 STAMP 2 to PC net monitor.

```
'bsnet1.bs2
' this program sends the word "Yin" followed
'by a linefeed and a carriage return. It then
waits to hear
'the word "Yang" plus LF/CR and then pauses 2
seconds and loops.

B_mode    con    32852

again:
   Serout 0,b_mode,["Yin",10,13]
   Serin   0,b_mode,[wait ("Yang",10,13]
   pause 2000
goto again
```

Listing 2-1 BSNET1.BS2 program listing.

see the words "Yin" and "Yang" appear on the screen at 2-second intervals, showing that the pair of STAMP 2's are sending and receiving on the same lines. This arrangement could be expanded to dozens of STAMP 2's if desired, with the right programming.

STAMP 2—RS-485 COMMUNICATIONS NETWORK

The RS-485 serial network protocol is vastly superior to the older RS-232 protocol, and is easy to implement. You can readily create a 32-node, differential 2-wire half-duplex network for distances up to 4000 ft using the RS-485.

```
'bsnet2.bs2
'this program waits for the word "Yin" plus
LF/CR
'then pauses 2 seconds and sends the word"Yang"
'and LF/CR, and then loops

b_mode   con 32852

again:
   Serin 0,b_mode,[wait("Yin",10,13)]
   Pause 2000
   Serout 0,b_mode,["Yang",10,13]
goto again
```

Listing 2-2 **BSNET2.BS2 program listing.**

The RS-485 standard allows the user to configure inexpensive local networks and multidrop communication links using twisted-pair wire. A typical RS-485 network can operate properly in the presence of reasonable ground differential voltages, withstand driver contention situations, and provide reliable communications in electrically noisy environments.

The Texas Instruments SN75176 differential bus transceiver chip and a Parallax BASIC STAMP 2 microprocessor will allow you to create a bidirectional RS-485 network using the RS-485 serial communication protocol, which can be used to provide strong serial signals up to 4000 ft at high baud rates in potentially noisy electrical environments. Two wires carrying an RS-485 signal (the A and B lines) provide a signal base from which many devices can communicate. Twisted-pair wire is recommended for long distances, but for short distances 24-gauge wire will work. Up to 32 devices can be connected to an RS-485 data line with the SN75176 and communicate by using a data protocol. This is referred to as an RS-485 network, or an RS-485 drop network. The heart of this communication is the Texas Instruments SN75176 differential bus transceiver chip. This chip converts RS-485 signals to RS-232 TTL-level signals, allowing devices that traditionally communicate over standard RS-232 serial connections to communicate over a single two-wire RS-485 network.

The Texas Instruments SN75176 chip has only 8 pins, and it can be easily connected to the BASIC STAMP, as shown in Fig. 2-16. The SN75176 requires 30 mA; therefore, it is recommended that only one of these chips be driven from the STAMP's V_{dd} pin, provided no other components require current. Setting $\overline{RE}$ (Not Receiver Enable) to *low,* the R (Receiver) pin is enabled, allowing the STAMP to receive any data coming over the A and B RS-485 network lines. Setting DE (Driver Enable) to *high* allows the STAMP to transmit data over the RS-485 network. V_{cc} and V_{dd} are electronic terms with the same meaning.

Setting up the RS-485 network for transmitting serial data using the BASIC STAMP would begin by setting P0 high and using Serout on P1 to transmit the data. In order to receive data, you would need to set P0 to low and using Serin to receive serial data on P1.

An RS-485 network operates with a controller/slave communication protocol. One controller commands everything, and one or more slaves respond to the commands. This example uses one BASIC STAMP as the controller and one BASIC STAMP as the slave. The circuit shown in Fig. 2-16 can easily be expanded to have up to 32 BASIC STAMPs communicating on one RS-485 network by connecting additional slaves to the network A and B lines.

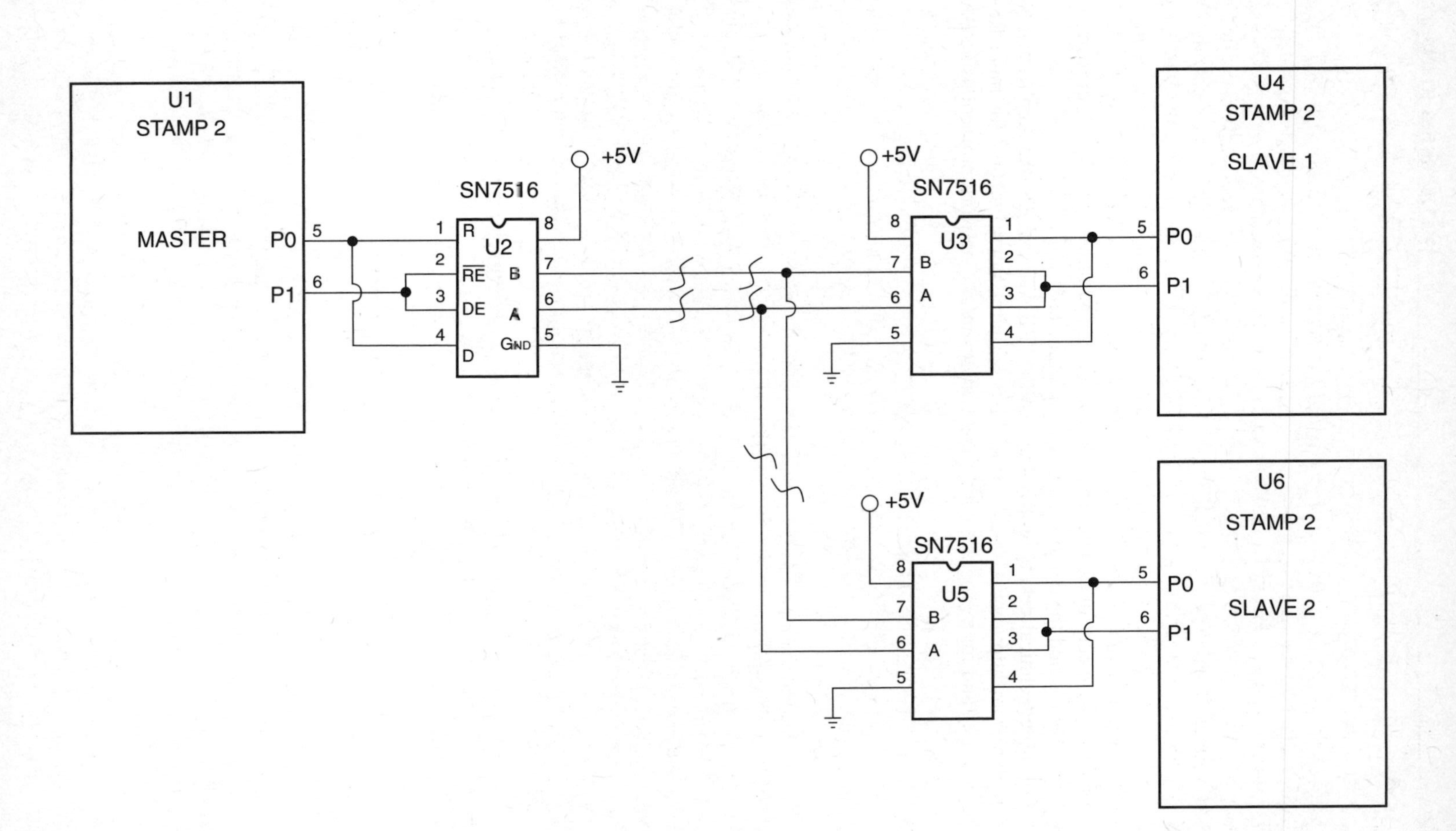

Figure 2-16 STAMP 2 to RS-485 network setup.

Construct the demo RS-485 network by first building the circuit as shown. Designate one BASIC STAMP to be the controller and the other to be the slave. On the controller, connect the supplied switch between the BASIC STAMP's P2 pin and ground, and the 10-kΩ resistor between P2 and V_{cc} as a pull-up resistor. This makes the line +5 V unless a button is pushed, in which case it goes to 0 V. On the slave side, connect the supplied LED and resistor in series between the BASIC STAMP's P2 pin and ground. Locate the WINDOWS STAMP 2 editor program titled STAMPW.EXE, and run it. Next, program the controller STAMP 2 with the supplied CONTROL.BS2 program (Listing 2-3). Finally, program the slave BASIC STAMP with the supplied program SLAVE.BS2 (Listing 2-4). Flick the switch on the controller. The slave LED will turn on or off depending on the controller's switch position, and the demo should start up immediately.

For more technical details on RS-485, see the "RS-422 and RS-485 Application Note" available from http://www.bb-elec.com. Also see http://www.lvr.com/ for more RS-485 information, and for a program listing that demonstrates a network protocol using BASIC see http://www.jdrichards.com.

In the process of developing STAMP 2 hardware and software projects, you will often need to test serial communication to and from the BASIC STAMP 2. If for example, you need to have the STAMP 2 send data to a terminal for display or data logging, a terminal program will make your life a lot easier. There are a number of ways to solve this problem. One solution is to use the Hyper Terminal program found in most Windows products under Accessories. Hyper Terminal is Windows-based and nice to look at, but can be a bit difficult to configure. Another solution is to use a full-featured terminal program such as Procomm, which is included on the CD-ROM. The provided Procomm is an older DOS version that is simple to understand and use, and also provides data logging capabilities for saving and manipulating data, i.e., through a spreadsheet. Another solution is to use QBASIC, which is also supplied on the CD-ROM. QBASIC is a DOS-based program. It is relatively easy to use, but QBASIC requires that your STAMP 2 send specific commands that QBASIC will understand, in order to open the serial port. So there are a few steps in getting the communication working. There are many terminal programs available that will work for communicating with the STAMP 2.

```
'control.bs2
1 (BS2 to RS-485 interface via a SN75176 chip)

' This program interfaces a BASIC Stamp 2 to an RS-485 network
' using the SN75176 Differential Bus Transceiver chip from
' TI. This program is meant to operate with another STAMP 2
' connected to the same RS-485 network and is running the
' slave.bs2 program.
' Pins 2 and 3 of the SN75176 chip are connected to pin 0 of
' the Stamp 2. Pins 1 and 4 of the SN75176 chip are
' connected to Pin 1 of the Stamp 2.
' This program expects an active-low button or switch to be
' connected to pin 2 of the Stamp2. When the button or switch
' sets pin 2 low, the character "L" is sent over the RS-485
' network. When the button or switch sets pin 2 high then
' the character "H" is sent over the network.
' Note. Setting pin 0 on the Stamp 2 high puts the SN75176 into
' transmit mode. So any serial data transmitted out of pin 1 on
' the Stamp 2 will be transmitted over the RS-485 network.

btnWk var byte 'Workspace for the BUTTON command
input 2 'Make pin 2 the button input
output 0 'Make pin 0 an output
high 0 'Put the SN75176 into transmit mode
if(in2<>1)then loop1: 'If pin 2 is initially High, send...
SEROUT 1,16468,1,["H"] '... an "H" over the RS-485 network
loop1:
BUTTON 2,0,255,0,btnWk,1,preloop2 'Wait till pin 2 is low
goto loop1
preloop2:
SEROUT 1,16468,1,["L"] 'Send a "L" over the RS-485 network.
loop2:
BUTTON 2,1,255,250,btnWk,1,loop_again 'Wait till pin 2 goes high
goto loop2
loop_again:
SEROUT 1,16468,1,["H"] 'Send an "H" over the RS-485 network.
goto loop1: 'Loop forever
```

Listing 2-3 **CONTROL.BS2 program.**

```
'slave.bs2
' (BS2 to RSS485 interface via a SN75176 chip)
' This program interfaces a BASIC Stamp 2 to an RS-485 network
' using the SN75176 Differential Bus Tranceiver chip from
' TI. This program is meant to operate with another Stamp 2
' connected to the same RS-485 network and is running the
' control.bs2 program.
' Pins 2 and 3 of the SN75176 chip are connected to pin 0 of
' the Stamp 2. Pins 1 and 4 of the SN75176 chip are
' connected to Pin 1 of the Stamp 2.
' This program expects an LED and resistor in series to be
connected to
' pin 2 of the Stamp 2. When an "H" comes across the RS-485
network
' pin 2 is set high, turning on the LED. When an "L" is received
' pin 2 of the Stamp 2 is turned off.
' Note: Setting pin 0 on the Stamp 2 low puts the SN75176 into
' receive mode. So any serial data received on pin 1 of the
' Stamp 2 will be read in with the SERIN command.

string var byte 'Used to hold the "H" or "L"
output 2 'Make pin 2 the LED connected pin.
output 0 'Make pin 0 an output pin.
low 0 'Put the SN75176 into receive mode.
loop1:
SERIN 1,16468,60000,loop1,[string]
'Read a byte of data coming in.
if(string<>"H")then is_low 'If H, then ...
high 2 '... set pin 2 high, turning on LED
goto loop1
is_low: 'If not an H, then turn off the LED
low 2
goto loop1 'Loop forever
```

Listing 2-3 **CONTROL.BS2 program (*Continued*).**

3

SERIAL BUILDING BLOCKS

CONTENTS AT A GLANCE

In this chapter, we will present a number of new and interesting special-purpose microprocessor integrated circuits used for timekeeping, video display, infrared (IR) decoding, temperature measurement, memory, remote control and data acquisition, etc. This chapter is not about a specific STAMP 2 project but rather is an introduction and/or idea generator for utilizing many new or special-purpose serial integrated circuits for creating enhanced STAMP 2 projects, including remote control and data acquisition. The special IC chips covered in this chapter will permit you to expand your horizons for creating future STAMP 2 projects. All of the integrated circuits in this chapter are dedicated microprocessor chips designed for a specific purpose. All of the IC chips covered in this chapter utilize either 1- or 2-wire serial control and are easy to interface to your STAMP 2 controller. Almost all of these support chips are readily available as support and add-on products from Parallax as well as third-party suppliers. These IC chips will add a unique and powerful dimension to your STAMP 2 projects. Most of these special-purpose chips are covered in detail on the CD-ROM in App. 1 as well as under application notes.

Dallas DS1302 Real-Time Clock Chip

The DS1302 is a real-time clock/calendar with 31 bytes of static RAM, all on an 8-pin DIP integrated circuit (see Fig. 3-1). The real-time clock counts seconds, minutes, hours, date of the month, month, day of the week, and year with leap year compensation. The DS1302 requires 5 V, and uses less than 300 nA at 2.5 V. The DS1302 communicates with a microcontroller such as STAMP through a three-wire serial connection. A temporary connection to a controller establishes the DS1302's time. Thereafter, the chip can operate as a standalone clock. The DS1302 has dual power supply pins for primary and backup, the latter of which may be powered by a supercapacitor input or rechargeable battery. The project relies on the chip's primary power supply input (Vcc2).

The DS1302 interfaces with controllers through a three-wire connection, consisting of a serial clock (SCLK) for data input, input/output line (I/O) for connection to the clock input,

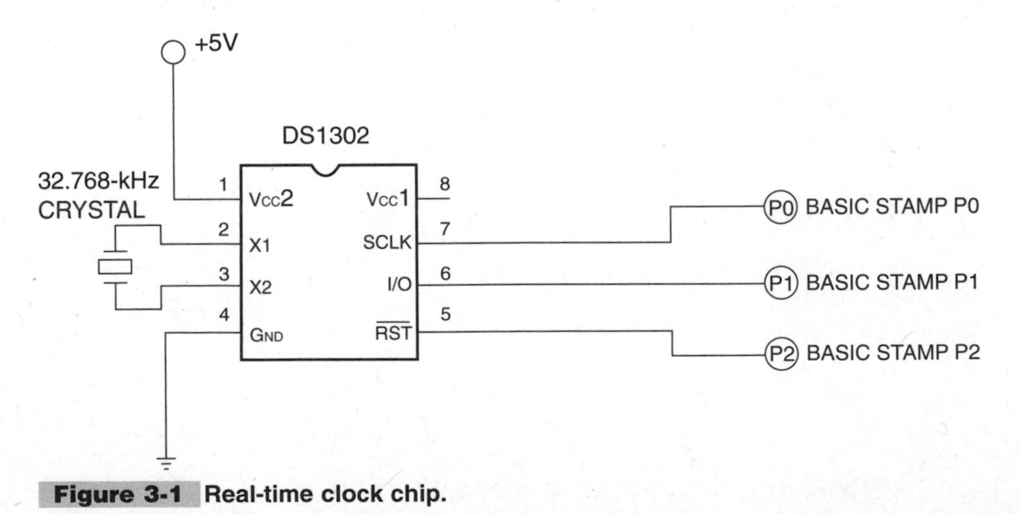

Figure 3-1 Real-time clock chip.

and reset (RST) for turning on control logic, which accesses the shift register and provides a method of terminating either single-byte or multiple-byte data transfer. The DS1302's X1 and X2 pins are connected to the leads of the 32.768-kHz crystal. In order have a STAMP communicate with the DS1302, you will need to first identify clock starting time by using different variable time registers. Next, you must reset the chip and send it an instruction telling it the starting time. Then, you need to read the time from the chip and debug it to the PC. Finally, you will need to deactivate RST after each step by taking it low.

Dallas DS1620 Digital Thermometer/Thermostat

The DS1620 is a complete digital thermometer on an 8-pin DIP chip, capable of replacing the normal combination of temperature sensor and analog-to-digital converter in most applications. The digital thermometer chip is depicted in Fig. 3-2. It can measure temperature in increments of 0.5 degrees Celsius (°C) from −55°C to +125°C. On the Fahrenheit (°F) scale, it measures increments of 0.9°F over a range of −67°F to +257°F. Temperature measurements are expressed as 9-bit, 2's complement numbers. The DS1620 communicates with a microcontroller such as the PIC or STAMP through a three-wire serial connection. The DS1620 can also operate as a stand-alone thermostat, if desired. A temporary connection to a controller establishes the mode of operation and high/low temperature set points. Thereafter, the chip independently controls three outputs: T(high), which goes active at temperatures above the high temperature set point; T(low), active at temperatures below the low set point; and T(com), which goes active at temperatures above the high set point and stays active until the temperature drops below the low set point.

The DS1620 interfaces with controllers through a 3-wire connection, consisting of a data input/output line (DQ), a synchronizing clock line (CLK), and a reset/select line (RST). The figure shows how to connect the DS1620 to the PIC or STAMP for the demo

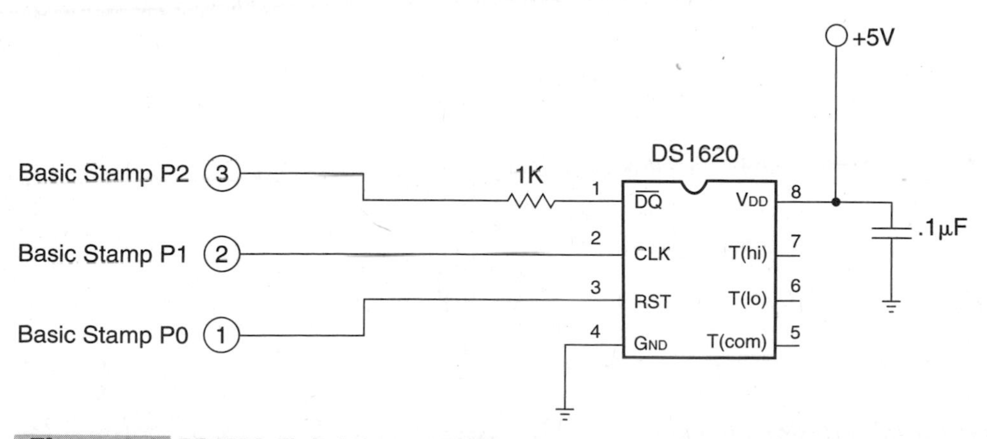

Figure 3-2 DS1620 digital thermometer.

programs. Do not omit the bypass capacitor—not even if you feel that your power supply is solid and well filtered. Locate that cap as close as practical to the supply leads of the DS1620. Although the 1-kΩ resistor is not strictly necessary as long as the firmware is functioning correctly, it's best to leave it in. In the event that both the controller (PIC or STAMP) and the DS1620 try to drive the data line at the same time, the resistor limits the amount of current that can flow between them to a safe value. In order to interface a STAMP to the DS1620, you will need to first activate RST by taking it high. Next you have to send an instruction (protocol) to the DS1620 telling it what you want to do. If you are reading data, shift it into the controller (PIC or STAMP). If you are writing data, shift it out to the DS1620. Finally, you need to deactivate RST by taking it low.

Linear Technologies LTC1298 12-Bit Analog-to-Digital Converter

The LTC1298 is a 12-bit analog-to-digital converter (ADC), illustrated in Fig. 3-3. It is internally referenced to the 5-V power supply, providing voltage measurements with 1.22-mV resolution. It has an internal sample-and-hold feature that prevents errors when it is used to measure rapidly changing signals. The 8-pin LTC1298 chip can be configured as a two-channel ADC or single-channel differential ADC. In two-channel mode, the selected channel's voltage is measured relative to ground and returned as a value between 0 and 4095. In differential mode, the voltage difference between the two inputs is measured and returned as a value between 0 and 4095. Supply current is typically 250 μA when the unit is operating and 1 nA (one-billionth of an ampere) when it is not. The maximum clock rate for the LTC1298's three-wire serial interface is 200 kHz, permitting up to 11,100 samples to be taken per second. The LTC1298 interfaces with controllers through 4 pins: chip select (CS), clock (CLK), data in (Din), and data out (Dout). Since PICs can readily switch between input and output on the fly, the interface shown in the figure combines Din and Dout into a single connection. As shown in the figure, the manufacturer recommends a 1- to 10-μF bypass capacitor with good high-frequency qualities—a ceramic unit at the low end of the scale, tantalum at the higher end. Since the supply voltage also serves as the ADC's voltage reference, it must be well regulated to avoid measurement errors. Since the LTC1298 draws very little current, you may use a precision voltage reference zener diode (such as the LM336) as a regulator. In order to animate the LTC1298, you will first need to activate CS by taking it low. Next you will have to send (shift out) configuration bits to the LTC1298. Now read (shift in) the 12-bit measurement from the LTC1298. Finally deactivate CS by taking it high.

Xicor X25640 8-kbyte EEPROM

The X25640 is an electrically erasable, programmable, read-only memory (EEPROM) device with 8192 bytes (8-kbytes) of storage, in an 8-pin DIP package. Like all EEPROMs,

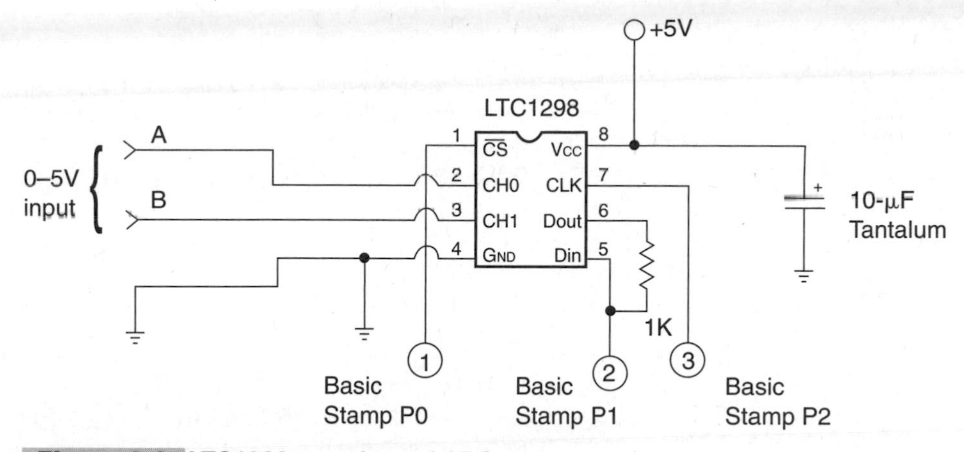

Figure 3-3 **LTC1298 two-channel ADC converter.**

it retains data with power off. It is intended for applications in which data is read often and written infrequently, since write operations gradually wear it out. Xicor says that the X25640 will survive a minimum of 100,000 writes. Data stored in the EEPROM should remain intact without power for 100 years or more. EEPROMs are typically used to store calibration tables, control settings, programs, maintenance logs, and small databases (such as lists of telephone numbers in an autodialer) that change fairly infrequently, but must be retained when the power is turned off. The X25640's relatively large storage capacity, simple interface, and low power consumption (100 μA inactive) make it well suited for data logging (see Fig. 3-4).

The X25640's interface is compatible with Motorola's serial peripheral interface (SPI) bus. The basic connections are chip select (CS), which activates the device; serial clock (SCK), which shifts data into or out of it; and data, which consists of the serial in (SI) and serial out (SO) lines tied together. There are two other control lines, hold (HOLD) and write-protect (WP), which are not used in these example applications. HOLD allows a busy, interrupt-equipped processor to pause a transaction with the X25640 while it tends to other business on the SPI bus. WP locks out any attempts to write to the device. Both of these lines are active low, so they are tied to V_{cc} to disable them. The figure shows how to connect the X25640 to the PIC or STAMP for the demo programs. Do not omit the bypass capacitor—not even if you feel that your power supply is solid and well filtered. Locate that cap as close as practical to the supply leads of the X25640. If you omit the bypass cap, you are almost certain to have intermittent communication with the X25640, especially during and after write operations. In order for a STAMP to communicate with the X25640, you will need to first activate CS by taking it low. Next you must send an instruction (opcode) to the X25640 telling it what you want to do. If you are reading or writing a memory location, send the address. If you are reading data, shift it into the controller (PIC or STAMP). If you are writing data, shift it out to the X25640. Finally, you must deactivate CS by making it high.

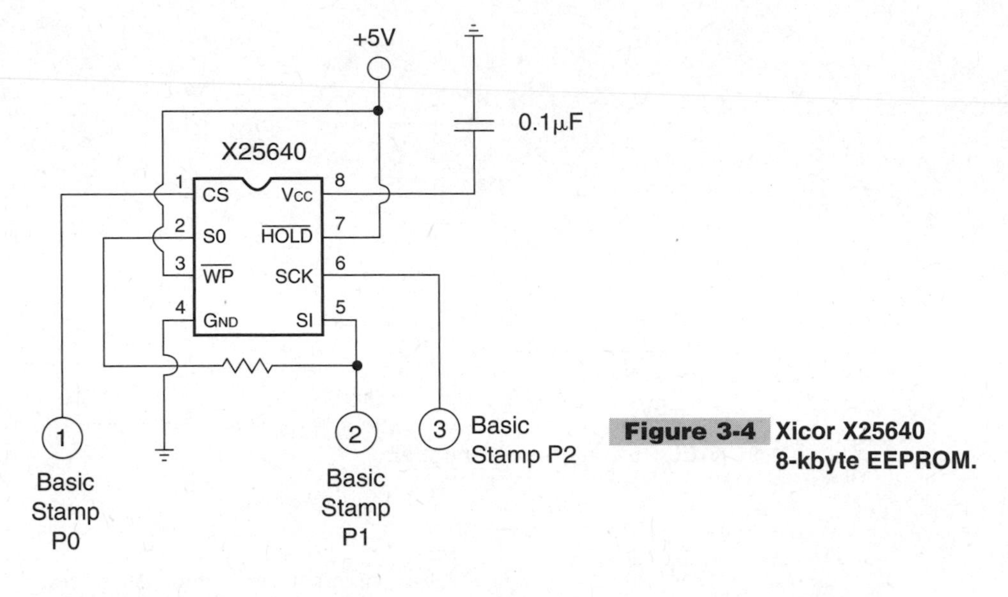

Figure 3-4 Xicor X25640 8-kbyte EEPROM.

K1EL IBM AT Keyboard to RS-232 Serial Converter

The K1EL PS/2 keyboard encoder chip is shown in Fig. 3-5. This specially programmed, 8-pin DIP package integrated circuit will allow you to connect up a full PS/2 keyboard to any of your STAMP 2 projects and is ideal for remote control. The keyboard encoder is an 8-pin microprocessor which requires only a few connections to produce a serial output which can be readily interface to a STAMP 2 microprocessor. Pin 1 is the V_{cc} input, while ground is applied to pin 8. The reset input at pin 2 is not used in our applications. The serial output is present from pin 3. The baud rate is set by pin 4; when this pin is set high, or at 5 V, then 9600 baud is selected. When the Baud pin is grounded, the output will be at 1200 baud. Pin 5 is the CLK pin, which connected directly to pin 1 of the PS/2 keyboard connector. The output sensing pin can be configured to either provide a high true marking output or a low true marking output. High true is the standard and in most cases would be used to connect directly to a microprocessor. Pin 6 sets the mode of operation. When pin 6 is high, then high true marking is selected, while grounding pin 6 allows the output to be low true marking. The DATA pin is fed directly from the PS/2 keyboard connector. Whenever the CTL, ALT, and DELETE keys are pressed at the same time, the Katbd will output a low true reset pulse on pin 2 of the encoder. The keyboard encoder is a great accessory chip for the STAMP 2 processor and will allow you construct many different types of remote control applications from robotics to model trains. All of the building block ICs are shown in their respective data sheets in App. 1.

SONY Infrared Decoder Chip

Our next building block IC is the 8-pin Sony IR decoder, which is a specially dedicated programmed microprocessor, shown in Fig. 3-6. The Sony IR decoder is from Cricket Robotics

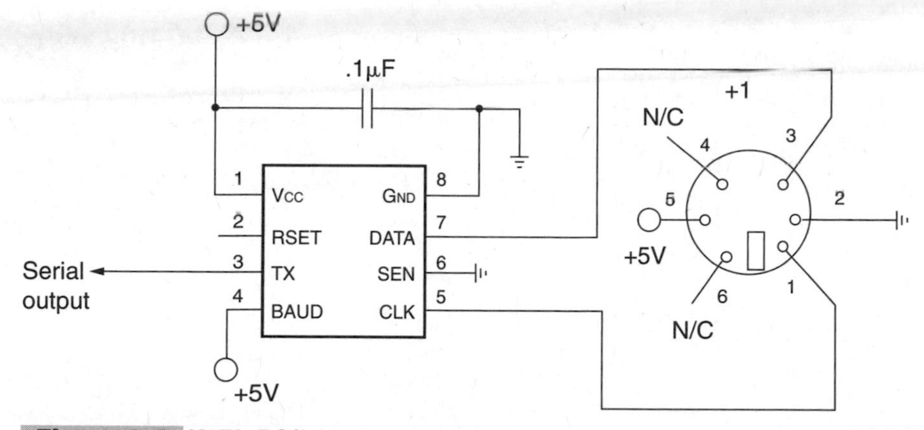

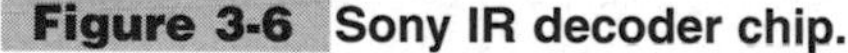
Figure 3-5 K1EL PS/2 keyboard encoder.

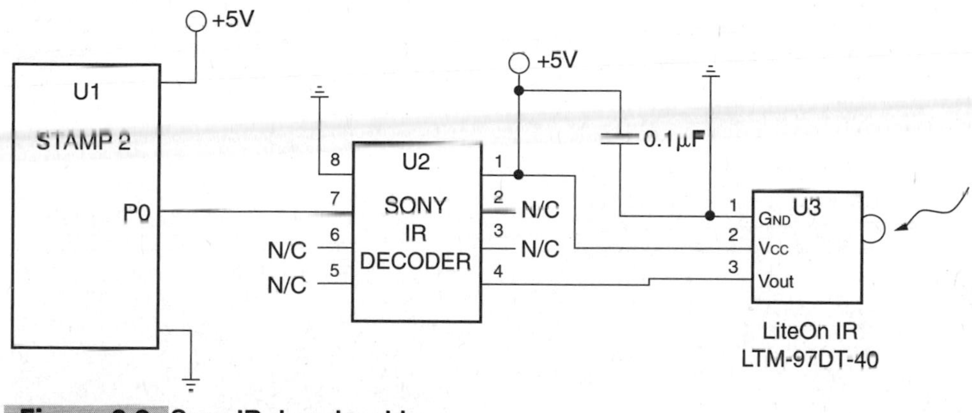

Figure 3-6 Sony IR decoder chip.

Inc. The Sony decoder IC takes a serial stream of bits from an IR receiver and converts the 12-bit frames of data to straight 8-bit ASCII. The data is converted in a continuous mode, which means the host processor must sync with the START bits of 2400-baud, 8-bit, no-parity frames being sent from a standard Sony IR remote. The Sony IR decoder chip is well suited to sending serial signals to a STAMP 2 processor. A Sony TV remote control outputs modulated infrared to communicate with a TV set. The IR receiver demodulates the IR signal and outputs a 12-bit serial word representing the key hit on the IR remote. The standard Sony IR signal outputs a 2-ms start bit. The Sony IR decoder IC watches for a valid START bit and then measures the high time of each bit of the 12-bit frame. The Sony IR decoder will decode each frame and then send the converted byte to the ASCII serial output pin 7. In operation, a LiteOn IR detector model LTM-97DT-40 is fed directly to the Sony IR decoder on pin 1. The Sony IR decoder is powered from a 5-V source at pin 8. The serial output of the Sony IR decoder is provided at pin 7, which can be readily interfaced to the BASIC STAMP 2. The key codes for the Sony IR decoder are shown in Table 3-1. A sample BASIC STAMP 2 application code is shown in the IR decoder demo program.

TABLE 3-1 SONY INFRARED REMOTE DECODER IC KEY CODES

REMOTE KEY	DECODER OUTPUT	REMOTE KEY	DECODER OUTPUT
0	137	Volume up	146
1	128	Volume down	147
2	129	Channel up	144
3	130	Channel down	145
4	131	Rewind	91
5	132	Play	92
6	133	Stop	88
7	134	Pause	89
8	135	Previous channel	187
9	136	TV-VCR	106
		Mute	148

SVID Single-Chip Video Controller

MVS Corporation's single-chip serial video controller (SVID) is a very interesting, low-cost way to display information and or graphics using a small microprocessor such as the STAMP or a PIC. The SVID video controllers feature low power consumption and no external memory chips, since the internal SRAM and EEPROM are used for display memory. High-speed serial and parallel interfaces are built in. Since the internal memory is nonvolatile and the SVID operates with or without a micro, it's perfect for stand-alone displays.

Applications include "smart home" systems, factory status panels, lab instrument readouts, vu-meters, portable sales terminals, etc. The NTSC version will drive a TV, VCR, or any other composite display. The new 4- to 6-in composite-type LCD panels are compact, low power, and inexpensive. CRT and LCD monitors (15, 17, 21 in, etc.) can be driven by the VGA version.

The diagram in Fig. 3-7 illustrates the VGA/CRT version of the SVID controller, while Fig. 3-8 depicts the NTSC video controller pinouts (http://star.net/people/nmvs)

In the character generator (CG) mode, 48 digits (8 × 6) are displayed from RAM with sixteen 5 × 7-character font tables held in EE. Graphics mode has 32 × 22 pixels total with 32 × 6 bit-mapped from RAM and the rest (32 × 16) from EE. Both RAM and EEPROM (electrically erasable read-only memory) memory can be updated by the user via the serial/parallel interface. The SVID is intended for low-resolution applications and educational use in the area of embedded controllers. For high-resolution applications, there's the MVSVGA2, a 640 × 480 VGA controller that drives color LCD and CRT VGA monitors.

Minimum components needed for all the video controllers are a 16-MHz crystal wired between pins 4 and 5 with two 47-pF caps to ground. A 5-Vdc supply or three alkaline cells

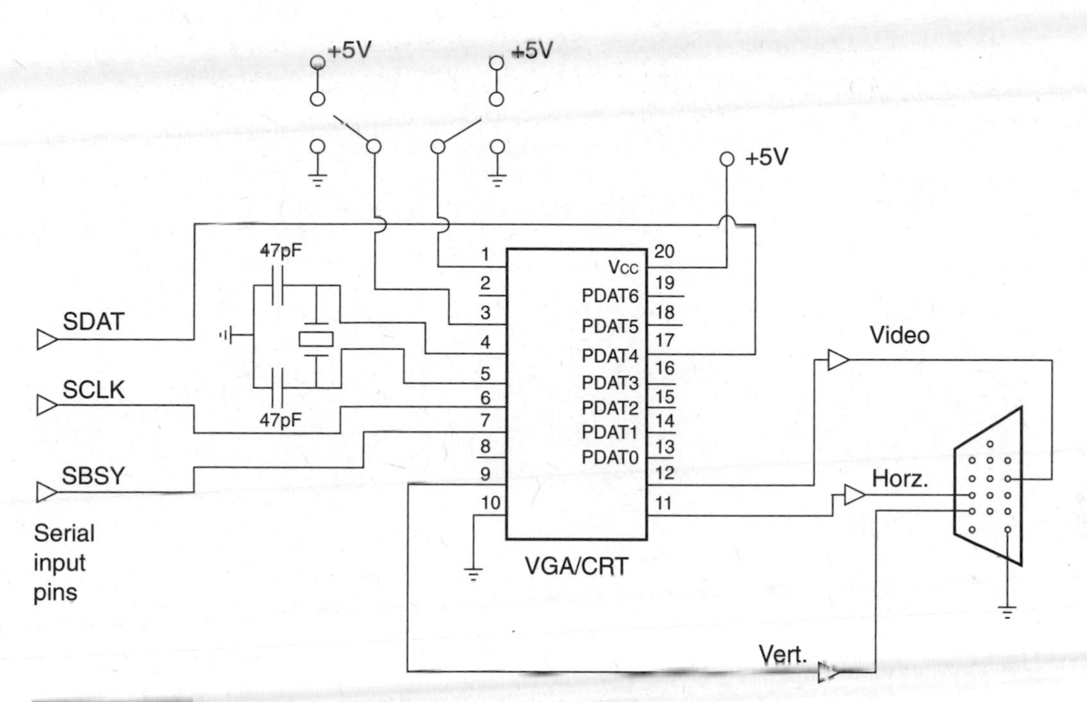

Figure 3-7 Single-chip serial video controller (SVID)—VGA/CRT version.

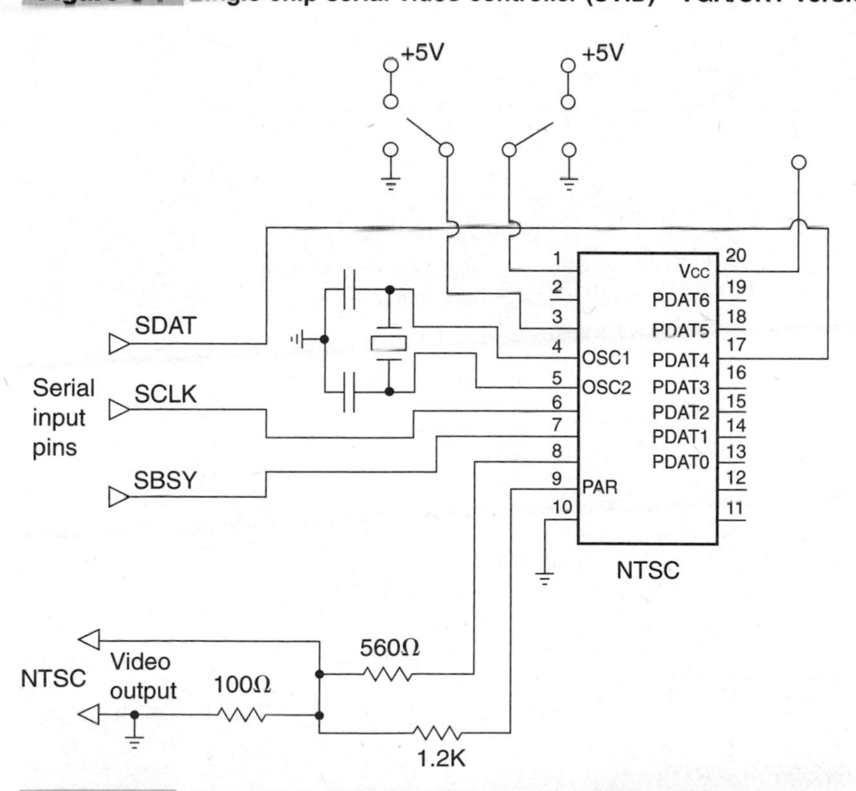

Figure 3-8 Single-chip serial video controller (SVID)—NTSC version.

are typical sources of power. When powered from batteries, the chip draws virtually no current when off, yet internal RAM retains data. The chip can be powered down by grounding the OFFL pin and powered up again by momentarily grounding ONL. (*Note:* if you are using pin 1, put a 0.01-μF capacitor to GND.) CRT and LCD monitors are driven directly from the VGA version as follows: generating composite from the NTSC version requires three resistors—100 Ω with one end on ground and the other end to 1.2 kΩ and 560 Ω. The 560-Ω resistor goes to pin 8; the 1.2-kΩ resistor goes to pin 9. The signal is across 100 Ω. There are four versions of the serial video controller:

SVID-cn	Character generator, NTSC out (TV, VCR, composite LCD)
SVID-cv	Character generator, VGA out (15-kHz CRT or LCD monitor)
SVID-gn	Graphic mode, NTSC out
SVID-gv	Graphic mode, VGA out

SN7516 Differential Bus Transceiver

The Texas Instruments SN7516 differential bus transceiver can be readily used to form a bidirectional RS-485 network when combined with the BASIC STAMP 2. The RS-485 serial protocol is commonly used to provide strong serial signals able to withstand long cable distances (up to 4000 ft) at high baud rates in potentially noisy electrical environments. Two wires carrying an RS-485 signal (the A and B lines) provide a signal base from which many devices can communicate. Twisted-pair wire is recommended for long distances, but for short distances 24-gauge wire works quite well. Up to 32 devices can be connected to an RS-485 data line with the SN75176 and communicate using a data protocol.

This is referred to as an RS-485 network, or an RS-485 drop network. The SN7516 chip converts RS-485 signals to RS-232 TTL-level signals, allowing devices that traditionally communicate over standard RS-232 serial connections to communicate over a single 2-wire RS-485 network. The SN7516 bus transceiver is an 8-pin in-line DIP chip that runs on +5 Vdc. A typical connection is illustrated in Fig. 3-9.

HA7S ASCII to 1-Wire Host Adapter

The ASCII to 1-Wire host adapter from Point Six, Inc. is a very powerful special-purpose chip. The HA7S is a TTL to 1-Wire interface in a very small 6-pin SIP designed to provide an ASCII command set for embedded controller and battery-operated applications that need to accommodate Dallas Semiconductor iButton and 1-Wire devices. The HA7S provides support for all 1-Wire devices using a serial protocol. Block mode commands support all 1-Wire device functions. The HA7S relieves the host of the burden of generating the time-critical 1-Wire communication waveforms while supporting all 1-Wire devices with simple ASCII commands that can be easily generated. The HA7S does all the hard work of interfacing 1-Wire networks.

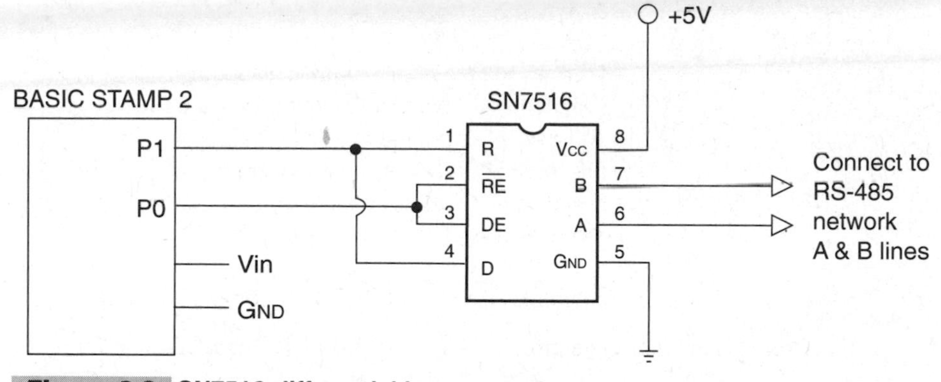

Figure 3-9 **SN7516 differential bus transceiver.**

Small size and very low power consumption as well automatic power-up and power-down features allow the HA7S to operate in low-power and battery-operated applications with no need for power control signals from the host device. The only interface signal required is the TTL-level TX and RX from a UART or microprocessor. The HA7S powers up into a very low power state. When a serial ASCII command is sent, the HA7S will wake up, process the command, send the response, and power itself down. The 1-Wire bus can be left in a powered-up or a powered-down state while the HA7S is powered down. While in power-down mode, the HA7S has a very low quiescent current requirement of about 5 μA.

The HA7S can perform search and family search functions, making it easy to acquire the unique 64-bit serial numbers of all connected devices. Many sensor devices require that extra power be delivered during periods of data conversion (DS1920 and DS1820 temperature sensors, for example). The HA7S automatically provides the extra current these devices require with a built-in smart strong pull-up. Dallas Semiconductor iButtons, which store data in TMEX touch memory file format, can be read or written with simple ASCII commands. The HA7S will automatically generate and check the CRC16 error checks from touch memory file records. The HA7S supports analog, digital, and temperature 1-Wire devices and all Dallas Semiconductor iButtons. The HA7S communicates at 9600 baud, 8 bits, no parity, and 1 stop bit.

Eight-Channel Infrared Remote Control System

The diagrams in Figs. 3-10 and 3-11 illustrate an eight-channel infrared remote transmitter and receiver system for remote control. This powerful remote control system can be used to control robots or devices around your home and office and can be readily interfaced with the STAMP 2 processor.

The eight-channel IR transmitter is centered around a specialty PIC at U1; this processor is designed to act as a carrier wave generator that generates a 38- or 40-kHz constant carrier wave signal, which activates driver transistor Q1, shown in Fig. 3-10. Transistor Q1

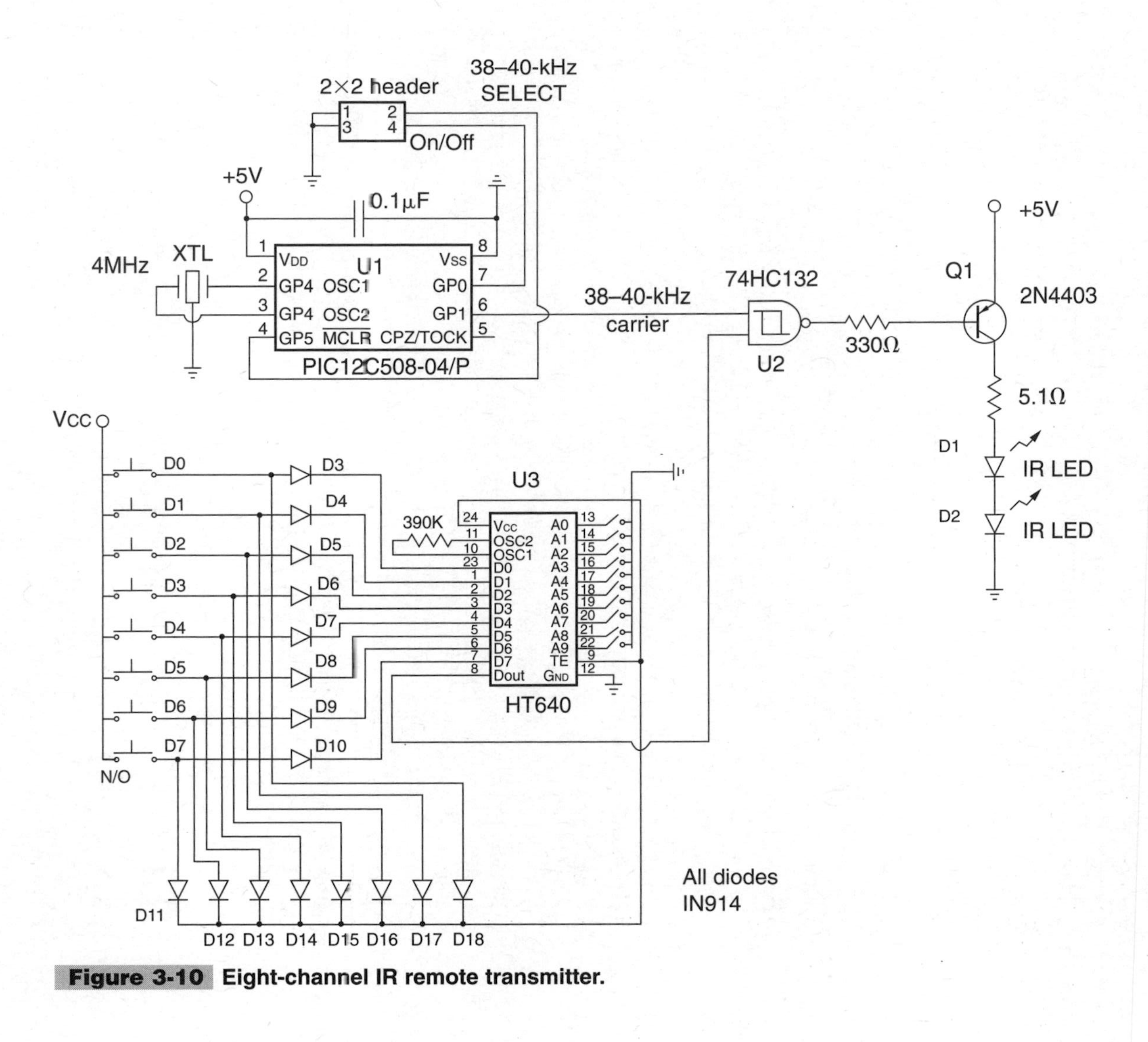

Figure 3-10 Eight-channel IR remote transmitter.

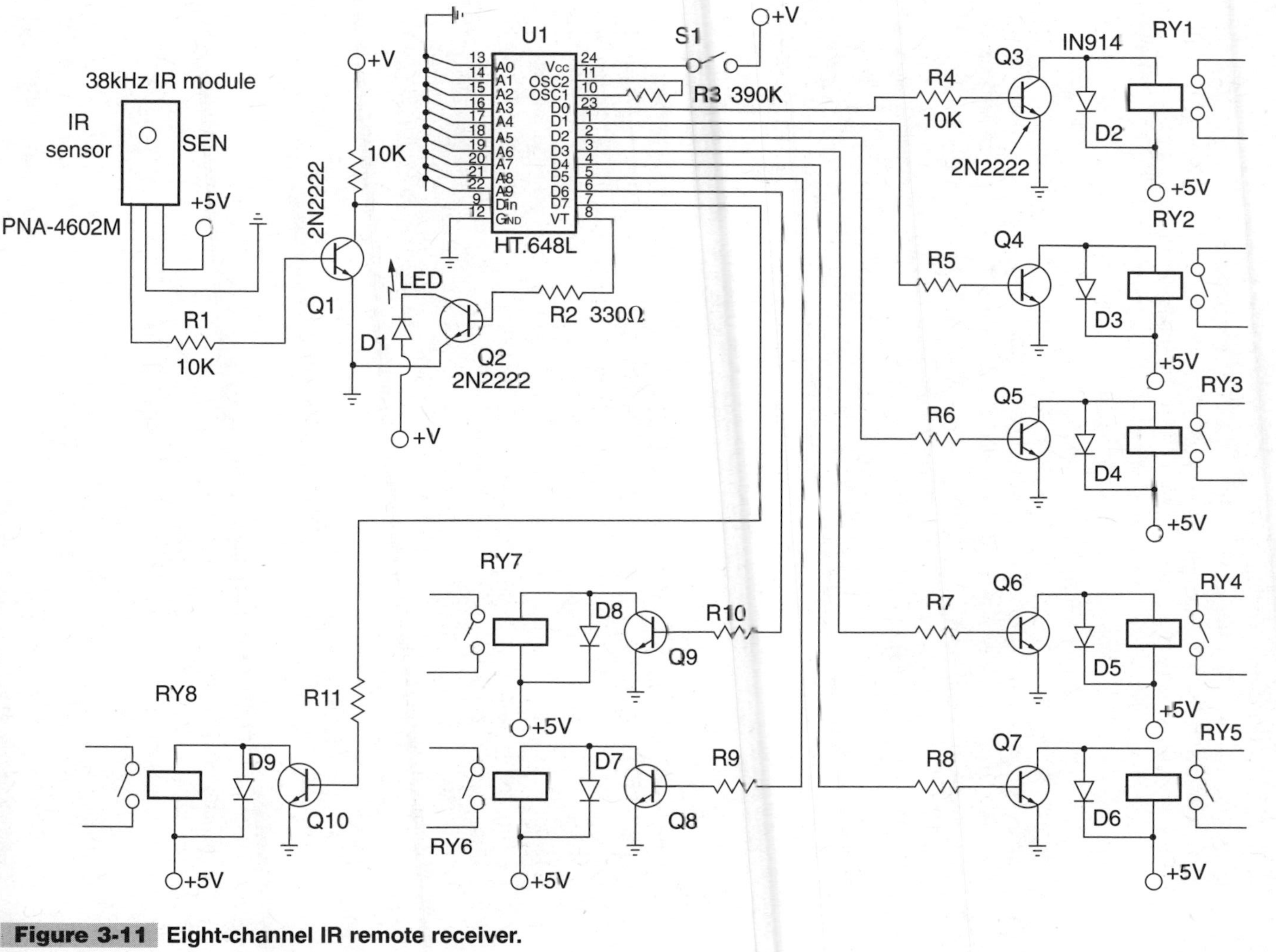

Figure 3-11 Eight-channel IR remote receiver.

drives both IR LED 1 and IR LED 2, which act as IR transmitters. When no jumpers are programmed at the header—i.e., the default condition exists as shown—then a 38-kHz carrier is generated. The HT-640 encoder sends a 4-word transmission cycle upon receipt of a transmission enable signal, i.e., a low signal at pin 9. The Holtek HT-640 digital encoder is an 8-channel input device with an 8-bit address programming. In operation, both transmitter and receiver addresses must be set exactly the same for the system to work. The output from the Holtek HT-640 encoder, in effect, modulates the 38-kHz carrier wave when it is combined with the signal from the carrier wave generator at U3. The IR remote control transmitter circuit is powered from a 5-Vdc regulator, such as an LM7805. Once the transmitter is built, an address is programmed into the encoder by grounding the address pins. In operation, once the receiver is set with the same programming configuration you are ready to remotely control devices. Pressing D0 on the transmitter encoder activates the D0 output on the corresponding receiver/decoder. Pressing D1 on the transmitter's encoder will thus produce an output at receiver/decoder's output D1. This remote control system is ideal for robotics and model railroad applications.

The corresponding 8-channel IR receiver/decoder is shown in Fig. 3-11. A 38-kHz PNA-4602M IR module from Rentron was used for the IR detector. The 3-pin device produces a demodulated serial data signal at its output. The IR detector is powered by 5 V, as is the HT-648L decoder. The output from the IR detector is coupled via a 10-kΩ resistor to Q1. Transistor Q1 drives the Holtek HT-648L decoder IC, on pin 14. A system status indicator LED is driven by transistor Q2 at pin 17. The decoder outputs on pins 10 through 13 are used to drive the four 5-V relays. Each decoder output is driven by a 2N2222 transistor, which in turn activates a relay when the correct output is selected at the remote encoder IC. Silicon diodes are used for protection at each of the relays to protect the transistors. In operation, as noted, the HT-1648L decoder addresses must be programmed by jumpering the address pins to ground. The address coding must be the same for the transmitter and corresponding decoder setup. This address coding scheme (Listing 3-1) allows many different encoders and decoders to operate in the same space and not disturb or interfere with each other. Complex system schemes can be developed with no interference between systems. Also note that the Holtek encoder/decoder series works well with small RF transmitters and receivers for remote control applications. The Holtek encoders/decoders make constructing remote control systems quite simple!

Addressable Serial-to-Parallel Converter

National Control Devices has a number of very interesting specially programmed microprocessor chips for simple serial control applications. The first device in our discussion is the addressable serial-to-parallel converter. The NCD-101 makes for a powerful control circuit with many potential applications. The NCD-101, shown in Fig. 3-12, was designed to perform high-speed serial-to-parallel conversions. The NCD-101 can be commanded from any serial source—either from a terminal program, from an RF system, from a STAMP 2, or from IR control. By simply sending three commands, you can readily control eight devices, and, by cascading seven of the NCD-101s, you can control up to 56 devices. The NCD-101 can be made to operate at three different baud rates, from 1200 to

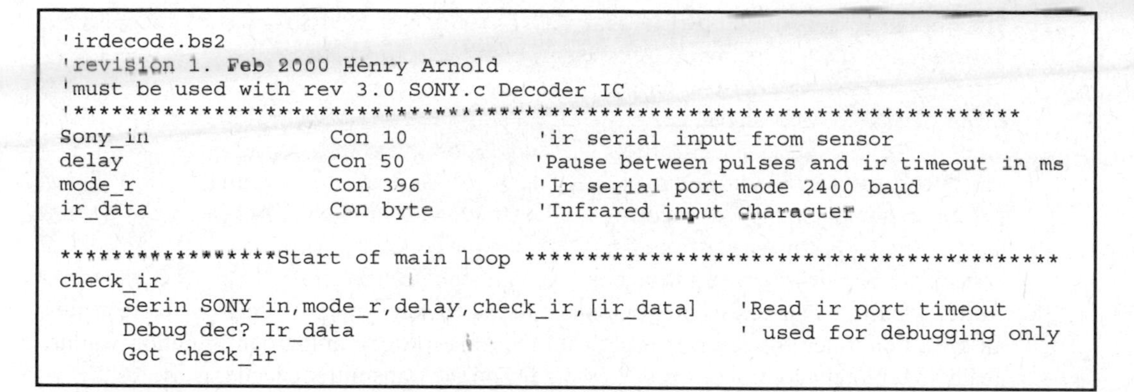

```
'irdecode.bs2
'revision 1. Feb 2000 Henry Arnold
'must be used with rev 3.0 SONY.c Decoder IC
'*********************************************************************************
Sony_in          Con 10          'ir serial input from sensor
delay            Con 50          'Pause between pulses and ir timeout in ms
mode_r           Con 396         'Ir serial port mode 2400 baud
ir_data          Con byte        'Infrared input character

****************Start of main loop ******************************************
check_ir
     Serin SONY_in,mode_r,delay,check_ir,[ir_data]     'Read ir port timeout
     Debug dec? Ir_data                                 ' used for debugging only
     Got check_ir
```

Listing 3-1 IR decoder demo program.

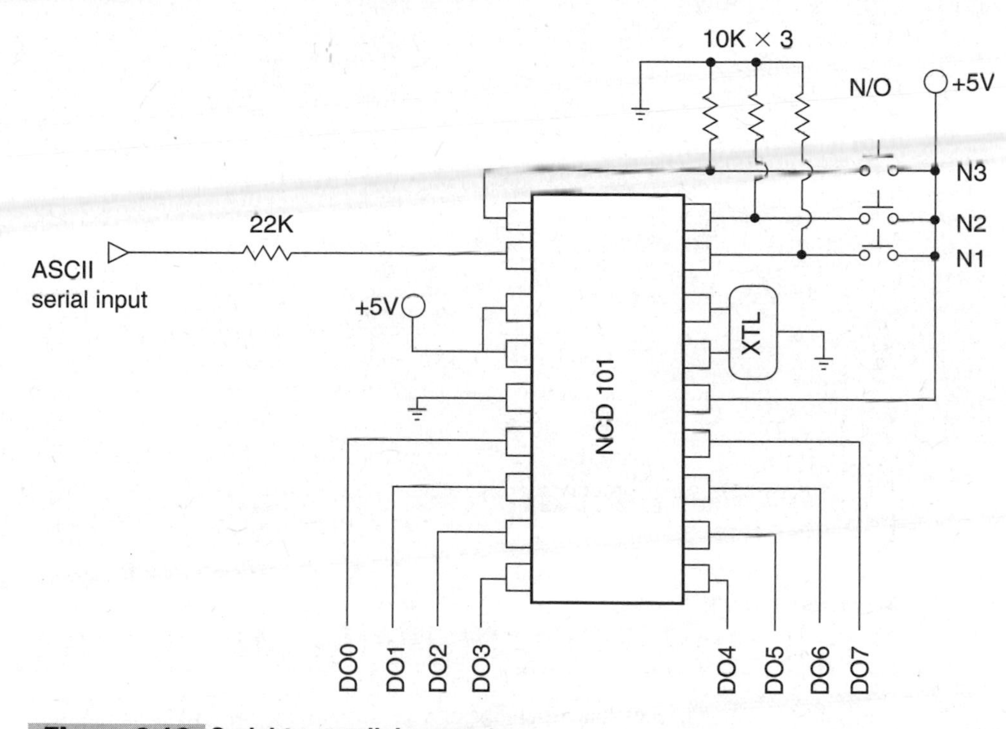

Figure 3-12 Serial-to-parallel converter.

9600 baud, simply by substituting different ceramic resonators. The NCD-101 is an addressable device that is programmed by the three switches, N1, N2, and N3, at pins 1, 17, and 18. Each converter chip has its own specific address, and when linked together forms its own network; see the address programming chart in Table 3-2.

A serial signal is fed to the converter on pin 2, and with simple commands you can control each of the eight outputs on each of the converter chips independently. The NCD-101

TABLE 3-2 NCD-101 ADDRESSABLE SERIAL-TO-PARALLEL CONVERTER

NAME/ADDRESS PROGRAMMING			
NAME	N3	N2	N1
0	0	0	0
1	0	0	1
2	0	1	0
3	0	1	1
4	1	0	0
5	1	0	1
6	1	1	0

requires only five external components for the system. Three 10-kΩ resistors are used for the address pins, and three jumpers or switches can be used for programming the converter chip. A crystal resonator is used to establish the speed or baud of the device, and a series resistor is used to accept the serial signals on pin 2. The serial-to-parallel converter is powered from a 5-Vdc source at pins 3 and 4.

In order to chain a number of converter chips together, you simply have to link each of the chips together via the serial input pins. When combining more NCD-101s together to form a larger control system, you must note that the RS-232 ground is shared among the converter chips. In order to "talk" to the NCD-101, you will need to send it simple serial commands, which are shown in Table 3-3.

Addressable Parallel-to-Serial Encoder

The NCD addressable parallel-to-serial encoder is a very powerful special-purpose serial control device. The NCD-102 can monitor 8-bit parallel input from a single 8-bit word or it can monitor eight different single binary input signals, and when queried with a serial word will output a status word indicating the state of the eight inputs, as shown in Fig. 3-13. The NCD-102 requires only three resistors and a ceramic resonator at pins 15 and 16. The NCD-102 can run at four different baud rates, from 1200 to 9600. Up to four NCD-102s can be attached to a single RS-232 serial port connection, providing a total of 32 input bits or channels. The name/address programming pins at 17 and 18 are used to set up each of the encoders in a larger system. Table 3-4 indicates the programming configurations. The NCD-102 is a polled device, operating in a "speak when spoken to" mode. The name jumpers determine the input's identity on the serial port. When a valid name and carriage return are received by the NCD-102, the device reads the current status of the eight inputs and replies with a byte of data. When four NCD-102s are connected together, the ASCII outputs must be ORed together. The NCD-102 is queried via the serial input connection at pin 2 through a 22-kΩ resistor, which

TABLE 3-3 NCD 101 ADDRESSABLE SERIAL-TO-PARALLEL CONVERTER

SERIAL CONTROL COMMANDS	
Turn a pin on:	<NAME>N<PIN NUMBER>RETURN>
Format:	0N3<RETURN>
Example:	Tell NCD-101 named 0 to turn on pin 3
Action:	
Turn a pin off:	<NAME>F<PIN NUMBER>RETURN>
Format:	6F3<RETURN>
Example:	Tell NCD-101 named 6 to turn off pin 3
Action:	
Output a byte:	<NAME>F<PIN NUMBER>RETURN>
Format:	3P1<RETURN>
Example:	Tell NCD-101 named 3 to send byte 1
Action:	
Example:	3P0<RETURN
Action:	Tell NCD-101 named 3 to send byte 0

protects the NCD-102. The ASCII output reporting, once queried, is sent out the ASCII output port on pin 1.

Addressable A/D and Digital Input/Output Control Device

The NCD-106 serial addressable A/D converter with digital I/O is a very interesting and useful device (see Fig. 3-14). The NCD-106 features dual unipolar 8-bit analog-to-digital converters with three additional digital inputs and four digital outputs. This converter/processor can be used for a number of different control, experimental, or benchtop applications in your home or shop. The NCD-106 is queried serially from pin 6 via a 22-kΩ protection resistor. The output status of the device is sent out from pin 1. The NCD-106 requires a ceramic resonator to establish the four baud rates. The NCD-106 only

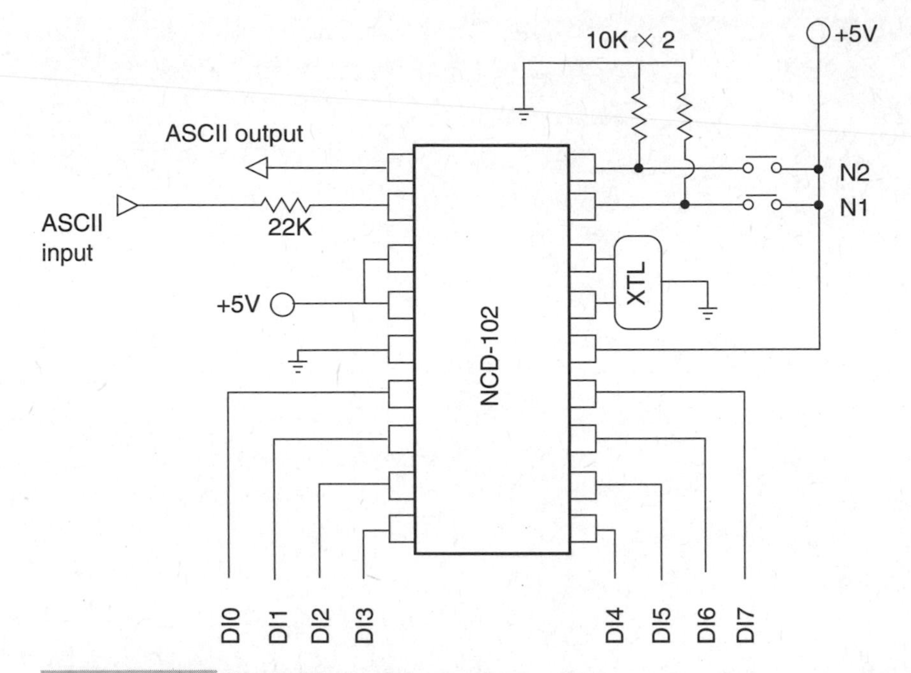

Figure 3-13 Parallel-to-serial encoder.

TABLE 3-4 NCD-102 ADDRESSABLE PARALLEL-TO-SERIAL ENCODER

NAME/ADDRESS PROGRAMMING

NAME	N2	N1
A	0	0
B	0	1
C	1	0
D	1	1

requires five resistors. One resistor is used to protect the serial input pin, while two resistors are used to protect the A/D pins and two resistors are used to address the device. Up to four NCD-106s can be connected together to form a system, which provides up to eight analog-to-digital inputs for measuring signals and voltages as well as 12 input channels and 16 output channels. This NCD-106 device could be used as a burglar alarm monitoring 12 binary sensor switches or the device could be used as a remote control device which can be used to turn on 16 individual devices. The NCD-106 controller could also be used in a feedback/control configuration. The system could monitor voltages via the A/D channels and/or 12 digital inputs and then control up to 16 outputs based on signals received from the input channels. Table 3-5 shows how to set the addressing pins on the NCD-106. The NCD-106 is commanded serially with simple commands, which can be seen in Table 3-6.

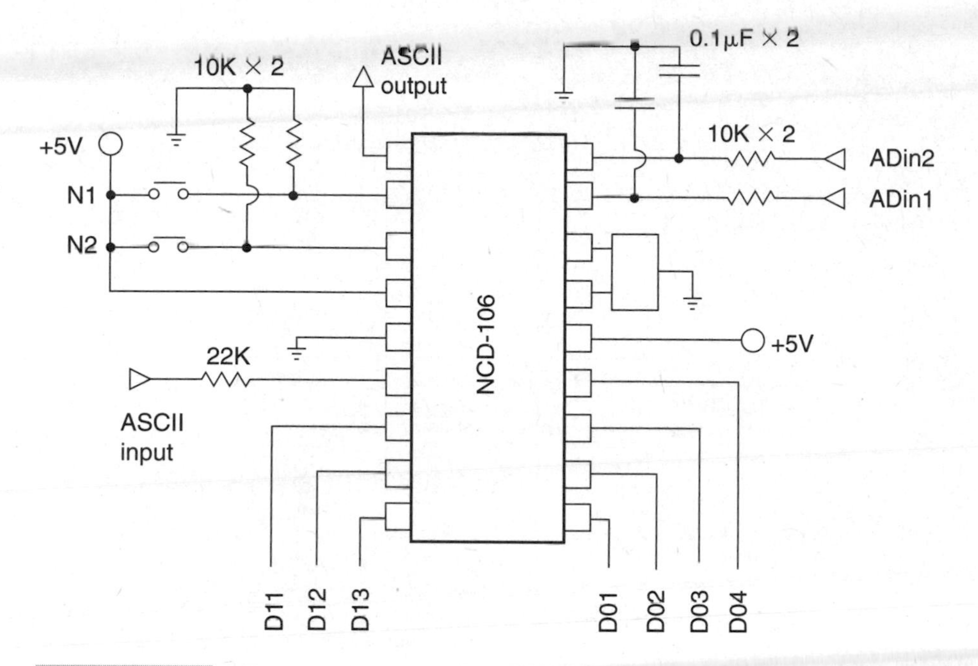

Figure 3-14 A/D and digital I/O converter.

Fifty-Six-Channel Radio Remote Control System

You can easily create a 56-channel radio remote control system by implementing the NCD-101 serial-to-parallel converter. The unique 56-channel radio remote control is ideal for model train or robot enthusiasts. The radio remote control system begins with a PS/2 keyboard encoder at U2, which serially drives an Abacom AM-RT5 data transmitter, as shown in Fig. 3-15. The keyboard controller and RF transmitter are both powered from a 5-V regulator at U1.

The receiver and controls section of the 56-channel remote control system are shown in Fig. 3-16. An Abacom HRR-3 superregenerative receiver is coupled to two sections of a MAX232 serial interface chip. The output of the MAX232 is next fed to seven NCD-101 serial-to-parallel converters to form a 56-channel radio remote control system. Each of the NCD-101 outputs would then be fed to a transistor relay driver for control applications. The 56-channel radio remote control receiver can be powered from a 5-Vdc power supply.

In order to activate a particular channel of NCD-101 converter, you just need to send <NAME>N<>PIN NUMBER><RETURN>. For example, if have a single NCD-101, and you wanted to activate output D03, you would press 0N3 on your keyboard. This sequence would tell NCD 101 named 0 to turn on pin 3. Recall that each NCD 101 has an address. An NCD-101 named 0 would be programmed for each address as 0,0,0 on the programming pins. Now, if you wanted to turn off pin 3 of NCD named 0, you would send 0F3. For more details on the NCD-101, check App. 1.

TABLE 3-5 NCD-106 ADDRESSABLE A/D DIGITAL INPUT/OUTPUT CONTROL DEVICE

NAME/ADDRESS PROGRAMMING

NAME	N2	N1
p	0	0
q	0	1
r	1	0
s	1	1

Thirty-Two-Channel Serial Input Scanner Alarm System

The 32-channel serial input scanner/alarm system is a unique system that can be used to monitor up to 32 channels—i.e., inputs or on/off conditions—over a single serial wire or wireless link. You could monitor 32 alarm conditions via a low-power wireless link and report this information back to a STAMP 2 set up as an alarm controller. The STAMP could then be used to display each channel via an LCD panel and sound an alarm if desired.

The heart of the 32-channel monitor system is an NCD-102 addressable parallel-to-serial converter chip, shown in Fig. 3-17. The NCD is a preprogrammed microcontroller designed to perform high-speed parallel-to-serial encoding from a standard serial source. The NCD-102 can be chained to four other NCD-102s to form a 32-channel monitoring system. The NCD-102 is a polled device operating in a speak when spoken to mode. Name jumpers are used to determine the input's identity in the system or network. When a valid name and carriage return are received by the NCD-102, the device reads the current status of each of the eight pins and replies with a byte of data.

The unique NCD-102 only uses four external parts per chip. Two 10-kΩ resistors are used for programming. Jumpers are set up to four possible combinations. A ceramic resonator is used to set the speed or baud rate of the NCD-102. Four speeds can be set, from 1200 to 9600 baud. The fourth component is a 22-kΩ series resistor at pin 2. The converter is powered from a 5-Vdc source at pin 3 and 4. System ground is found on pin 5.

Four converter chips can be connected together to form a 32-channel system. Note that the outputs from the converters must be ORed together as shown, and each NCD-102 should have its own name in the system via the programming pins. The RS-232 ground system must be shared with the ground of the circuit. This system is a three-wire system; i.e., it has ASCII input for serial inquiry and ASCII output for information of received data and ground.

In order to turn on a pin, you would use the format <NAME><RETURN>. An example would be A<RETURN>; this would tell the NCD converter named A to read pins and reply with a byte of data (see Table 3-4 and App. 1). By combining four NCD converters together, you can develop a powerful monitoring system for watching various on/off sensors such as doors, windows, and comparator outputs. These converter devices are a great building block in creating a complex control system.

TABLE 3-6 NCD-106 ADDRESSABLE A/D DIGITAL INPUT/OUTPUT CONTROL DEVICE

SERIAL CONTROL COMMANDS	
Send analog input commands:	
Format:	
Example:	<NAME>A<CHANNEL<>RETURN>
Action:	Tell NCD-106 named P to read A analog input 1—reply with byte of data
Send digital input status:	
Format:	<NAME><RETURN
Example:	p<RETURN>
Action:	Tell NCD-106 named p to read digital inputs
Turn on a digital output:	
Format:	<NAME>N<RETURN>
Example:	pN1<RETURN>
Action:	Tell NCD-106 named p to turn on output 1
Turn off a digital output:	
Format:	<NAME>F<RETURN>
Example	pF1<RETURN>
Action:	Tell NCD-106 named p to turn off output 1
Output a byte to digital outputs:	
Format:	<NAME>P<BYTE><RETURN>
Example:	pP0<RETURN>
Action:	Tell NCO-106 named p to send a byte ASCII

Multichannel Remote Serial I/O System

The multichannel remote serial I/O system is a an extremely powerful system building block in forming a complex monitoring system. This multichannel control and reporting system centers around an NCD-106 preprogrammed microprocessor. The NCD-106 combines two-channel, 8-bit A/D converters, three input and four output channels all in one chip. The NCD-106 can be interrogated via a standard serial port and will monitor analog

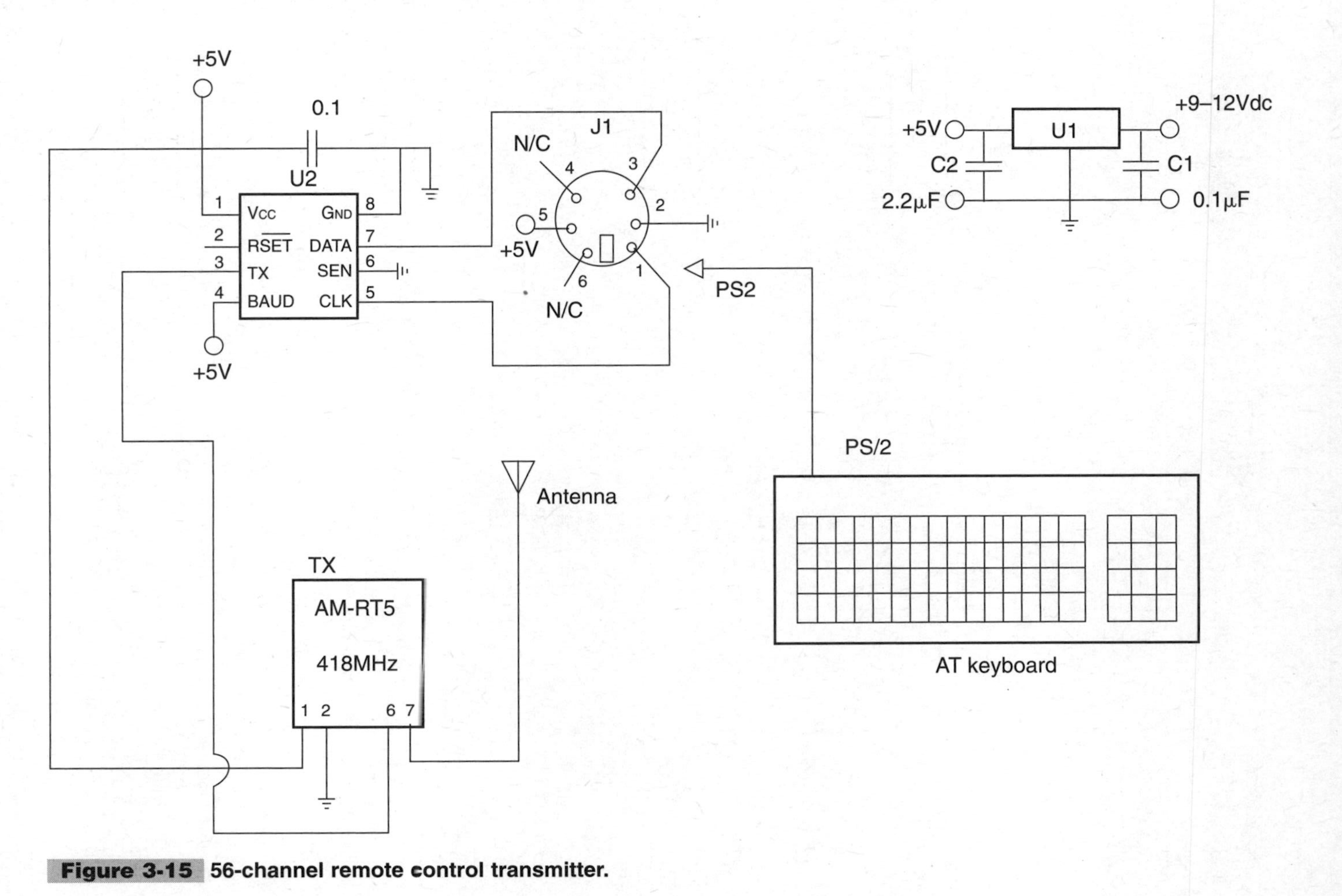

Figure 3-15 56-channel remote control transmitter.

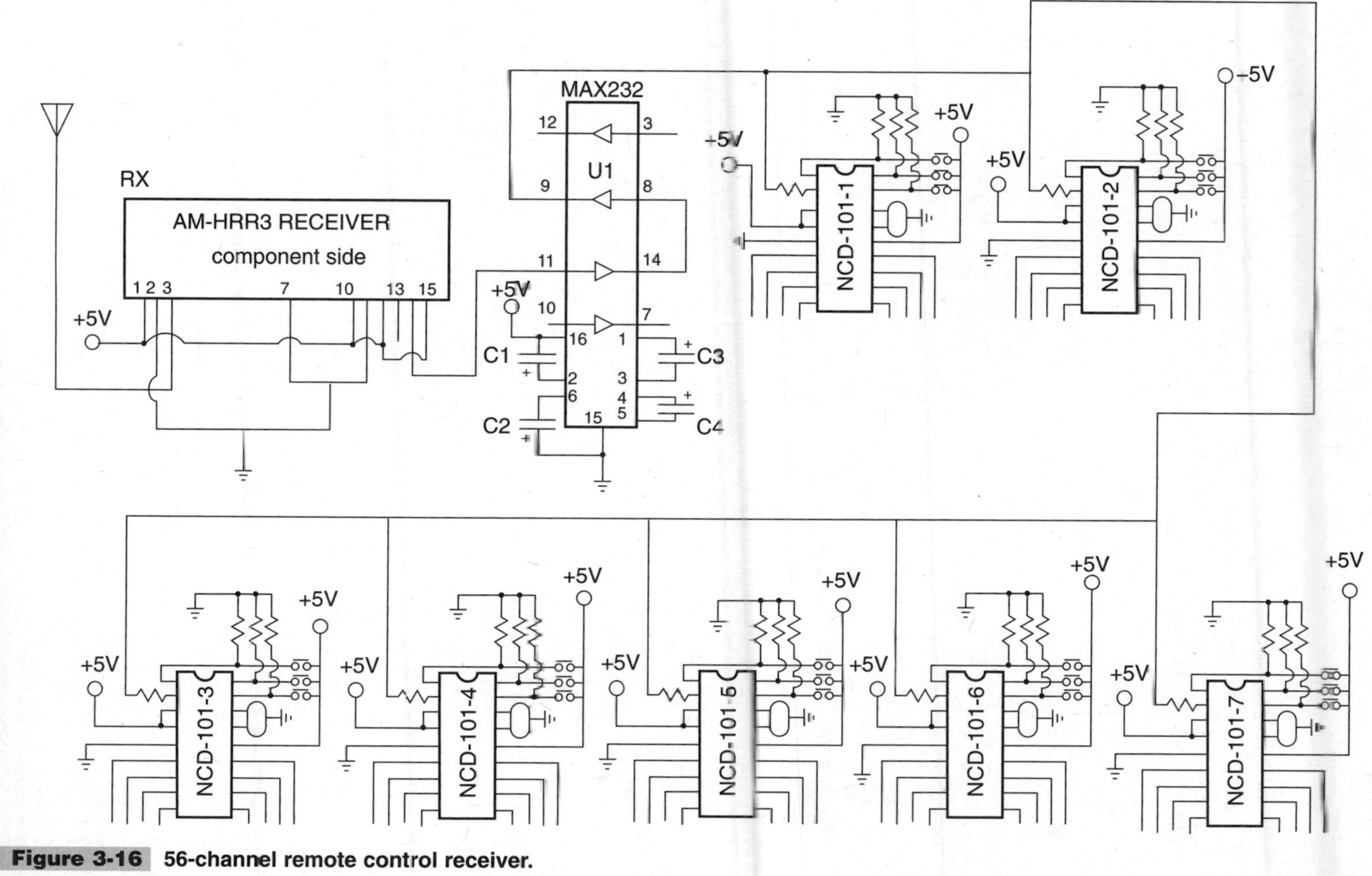

Figure 3-16 56-channel remote control receiver.

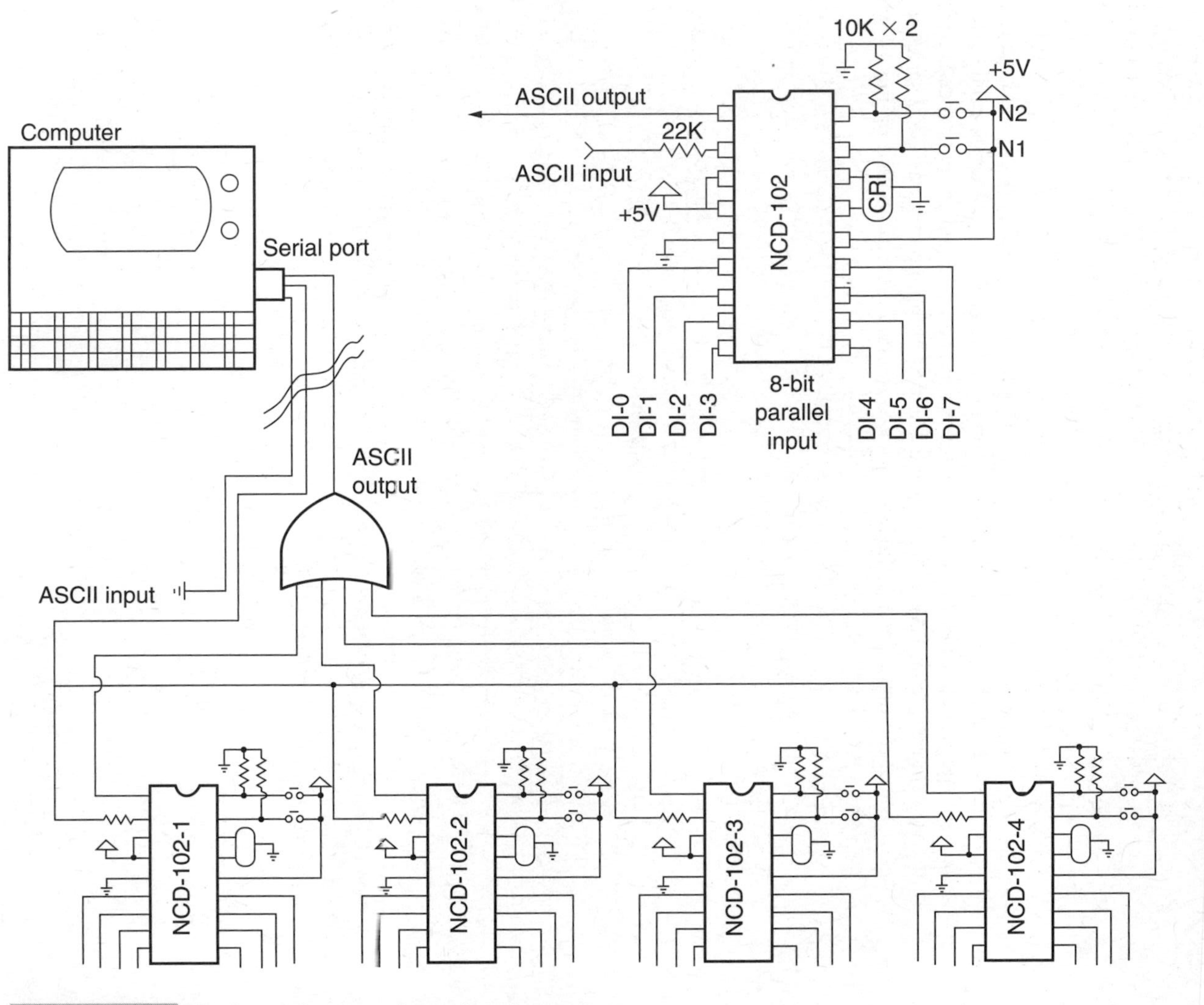

Figure 3-17 32-channel remote input inquiry system.

as well as digital signals. This multichannel chip can be combined with up to four NCD-106s to form a network capable of 8 analog-to-digital converters, 12 input monitor channels, and 16 output channels.

The NCD-106 processor can be used in a wide variety of control applications and would prove itself very useful on the benchtop or for experimental purposes. The NCD-106 can be interrogated by STAMP 2, IR remote, a PC with terminal program, or even a Palm Pilot.

The heart of the multichannel RS-232 A/D I/O is the NCD-106 processor, shown in Fig. 3-18. This 18-pin processor uses a single serial input on pin 6 to interrogate the device and a single serial ASCII output at pin 1, which displays the inquiry results. Two address programming inputs are provided at pins 2 and 3; both use a 10-kΩ resistor tied high to 5 V. The address input pins have four programming combinations, as shown in Table 3-4. The processor speed is controlled by a ceramic resonator at pins 15 and 16. Four speeds or baud rates are achieved, from 1200 to 9600 baud. The ASCII input pin has a 22-kΩ series resistor to protect the input from overload. Power to the processor is applied on pin 14 while the system ground is at pin 5.

Two analog-to-digital inputs are provided at pins 17 and 18. The two 8-bit A/D inputs will accept 0- to 5-Vdc inputs through 10-kΩ series resistors. Each analog input has a 0.1-μF capacitor across it to ground. A powerful two-way radio control multichannel data acquisition and I/O control system can be formed by using four NCD-106 chips. With this system you can monitor voltages remotely and then control functions based on your analog-to-digital readings. This system's functions range from monitoring switches, sensors, and voltages to controlling fans, motors, and appliances.

Two wireless Abacom BIM418 transceivers are used in this system. One of the wireless transceivers is used at the operator end of the system and is connected to a K1EL PS/2 keyboard controller to send commands to the receiver control side of the system. A serial LCD is connected to the operator side transceiver to monitor status signals returning from the NCD-106. On the NCD-106 controller end of the system, a BIM418 transceiver is used to relay the commands from the operator end of the system.

Controlling the NCD-106 is simple once you get the picture. Basically, you can send an analog input status command to check the status of the analog-to-digital converter. You can send a digital input status command to check the status of the digital inputs or you can turn on/off a digital output. In addition you can control an entire output byte on each NCD-106 at once, if desired; see Table 3-5.

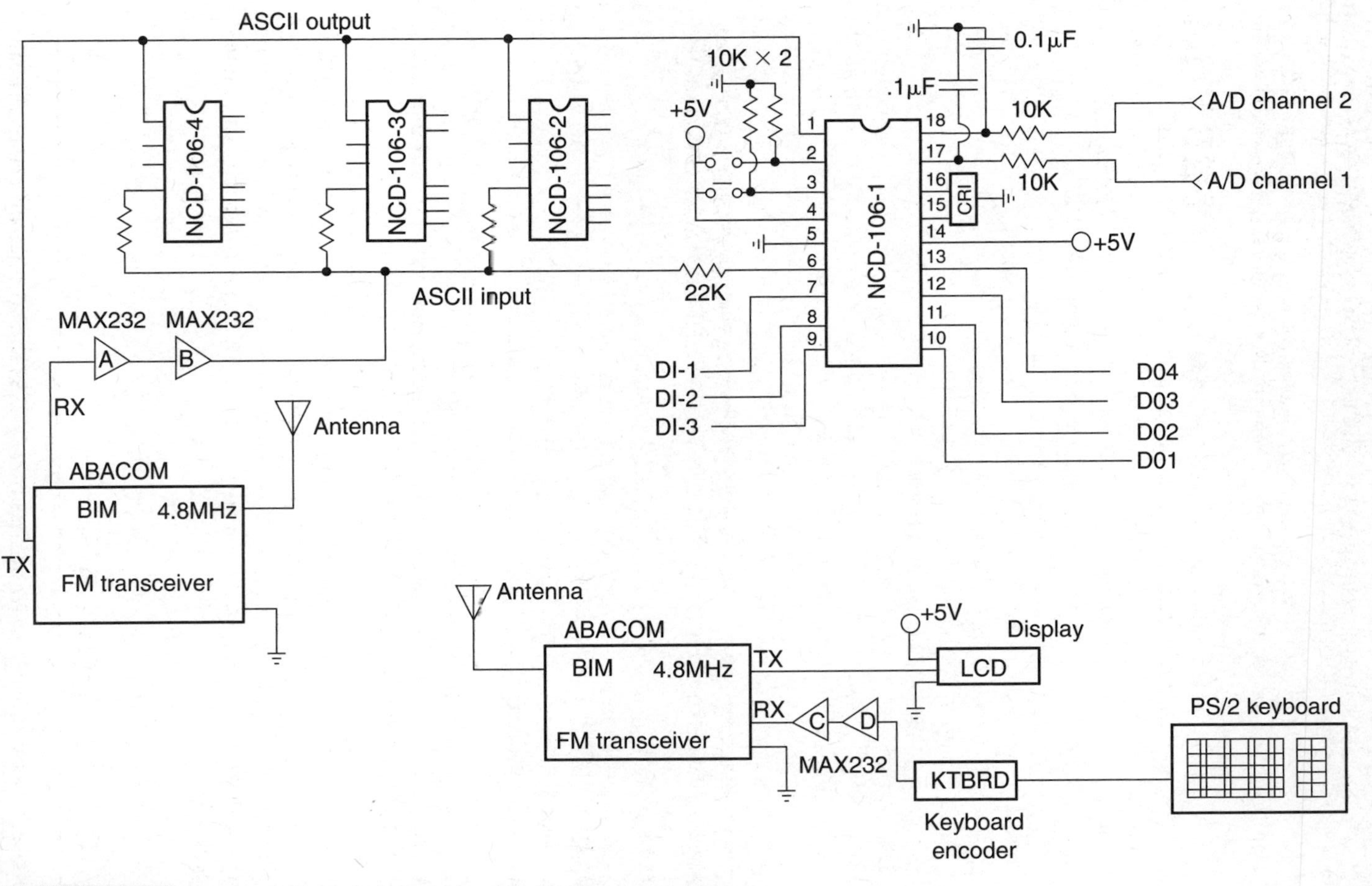

Figure 3-18 Remote A/D and digital input radio inquiry system.

4

10BASE-T NETWORK CABLE TESTER

CONTENTS AT A GLANCE

The 10Base-T network cable tester is a very useful gadget that will save lots of time and energy. If you've been involved or anticipate being involved in wiring a network either at home or at work you'll find this project a real lifesaver. Wiring a 10Base-T network becomes a tedious job in a short time and is therefore prone to mistakes. Keeping track of 8 wires per station, or 16 wire connections per computer, increases the chances of a wiring error, either from an open or short or from incorrect wire sequence in a connection. Building the 10Base-T cable tester will allow you to get it right the first time, with a minor investment.

Basic Components

The 10Base-T Network cable tester utilizes a BASIC STAMP 2 microprocessor in a simple but ingenious way. The STAMP 2 initializes all lamps, and then sequences seven LED lamps one after the other, from LED 1 through LED 7. On reaching the last LED, the STAMP 2 sequences back down from LED 7 to LED 1. The STAMP 2 then shuts down all lamps, waits a few seconds, and repeats the sequence over and over again.

The 10Base-T network cable tester is actually a two-piece, or two-part, testing system. The sending unit or transmitter part of the cable tester is shown in Fig. 4-1. The first seven I/O pins, P0 through P6, are connected to an 8-pin RJ-45 10Base-T cable connector through an 8 pin header on a perfboard or circuit board. The STAMP 2 LED sender portion of the cable tester can then be plugged into an RJ-45 jack at one end of your cable connection. The receiver section of the cable tester is depicted in Fig. 4-2. It contains the seven LEDs with series resistors connected to a second RJ-45 connector at the opposite end of the cable under test.

Once the driver or transmitter section of the cable tester is plugged in and the receiver portion is plugged in at the other end of the same cable, you can begin the test. The 10Base-T network cable tester will check the cable by first sequencing the LEDs in one direction, then pausing for a few seconds and completing the scan by sequencing in the opposite direction. The tester will check for open-circuit conditions, since all the LEDs should light up in sequence. Any LED failing to light up would indicate an open condition in that particular wire. The cable tester also turns on all LEDs at once to allow you a second look at the wires. The cable tester allow will indicate if there is a short in the wiring by having one or more LEDs light with diminished output or brilliance. The sequencing feature built into the cable tester also indicates if your wiring order is correct in each connector, so the cable tester performs a number of tests all at once. The 10Base-T cable tester can then be moved to the next set of cables and the test repeated once again, until all the 10Base-T cables have been tested.

Construction of the Cable Tester

The 10Base-T network cable tester is simple and straightforward to build, and should take less than an hour to complete. You can quickly build the network cable tester sending unit on 3 by 3-in perfboard to save time or you could fabricate your own small printed circuit board if desired. Install a 24-pin DIP socket, if you will be using the original STAMP 2; if you choose the alternative STAMP 2, you will need to use a 28-pin IC socket. It is important to take a few extra minutes when you are constructing the 10Base-T cable tester to observe the correct orientation of the microprocessor as well as to observe the correct polarity of the capacitors, diodes, and LEDs. If you use the original STAMP 2 you can attach a 9-V transistor-radio-type battery in order to power the tester. If you will be using the alternative STAMP 2, then you will need an additional 5-V regulator to power the microprocessor. Switch S1 is used to apply power to the cable tester circuit, and switch S2 is provided as a reset button, in the event of a microprocessor lockup condition. A normally open PC-mounted pushbutton is used for the reset switch. The four programming pins on the STAMP 2 microprocessor, i.e., pins 1 through 4, are connected to a 4-pin 0.010-in male header mounted on the circuit board of the sending unit. A mating 4-pin female header is wired to a 9-pin female serial RS-232 connector, which is used to connect to your programming computer. Each of the STAMP 2 output pins P0 through P6 is coupled to 150-Ω series resistors, which in turn are connected to a chassis-mounted female 8-pin miniconnector.

The prototype 10Base-T cable tester sending unit is mounted in a small 3- by 5-in plastic box; a small 8-pin female miniconnector is mounted on the outside of the enclosure. This

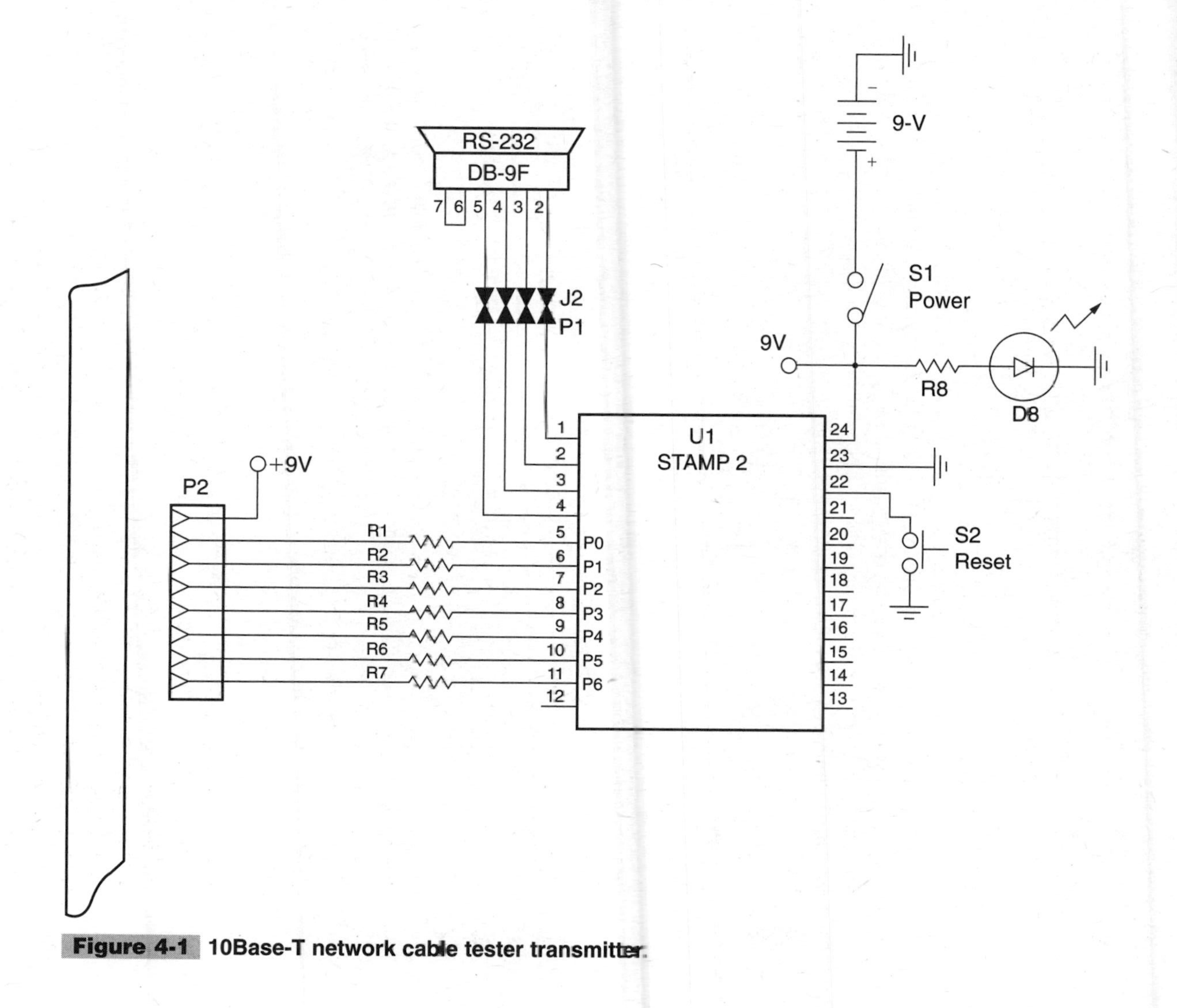

Figure 4-1 10Base-T network cable tester transmitter.

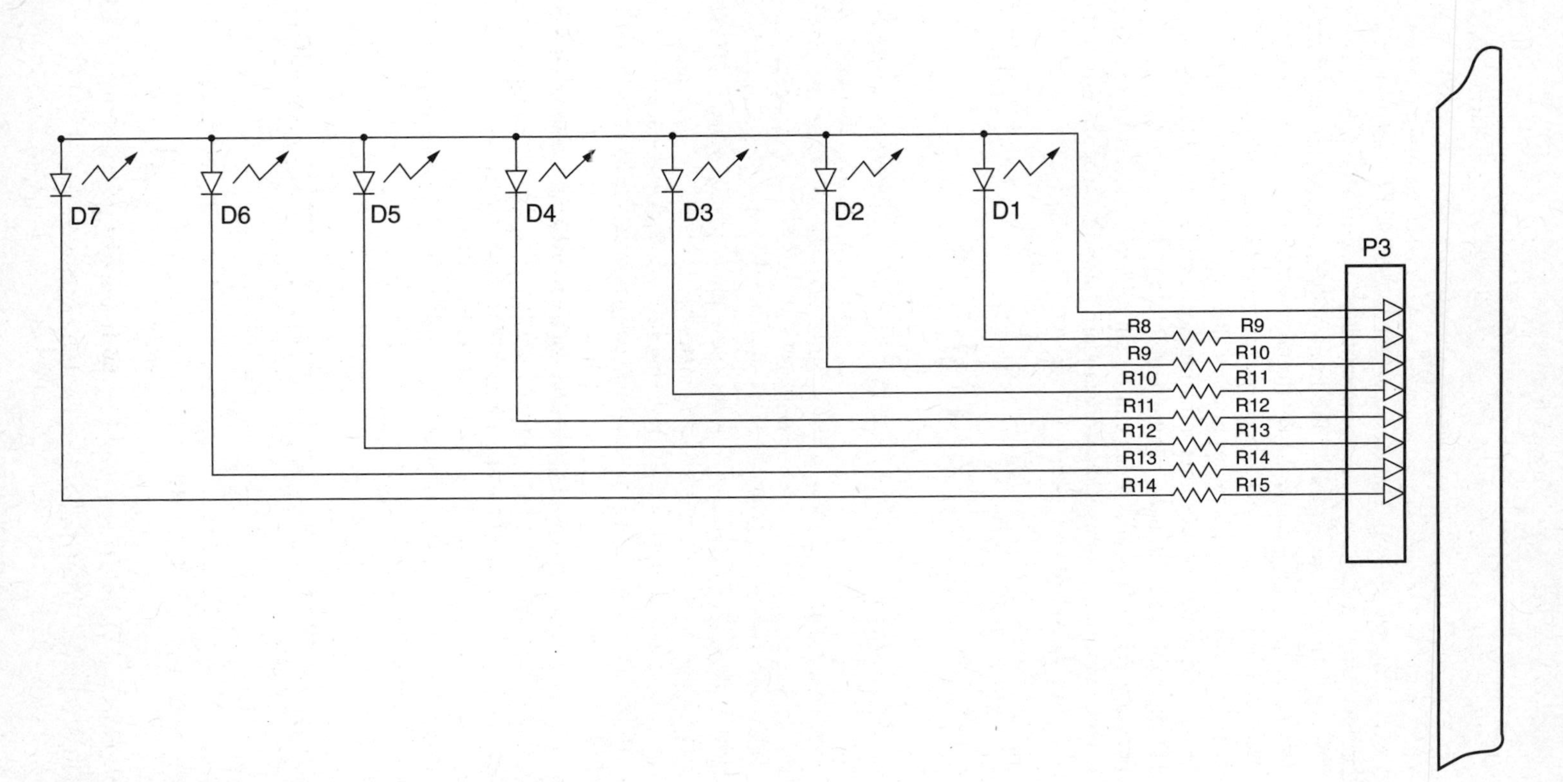

Figure 4-2 **10Base-T network cable tester receiver.**

connector is wired to the series resistors from the outputs of the STAMP 2. A mating 8-pin male connector is wired by an 8-ft length of CAT5/CAT8 cable to an RJ-45 Ethernet connector plug at the opposite end of the cable. A power switch and power indicator LED are mounted on the front of the chassis box. An alternative cable could be constructed with an 8-pin male connector at one end and, at the opposite end of the cable, each individual wire would be brought out to separate alligator clips, used to test circuits without an RJ-45 connector.

The prototype of the 10Base-T receiving unit is housed in a small 3- by 5-in plastic chassis box. Seven LEDs are mounted on the front of the enclosure. A series resistor is connected to each LED to prevent the LEDs from drawing excessive current and burning out prematurely. An 8-pin female miniconnector is mounted to the outside of the receiving unit chassis box. You will need to make up another cable, using a length of CAT5/CAT8 Ethernet cable. At one end of the cable, solder a mating 8-pin male miniconnector and at the opposite end of the cable you will need to attach an RJ-45 plug.

Constructing RJ-45 "patch" cables can be simplified by referring to Table 4-1, which illustrates the color pair combinations for standard "straight through" Ethernet cable. Table 4-2 depicts a color-coded wiring chart for "cross-over" ethernet cables.

Programming the Cable Tester

Once both the sender and receiver sections of the 10Base-T cable tester have been completed, you can move on to loading the software into the new 10 Base-T network cable tester. First, you will need to connect up your newly made programming cable between the sending unit and your personal computer. Next, you will need to locate the Windows editor for the STAMP 2, titled STAMPW.EXE. Once the editor is running, and your programming cable is attached, apply power to the STAMP 2 microprocessor. Next, locate and download the CABTEST.BS2 software (Listing 4-1 on p.69) needed to animate the 10Base-T network cable tester. Finally, run the program and you're ready to begin using your new network cable tester.

Most 10Base-T systems incorporate an RJ-45 patch panel at the computer switch/server side, so the above arrangement will work very well. In some instances, where no patch panel is used on the switch/server side, you could still use the cable tester but you would need to make a cable from the STAMP 2 sender with clip leads; the problem with this method is that you will have to pay strict attention to color codes used for each wire, since you are not using the standard RJ-45 connector with its wiring sequence and conformity.

The 10Base-T network cable tester will be a welcome addition to your test gear. You will wonder how you did wiring without it.

10Base-T Network Cable Tester Parts List

U1	BASIC STAMP 2 (original)
R1, R2, R3, R4, R5, R6, R7	50-Ω 1/4-W resistor
R9, R10, R11, R12, R13, R14, R15	150-Ω 1/4-W resistor

R8	330Ω 1/4W resistor
D1, D2, D3, D4	LEDs
D5, D6, D7, D8	LEDs
S1	SPST toggle power switch
S2	Normally open SPST pushbutton switch
J1	4-pin 0.010-in female header
P1	4-pin 0.010-in male header
P2, P3	RJ-45 8-pin 10Base-T connectors
Battery	9-V transistor radio battery
Miscellaneous	24-pin DIP socket, header, wire, PC board, battery clip, CAT5/CAT8 cable, alligator clips, etc.

TABLE 4-1 STRAIGHT THROUGH ETHERNET CABLE WIRING CHART

RJ-45 PIN NO.	COLOR (BOTH SIDES IDENTICAL)
Pin 1	White with orange
Pin 2	Orange
Pin 3	White with green
Pin 4	Blue
Pin 5	White with blue
Pin 6	Green
Pin 7	White with brown
Pin 8	Brown

TABLE 4-2 CROSSOVER ETHERNET CABLE WIRING CHART

RJ-45 PIN NO.	SIDE 1 COLOR	SIDE 2 COLOR
Pin 1	White with orange	White with green
Pin 2	Orange	Green
Pin 3	White with green	White with orange
Pin 4	Blue	Blue
Pin 5	White with blue	White with blue
Pin 6	Green	Orange
Pin 7	White with brown	White with brown
Pin 8	Brown	Brown

```
'Cabtest.bs2
DIRA=%1111
DIRB=%1111
DIRC=%0000
DIRD=%0000


Top:
OUTA=%0000
OUTB=%0000


pause 2000
OUT0=1
pause 300
OUT1=1
pause 300
OUT2=1
pause 300
OUT3=1
pause 300
OUT4=1
pause 300
OUT5=1
pause 300
OUT6=1
pause 300
OUT7=1
pause 300


OUTA=%0000
OUTB=%0000


pause 300
OUT6=1
pause 300
OUT5=1
pause 300
OUT4=1
pause 300
OUT3=1
pause 300
OUT2=1
pause 300


OUT1=1
pause 300
OUT0=1
Pause 3000
goto top
```

Listing 4-1 10Base-T network cable tester program.

5

IRDA COMMUNICATOR

CONTENTS AT A GLANCE

Infrared communication via Infrared Data Association (IrDA) is built into many portable computers and PDAs, such as the Palm, Windows CE machines, the old HP200LX, and many laptops. Infrared communication provides a convenient way to download data loggers and to communicate with other mobile platforms as well as for remote control without wires over short distances.

The IrDA protocol is very similar to standard RS-232, as illustrated by the diagram in Fig. 5-1. The standard RS-232 protocol uses voltage levels to transmit the data. From left to right on the time-line, the voltage is in a resting low state when a high start bit comes along, followed by 8 data bits, and last at least 1 low stop bit. A new character can start at any time after the conclusion of the previous stop bit(s). There are several conventions. RS-232 is an inverted logic, so that a logical 1, or "mark," is represented by a low level. This comes from old teletypewriter days when a pen was down to make a mark and lifted up by an electrical current in a magnet to make a space, or zero. Also, by convention, the data is transmitted least significant bit first, and a logical 1 is transmitted as a low level and a logical zero is transmitted as a high level. At 2400 baud, the length of each bit period is 1/2400 = 417 μs.

Conserving Power

Imagine transmitting the same data by light waves. You could leave the light on to represent the mark state, in which case space would be light off. But if characters are sent only infrequently, that could waste a lot of power, because the light would be on most of the time. Our IrDA communicator will be portable, battery-powered equipment, so we want to conserve power.

Alternatively, light off could represent mark and light on, space. But still that would consume more power than necessary. The designers of the IrDA protocol realized this and decided to make a flash of light represent the logical 0 and to make the flash brief, shorter than the

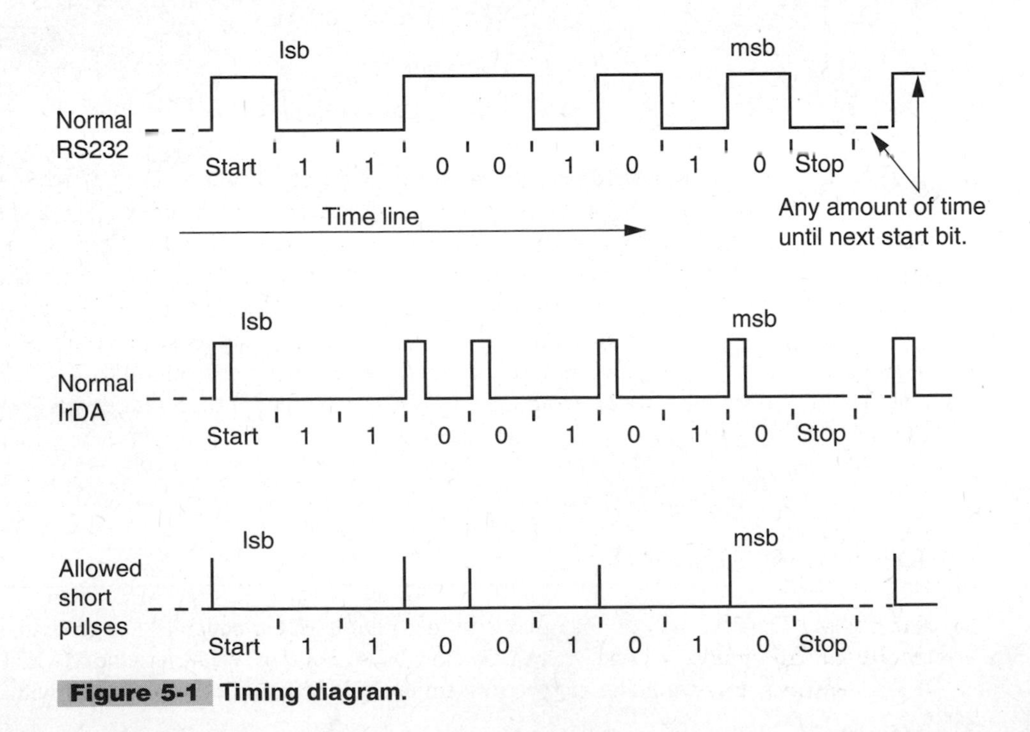

Figure 5-1 **Timing diagram.**

standard RS-232 bit period. This allows the flash to be very bright, and to transmit farther for a given input of battery power. The accepted length of a flash of light to represent a 0 is $^3/_{16}$ of the bit period. The flash of light for IrDA at 2400 baud would be 78 μs instead of 417 μs, but the bit period is still 417 μs. The $^3/_{16}$ rule holds for any baud rate, not just 2400.

The IrDA specification for low-speed (115K baud and under!) communications allows the length of the transmitted pulse to be much shorter than $^3/_{16}$ of the bit period. The pulse of light for any baud rate can be low as 1.7 μs in duration to represent a 0 (1.7 μs is $^3/_{16}$ of the bit period at 115K baud). This very short pulse allows a transmitter to create a very bright flash, and still achieve low battery consumption.

In practice, IrDA transmitters and receivers vary somewhat in the pulse width that they transmit and what they will receive correctly. This is something to check if you are having trouble getting one IrDA device to talk to another. The light is usually emitted at 880 nm. Infrared receivers usually have a fairly wide bandwidth to accept energy from any IR LED, 880 to 940 nm.

The IrDA transmitter circuitry must be able to turn normal RS-232 pulses into the short pulses of IrDA. This takes a clock synchronized with the baud rate. It is not possible to turn normal RS-232 into IrDA pulses simply by keying on the transitions in the signal. There are not necessarily transitions in the data; for example, the ASCII null character is all zeros, so the start bit plus the 8 zero bits form one long pulse 9 bits long with no transitions:

```
______- - - - - -______     Transmission of ASCII null
S00000000s                  Start and data bits are 0s
___|||||||||____            IrDA version, 9 short light pulses
___-____________            Transmission of ASCII $ff
S11111111s                  Data and stop bits are 1s
___|____________            IrDA version, 1 short light pulse
```

The IrDA receive circuitry must be able to detect the short pulses of light and stretch them to fill the standard bit width at the current baud rate for input to standard RS-232 serial ports. The above describes what is called the *physical layer* of IrDA. There are other layers that implement protocols for such things as sending packets of error-free data and for automatic negotiation of baud rate. We will not discuss these other layers in this project. The STAMP's limited brain would have trouble implementing those protocols. We are dealing here just with the physical layer, where IrDA is simply a wireless RS-232 substitute. The link must be tight to avoid errors, or you must implement some ad hoc error checking in software at both ends.

Basic Components

The STAMP itself has an RS-232 port, but it does not have the capability to generate and receive the short pulses of either normal or short IrDA. For that, we turn to the MAX3100. It is a universal asynchronous receiver-transmitter (UART) that communicates with the

BASIC STAMP over the SPI interface, and sends and receives the data to and from the outside world in the form of RS-232 or, optionally, sends and receives the data directly as IrDA. All it takes is an IR emitting diode and an IR photodetector. The MAX3100 requires a 3.6864-MHz crystal, on which it bases its baud rates.

The IrDA communicator circuit in Fig. 5-2 has a simple phototransistor as the receiver connected to the RX pin, which is fine for low speeds and short distances. For output from the TX pin, a high-efficiency IR LED is driven directly, sinking current to turn the light on. This is a very short range circuit, a fraction of an inch range. The alternative receiver depicted in Fig. 5-3 uses a Burr-Brown integrated photodiode amplifier to achieve greater sensitivity and speed. The output b from the OPT101 positive inverter circuit would replace the phototransistor input to the RX pin on the MAX3100. If more power is needed on the transmit side, it can be driven by a power transistor or a more capable current source.

The red LED D1 and the pushbutton S3, shown in the circuit connected to the RTS\output and CTS/input signal lines on the MAX3100, are for test purposes only, as illustrated in the program below.

The MAX3100 has additional desirable features. One of these is a buffer that can hold up to eight received characters until the STAMP gets around to reading them out. Another feature is a sleep mode, with automatic wake-up on detection of a start bit. (Note that the MAX3100 operates in noninverted mode, with a logical 1, or mark, state being a high level. For normal RS-232, MAX3100 would interface with the RS-232 line through an inverting RS-232 driver chip.)

Here is a simple getting-started demo of some of the main commands required to use the MAX3100. The demo first configures the MAX3100 for 2400 baud using the IrDA timing at the TX and RX pins. It then transmits "the quick brown fox jumps over the lazy dog" plus a control/line feed (CRLF). Then it goes into a loop to receive data from the IrDA port and print it on the debug screen. If the pushbutton on the above circuit is pressed, the program goes back and transmits the "fox and lazy dog" phrase again, so long as the pushbutton is down. The RTS line is made high to turn on the red LED when the MAX3100 is transmitting. You can readily change the output text to reflect your application.

To use this with an HP200LX, enter the HP COMM program and choose the connect menu. Choose the infrared port and 2400 baud. Now, with the two infrared ports coupled, you should see the fox message printed on the HP screen, and what you type on the HP200LX keyboard should be echoed on the STAMP debug screen.

The MAX3100 is one of those IrDA devices that generate an output light flash that is precisely $^{3}/_{16}$ of the bit width. And it expects the same for its input. It will reject incoming flashes if they are less than $^{1}/_{8}$ of the bit width. That is a problem with some devices that generate a shorter pulse. A pulse stretcher can be added to the circuit for use at a fixed baud rate.

There are some peculiarities in the Palm Pilot IrDA implementation. Note that the exact order of the IR commands used to open the IR serial port in NSBASIC makes a considerable difference in successful communication with Palm Pilot devices. Palm users should read the NSBASIC messages in Table 5-1 for additional help.

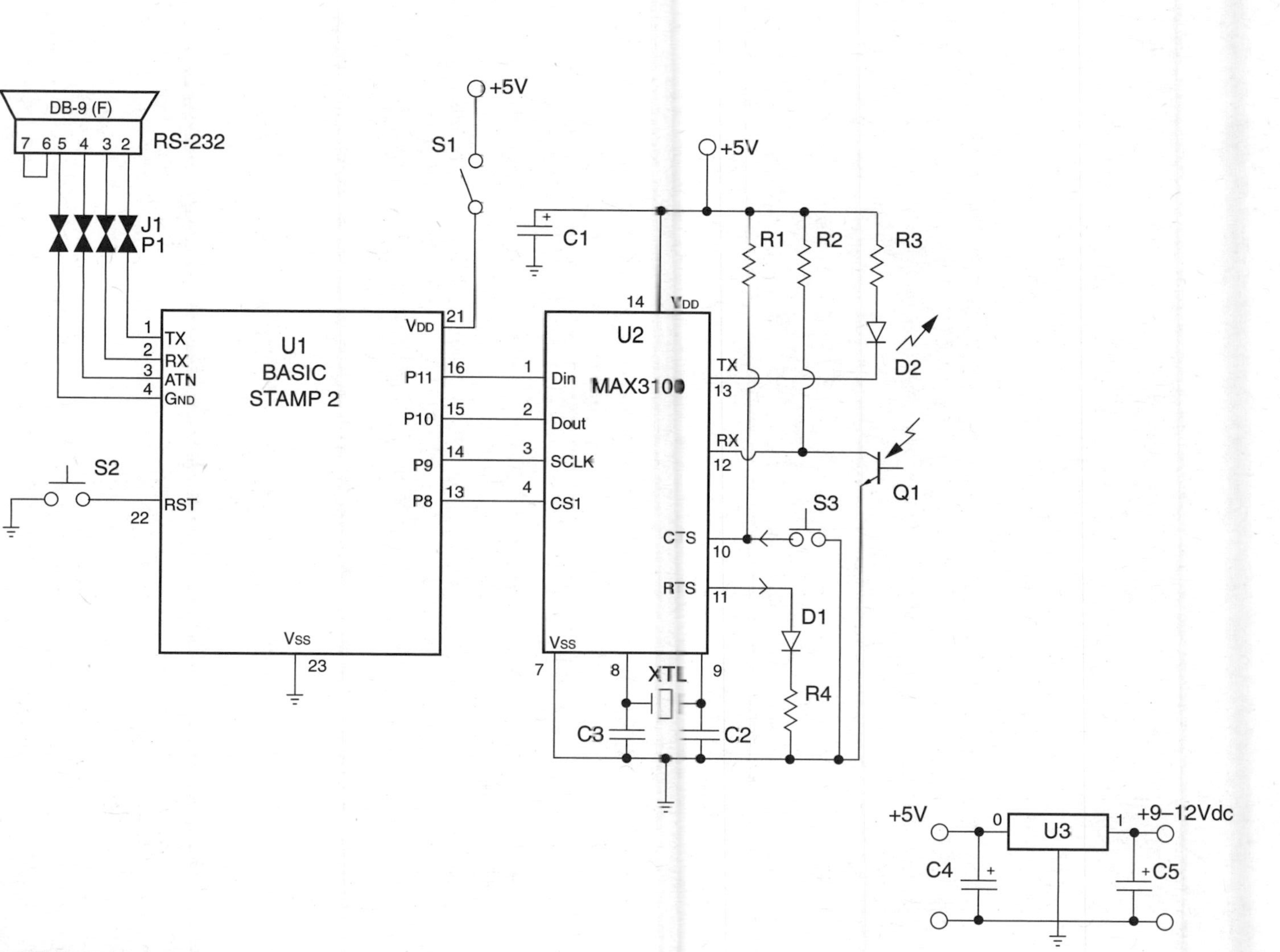

Figure 5-2 IrDA transceiver.

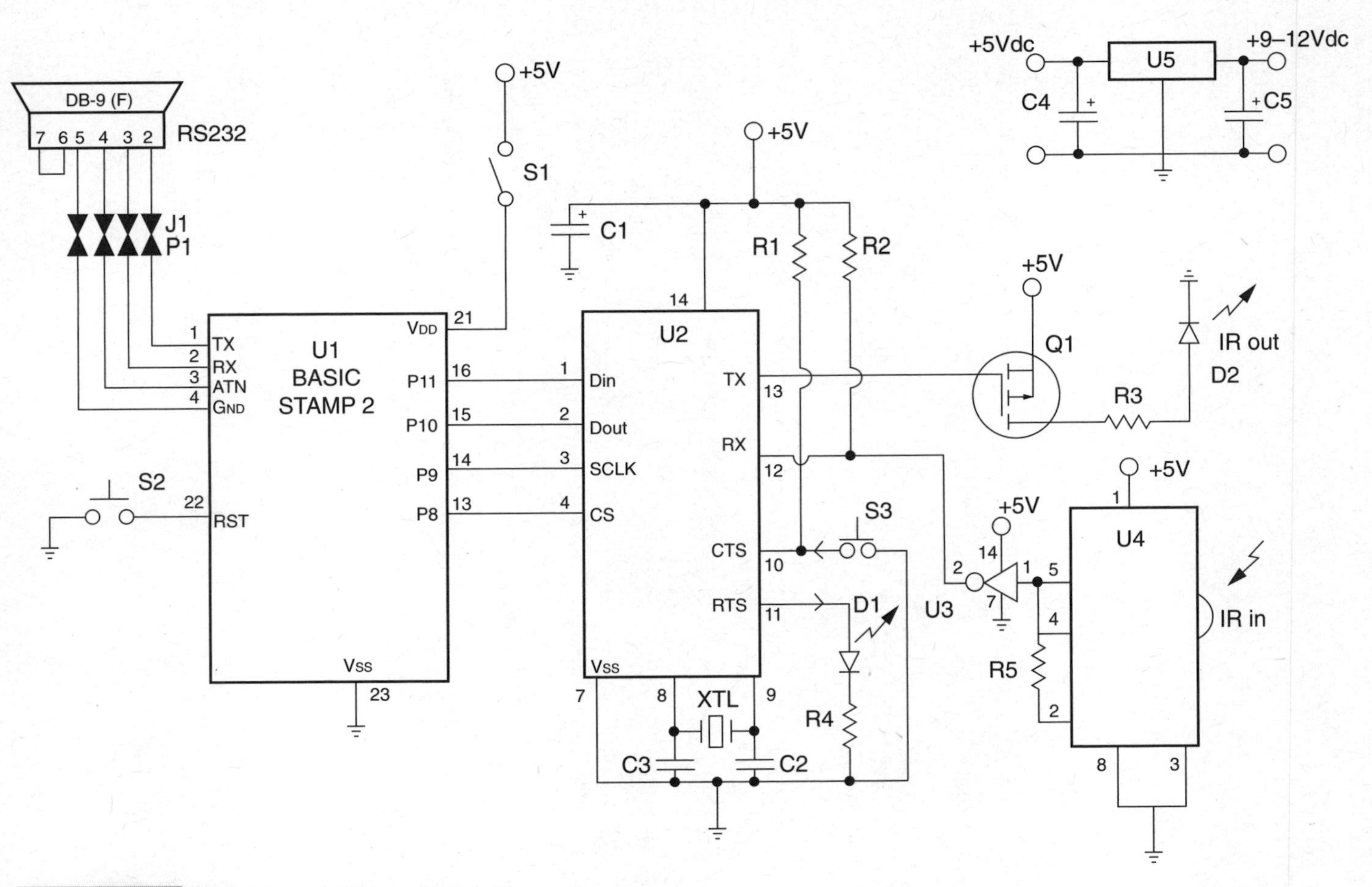

FIGURE 5-3 High-power IrDA communicator.

Construction of the Communicator

The IrDA communicator was built with a small 4- by 4-in glass epoxy circuit board. A circuit board will ensure the maximum speed and the shortest possible circuit traces. A PC board will also keep the system more reliable. STAMP 2 pins 1 through 4 are used to program the STAMP 2 for your particular application. A 4-pin 0.010-in male header is placed on the circuit board at P1, and a mating 4-pin female jack at J1 is wired to a 9-pin female RS-232 connector, which connects to your programming PC. A momentary normally open pushbutton reset switch is connected to pin 21 of the STAMP 2 in the event of a system lockup condition.

Be sure to use ICs for all of the circuits when constructing the circuit. Also be sure to observe the correct polarity when installing the capacitors and diodes as well as the IR LEDs. After assembling the circuit board be sure to take an extra moment to orient the integrated circuits properly. When installing ICs, be sure to locate the notch or cutout that is either on the top left of the IC or in the center of the top of the IC. This precaution will ensure that there will be no surprises when you initially apply power to the IrDA communicator. The IrDA communicator operates from a 5-V power source. A 5-V regulator is shown in the schematic; it can be fed with from a 9-V "wall wart" power supply, or you can elect to power the regulator from a 9- or 12-V battery pack if desired.

When you have completed the IrDA communicator, you can place it in a small plastic enclosure. The power switch S1, the reset switch S2, and the send-data pushbutton at S3 are all mounted on the front of the communicator enclosure. A power jack and the programming jacks are mounted on the rear of the case. If you use the IrDA communicator as an alarm or sensing device, you will need to make provisions for sensor input jacks on the case as well. You can use the communicator to sense a particular condition and have it report to another device via IR if desired. Once the IrDA communicator has been assembled, you will need to program the communicator for your particular application. First, you will need to locate the STAMP 2 Windows editor program titled STAMPW.EXE. Next, you will need to connect up your programming cable and apply power to the IrDA communicator. Last, you will need to download the IrDA.BS2 program (Listing 5-1) from the CDROM. The IrDA.BS2 program is basically a demo program to get you started using the communicator. Once you develop more applications you can rewrite or modify the program to suit your new application.

IrDA Communicator (Low-Power Version) Parts List

U1	BASIC STAMP 2 (original)
U2	Maxim MAX3100 UART IC
U3	LM7805 5-V regulator IC
R1	10-kΩ $^1/_4$-W 5% resistor

R2	2-kΩ $^1/_4$-W 5% resistor
R3	240-Ω $^1/_4$-W 5% resistor
R4	330-Ω $^1/_4$-W 5% resistor
C1	0.1-μF 35-V tantalum capacitor
C2, C3	20-pF disk capacitor
C4	4.7-μF 35-V electrolytic capacitor
C5	0.22-μF 35-V disk capacitor
D1	Red LED
D2	IR LED
Q1	IR detector transistor
S2, S3	Normally open momentary pushbutton
S1	SPST toggle power switch
Miscellaneous	PC board, 4-pin male header, 4-pin female header, DB-9 connector wire, enclosure, etc.

TABLE 5-1 ADDITIONAL PALM IrDA IMPLEMENTATION

NSBASIC MESSAAGE GROUP DISCUSSION FOR IRDA PALM USERS
http://groups.yahoo.com/group/nsbasic-palm/message/11025
http://groups.yahoo.com/group/nsbasic-palm/message/14341
http://groups.yahoo.com/group/nsbasic-palm/message/14332
http://groups.yahoo.com/group/nsbasic-palm/message/8625
http://www.markspace.com/online.html (FAQ for online terminal emulator for Palm)

```
' IRDA.BS2
' using BS2E
' program for the MAX3100 in IR mode
' with 3.6864mhz crystal
outs=$0100
dirs=$ffff
din    con  11      ' data sent from stamp to uart
dout   con  10      ' data out from uart to stamp
sclk   con   9      ' data clock
cs     con   8      ' active low chip select
fox    data         "the quick brown fox jumps over the lazy dog",13,10,0
char   var  word    ' will combine 8-bit command/status and 8-bit data
ix     var  byte
rts    var  char.byte1.bit1  'for data send
cts    var  char.byte1.bit1 ' for data read
```

Listing 5-1 IrDA communicator program.

```
tst    var  bit
Rbit  var     char.byte1.bit7


high cs
pause 100


' _______________
' set the mode to 2400 baud, IrDA
' this writes to the configuration register
writemode:
low cs
 shiftout din,sclk,msbfirst,[$C08D<\\>16]
high cs


' _______________
' transmits the message in data space "fox"
' "the quick brown fox...  +CRLF
sendfox:
 char.byte1=$80                ' make rts high, red led on
                               ' rts is bit 1, pin is made high when bit is low.
 ix=0
sendfox1:                      ' this loop gets fox, null terminated string.
 read ix,char.byte0
 if char.byte0=0 then sendfoxend
sendfox2:
 low cs
 pulsout sclk,2                  ' get one bit
 tst=dout                        ' test that transmit buffer is empty


 high cs
 if tst=1 then sendfox2          ' loop back until ready
 low cs                          ' now send the character
 shiftout din,sclk,msbfirst,[char<\\>16]
 high cs
 ix=ix+1
 goto sendfox1                   ' look for next char to send
sendfoxend:
 low cs                          ' now turn off RTS pin
                                 ' the TE<\\> bit=1, so no data is transmitted.
  shiftout din,sclk,msbfirst,[$8600<\\>16]
  high cs
  goto terminal                  ' back to receiving characters
                                 ' and testing cts


' _______________
' now enter a loop to receive data and print it on a debug screen
' jumps to sendfox routine when the hardware cts line is low
' by pressing the pushbutton in the above circuit
terminal:
low din                          ' sequence to receive data, both 8-bit status and
                                   8-bit data
 low cs
  shiftin dout,sclk,msbpre,[char<\\>16]    ' 16 bits total
 high cs
 if cts then sendfox                        ' Jump to sendfox is pushbutton down.
                                            ' cts bit is high when the pin is low
 if Rbit=0 then terminal                    ' skip printing if there is no character
                                            ' Rbit is the flag for this.
 debug char.byte0                           ' print the character
goto terminal  ' receive more bytes
```

Listing 5-1 IrDA communicator program (*Continued*).

IrDA Communicator (High-Power Version) Parts List

U1	BASIC STAMP 2 (original)
U2	MAX3100 UART IC, Maxim
U3	CD4069 inverter IC, National
U4	OPT101 IR receiver IC, Burr-Brown
U5	LM7805 5-V regulator, National
R1	10-kΩ $^1/_4$-W 5% resistor
R2	2-kΩ $^1/_4$-W 5% resistor
R3	49-Ω $^1/_4$-W 5% resistor
R4	330-Ω $^1/_4$-W 5% resistor
R5	100-kΩ $^1/_4$-W 5% resistor
C1	0.1-μF 35-V tantalum capacitor
C2, C3	20 pF 35-V disk capacitor
C4	4.7-μF 35-V electrolytic capacitor
C5	0.22-μF 35-V disk capacitor
D1	Red LED
D2	High-power IR LED (RS-276-1443)
Q1	p-channel MOSFET IRF9530
S2, S3	Normally open momentary pushbutton
S1	SPST toggle power switch
Miscellaneous	PC board, 4-pin male header, 4-pin female header, DB-9 connector wire, enclosure, etc.

6

MULTICHANNEL RADIO ALARM SYSTEM

CONTENTS AT A GLANCE

The multichannel radio alarm system is an extremely useful portable alarm system that has many applications around your home, office, or shop areas. The multichannel alarm system is a two-unit system, consisting of (1) a 12-channel alarm controller utilizing the BASIC STAMP 2 processor and an Abacom RT5 mini data transmitter that sends alarm activations and (2) a remote receiver/display unit. The remote receiver/display section of the system consists of an Abacom HRR-3 receiver, an LCD display, and a piezoelectric alarm buzzer. The piezo buzzer alerts you to an alarm condition; you can then look at the display and see which alarm channel has been activated.

The 12-channel radio alarm system will enable you to monitor many different types of sensors or alarm conditions in a single room or numerous sensors/switches in surrounding rooms, and it will send an alarm condition via radio to the receiver/display unit in a remote location up to 150 ft away, such as an adjoining room or building. Substituting an Abacom AM-RRS2 receiver will allow a 300-ft range.

Basic Components

Our discussion of the multichannel radio alarm begins with the alarm input conditioner circuit shown in Fig. 6-1. The input conditioner circuit is designed to accept both normally open and normally closed inputs. Switches S1, S2, and S3 can be normally open sensors or door switches. You can add any number of additional switches if desired. Normally closed or supervised switches or sensors are shown at S4, S5, S6, and S7. Additional switches or sensors can also be connected in series with these switches if desired. The input conditioner is powered from a 5-Vdc source. The diodes D1 and D2 apply a trigger signal to activate transistor Q1, which in turn is used to supply 5 V on activation of any alarm to the input of the STAMP 2. This circuit represents an "alarm zone." If desired, an input conditioner can be placed ahead of any of the STAMP 2 input channels. In operation the output of the input conditioner is connected directly to an input channel of the multichannel alarm input instead of through the switch, as shown on the system diagram in Fig. 6-2.

The heart of the 12-channel radio alarm is centered around the BASIC STAMP 2 microprocessor shown at U1 in Fig. 6-2. The 12 input channels are featured from P0 through P12. Each input is tied to ground through a 10-kΩ resistor. Each of the inputs is shown connected to normally open switches, but the switch could be any type of normally open switch or sensor, such as magnetic door or window switches or floor mat sensors. Note that the switches or sensors are tied to 5 V. The STAMP 2 controller is powered on pin 21 through toggle switch S1, from a 5-Vdc source from the U2. A system reset push button is connected to the reset pin 22, to allow resetting the processor in the event of a system lockup. A PC board normally open mini pushbutton was used for the reset button. A system status lamp D2 is driven by STAMP 2 pin P13. This indicator is used to show the system start-up and status conditions. A siren or local alarm output is provided at P14, which can be used to drive a flasher light, siren, or Sonalert via relay RY1. Note that a larger-current device can be accommodated if RY1 is used to drive a larger-current relay, which in turn drives a high-current output device. The STAMP 2 controller is programmed via the DB-9 connector. Pins 1 through 4 of the STAMP 2 are brought out to a 4-pin male header at P1 that mates to a female header at J1. A programming cable is then wired between J1 and the a DB-9 female connector.

The STAMP 2 is coupled to an Abacom AM-RT5 data transmitter at TX (see Fig. 6-3). The RT5 is a complete hybrid RF transmitter with a 70-m range. The transmitter can transmit data at up to 4 kHz from any standard CMOS or TTL source. Power consumption of the minitransmitter is about 4 mA. This low-cost 4-pin data transmitter measures only 10.1 by 17.7 mm. The RT5 is powered from a 5 Vdc source at pin 1. A ground connection is

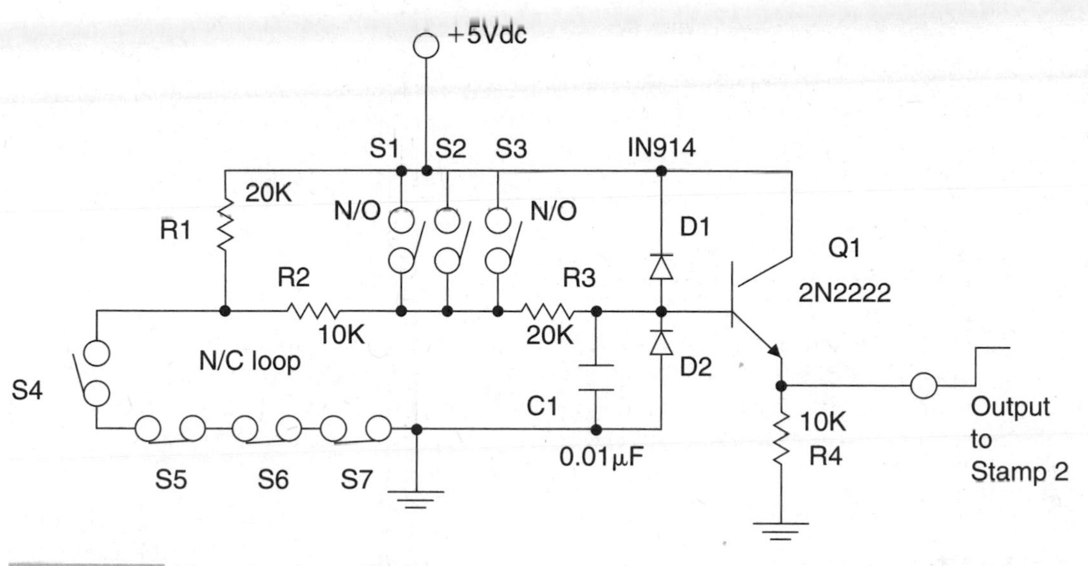

Figure 6-1 Alarm input conditioner.

provided at pin 2, while the input data is applied to pin 6 of the RT5. A quarter-wave vertical whip or half-wave dipole antenna can be connected directly to pin 7 for maximum range. A quarter-wave vertical whip is shown in Fig. 6-4, and a half-wave dipole antenna is shown in Fig. 6-5.

The receive/display section of the multichannel alarm begins with either the AM-HRR3-418 superregenerative or the optional AM-RRS2-418 superheterodyne receiver. The AM-HRR3 receiver is shown in Fig. 6-6. This AM-type mini data receiver measures only 38 by 13.7 mm. The AM-HRR3 receiver is a thick-film hybrid module with laser-trimmed inductor and requires no external adjustment. The AM-HRR3 consumes only about 0.8 mA and has a range of 150 ft with an external antenna, described above. The AM-HRR3 receiver is a 15-pin device; pins 1, 10, 12, and 15 are connected to 5-Vdc, while pins 2, 7, and 11 are connected to ground. An antenna is connected to pin 3, and the data output is shown at pin 14. A test pin is provided at pin 13.

The Abacom AM-RRS2 series receivers are AM superheterodyne-type receivers that have a 300-ft range with external antenna. The AM-RRS2 series receiver can be readily substituted in place of the AM-HRR3 receivers, but be aware of the pinout differences. The AM-RRS2 receiver is depicted in Fig. 6-7. This mini data receiver measures 30 by 20.3 mm and consumes about 3.7 mA. The AM-RRS2 receiver is also powered from a 5-Vdc source on pin 10, with a ground connection on pins 5 and 9 as shown. An antenna is connected to pin 1, while the data output pin is provided on pin 12.

The complete receiver/display section of the multichannel alarm system is shown in Fig. 6-8. The AM-HRR3 receiver is shown connected to an ILM-216L serial LCD display unit from Scott Edwards Electronics (www.seetron.com); see App. 1 on the CD-ROM for more details. The ILM-216L features a 2 × 16-type display, with 1200- to 9600-baud operation,

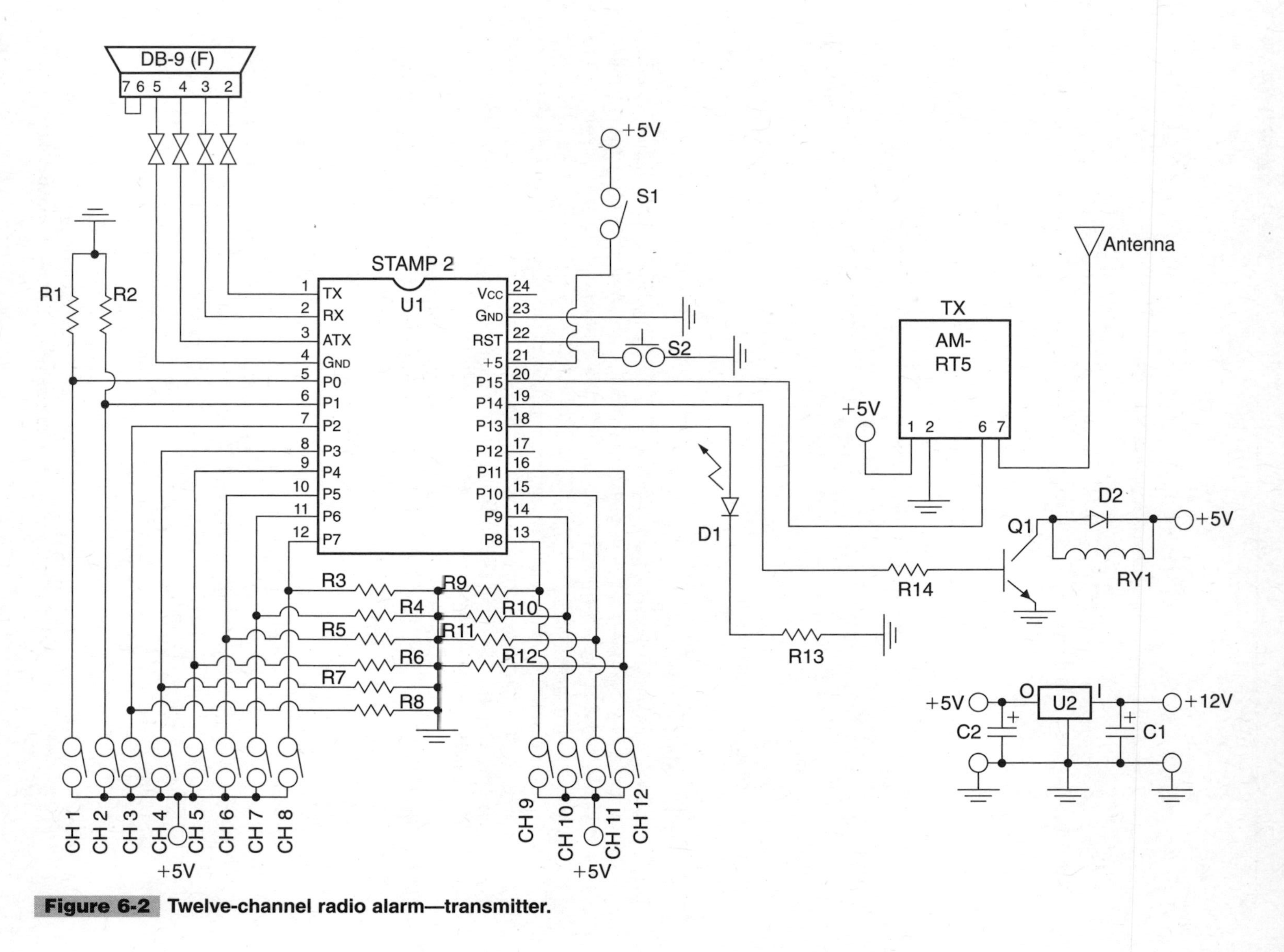

Figure 6-2 **Twelve-channel radio alarm—transmitter.**

four switch inputs, software-controlled backlight, custom characters, and buzzer output. The data output of the AM-HRR3 on pin 14 is fed directly to the serial ILM-216L display. The ILM-216L display unit provides 8 pins for interfacing. Pins 1 and 6 are tied to ground, and 5 Vdc is applied to pin 2 to power the display. Serial data input is applied to pin 3, while data output from the display is provided on pin 4. A piezo alarm buzzer can be connected to pin 5. The baud rate can be changed via pin 8, using the configuration/test input on pin 7. The receiver is coupled directly to serial input on pin 3 of the display. The receiver and display are both powered from the 5-V regulator at U1. A piezo buzzer is connected to pin 5 of the display unit to provide an audible indication that an alarm condition has occurred.

In operation, once the alarm controller is triggered, the STAMP 2 sends out serial data through the transmitter to the receiver at RX. The STAMP 2 sends out a serial word to activate the buzzer three times to get your attention; you can modify the program to preset a longer alarm sound if desired. Once the buzzer has been activated, the STAMP 2 begins sending out a serial data message to the LCD display unit, which indicates which alarm channel has been activated. The transmitter and receiver combination will operate up to 3 kHz, so the serial data stream from the STAMP 2 to the transmitter was set up to 2400 baud. Note that by using the Abacom FM transmitter/receiver pair the system is capable of 9600-baud operation.

Construction

The STAMP 2 alarm controller and transmitter were mounted on a small PC board measuring 4 by 4 in. It is advisable to use an integrated circuit socket for the STAMP 2 controller, in the event of a failure. Be careful to correctly orient the STAMP 2 when installing in the IC socket. When assembling the transmitter/controller, be sure to observe the correct polarity of the capacitors, diodes, and LEDs. A 4-pin 0.010-in male header at P1 is soldered to the PC board. A mating 4-pin 0.010-in female header at J1 is cabled to a 9-pin DB-9 female RS-232 connector, for programming the STAMP 2 processor. If you elect to use the

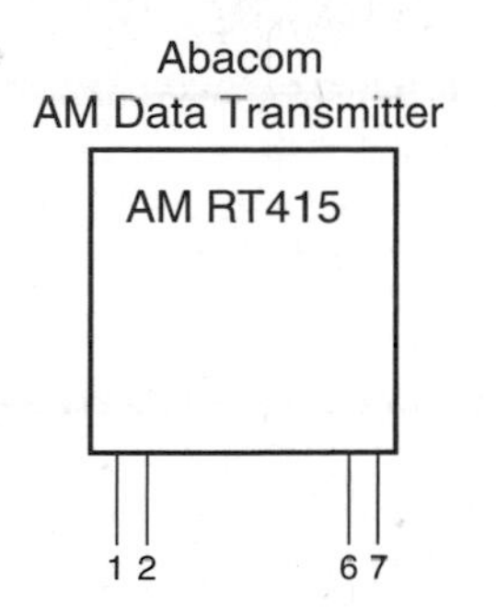

RT4	RT5*	Function
1	1	Vcc (+)
2	2	GND
3	6	Data in
4	7	Ant

Figure 6-3 AM data transmitter.

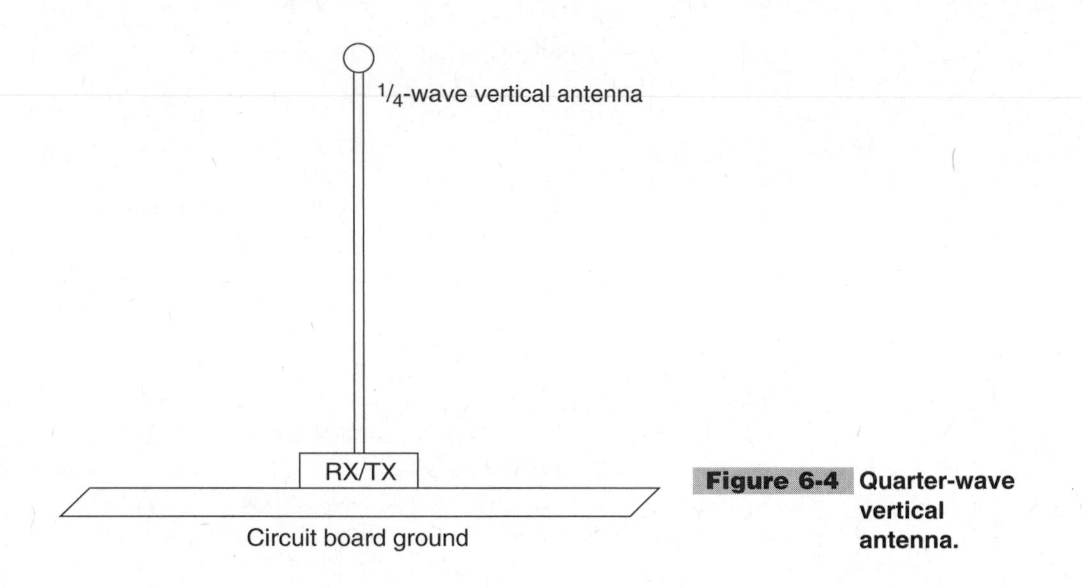

Figure 6-4 Quarter-wave vertical antenna.

input conditioner circuit, or more than one conditioner circuit, you will need to wire these to inputs to the main controller board. You will then need to select an enclosure that will house the STAMP 2 controller, the transmitter, and all the input conditioner circuits. An aluminum chassis box is recommended, in order to reduce interference to the transmitter circuit from outside noise sources. A two-circuit coaxial power jack can be mounted on the rear of the chassis box. A screw terminal bank was used to accept input sensor connections. An antenna jack was mounted on the rear of the chassis in order to connect to the external antenna. The transmitter ground and controller ground are connected together to form a common ground and are tied to the metal chassis box enclosing the alarm controller assembly. A quarter-wave vertical whip or a half-wave dipole could be used for the antenna. A 9-V "wall wart" power supply is used to power the 5-V regulator; also note the controller could be powered from a rechargeable battery for fail-safe and/or portable operation.

The receiver display unit is fabricated on a 3- by 5-in printed circuit board, with the receiver, display, buzzer, and regulator. A metal chassis box is used to house the receiver/display unit, to reduce the possible interference from outside noise sources. The most difficult part of the fabrication is mounting the display in the chassis box: The chassis box is marked up and the outline hole is outlined. A small drill bit is used to drill out holes along the perimeter of the display outline. Then a file is used to custom-fit the display into the enclosure. A wall wart power supply can be used to power the receiver/display unit or a 9- to 12-V rechargeable battery can be used for portable operation if desired. A small hole is drilled on the top of the case to allow the piezo buzzer to soundoff through the case. The buzzer is epoxied to the underside of the enclosure. A coaxial power jack is mounted on the rear of the receiver chassis, along with an antenna jack to accept an external antenna. A quarter-wave vertical whip or half-wave dipole can be used for the antenna on the receiver display unit. Be sure to use the same type and polarity of antennas between the transmitter and receiver units to ensure maximum range. For example, if the transmitter utilizes a vertical antenna, the receiver section should also use a vertical antenna. If you elect to use a half-wave dipole for the transmitter, then a half-wave dipole should be used at the receiver unit as well.

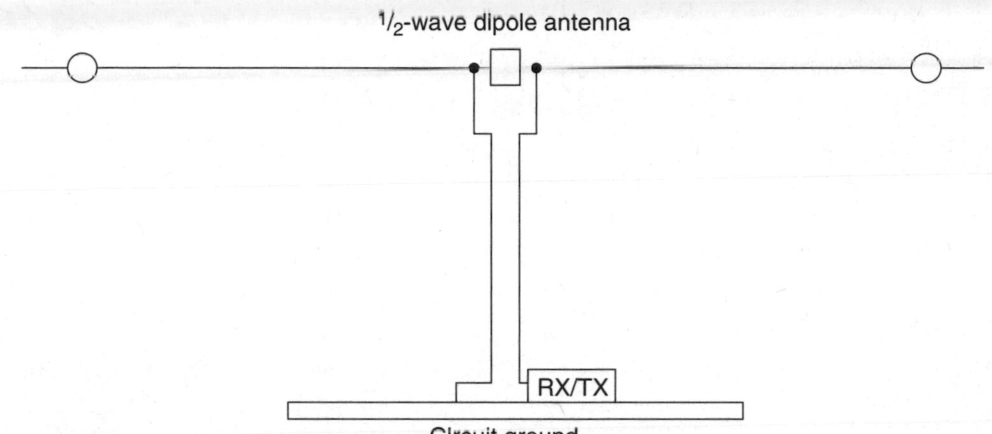

Figure 6-5 **Half-wave dipole antenna.**

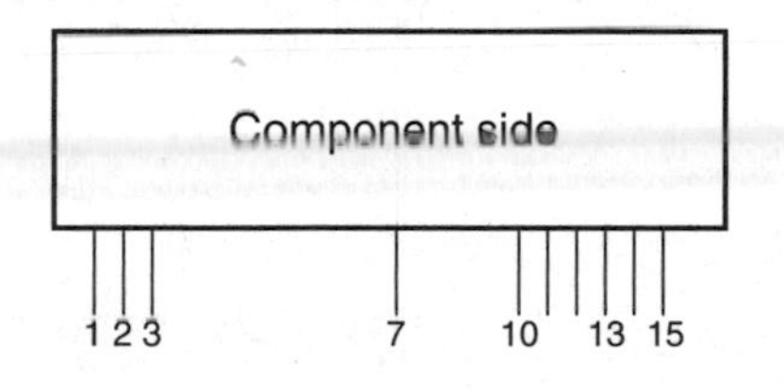

Pin #	Pin name	Pin #	Pin name
1	RF +VCC	8, 9	N/C
2	RF GND	10	AF +VCC
3	Data in	11	AF GND
4	N/C	12	AF +VCC
5	N/C	13	Test
6	N/C	14	Data out
7	RF GND	15	AF +VCC

Figure 6-6 **Abacom AM-HRR3 data receiver.**

Once the multichannel alarm system has been constructed, you will need to download the software into the STAMP 2 controller board. Connect your personal computer and the STAMP 2 via the programming cable you constructed. Now locate the STAMP 2 Windows editor titled STAMPW.EXE on the CD-ROM. Next apply power to the STAMP 2 controller transmitter board. Finally, locate and download the RADALARM.BS2 program (Listing 6-1) into the BASIC STAMP 2 controller. Your multichannel alarm system is now completed and ready to test. Turn on both the alarm controller transmitter and the alarm controller receiver. Now you will need to simulate an alarm condition to test the transmitter and receiver combination. Connect one of the input channels on the controller to 5 V to simulate an alarm condition. Once an alarm condition has been triggered, it should propagate through the system. The receiver

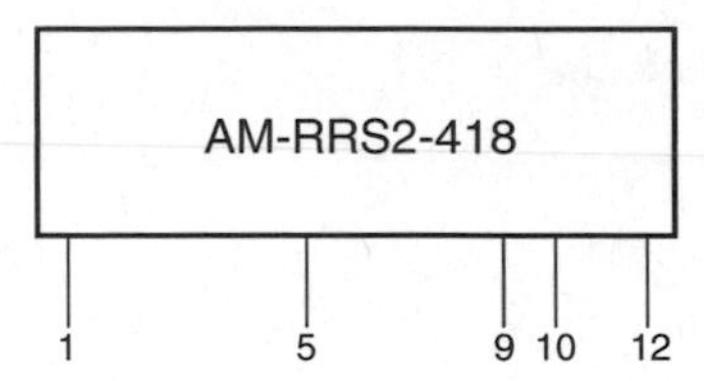

Pin #	Pin name
1	Data in (RF)
5	GND
9	GND
10	VCC
12	Data out

Figure 6-7 Abacom AM-RRS2 superheterodyne receiver.

should come to life after a few seconds and the piezo alarm buzzer should begin to sounds three times in a row. The LCD display will then begin to display the actual alarm channel that has been activated. Your multichannel alarm system is now ready to serve you!

Multichannel Alarm Transmitter Parts List

U1	BASIC STAMP 2 (original)
U2	LM7805 5-V regulator
TX	Abacom AM-RT5 transmitter module
R1–R12	10-kΩ 1/4-W resistor
C1	1-μF 35-V electrolytic capacitor
C2	4.7-μF 35-V electrolytic capacitor
R13, R14	1-kΩ 1/4-W resistor
D1	LED
D2	1N4002 silicon diode
Q1	2N2222 silicon transistor
RY1	5-V minirelay (RadioShack)
Miscellaneous	IC socket, PC board, wire, headers, DB-9, antenna, etc.

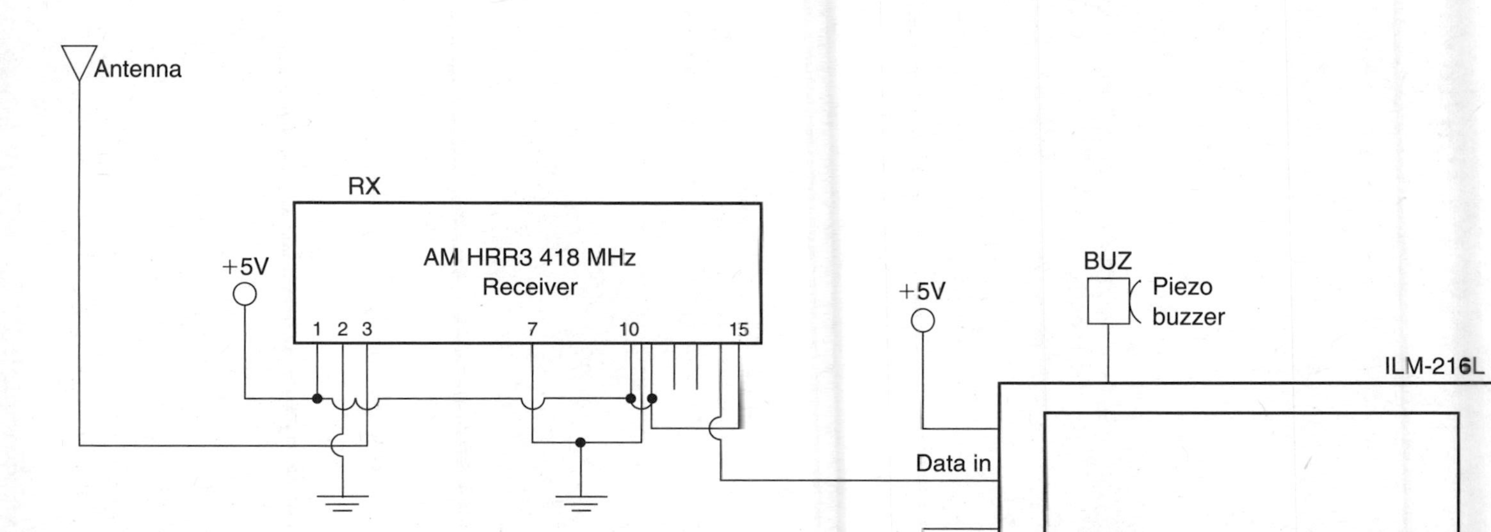

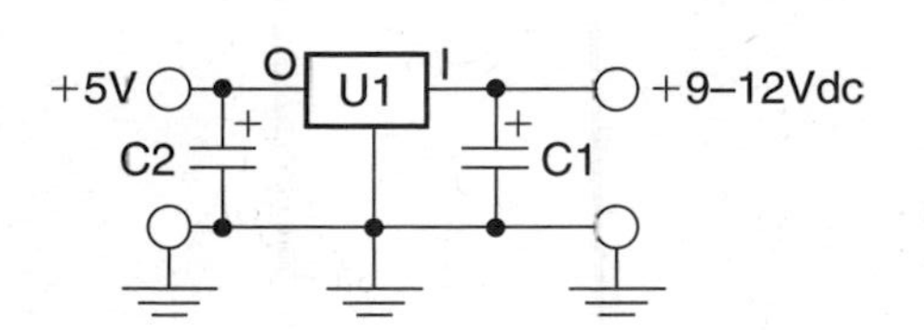

Figure 6-8 Radio alarm receiver/display unit.

Multichannel Alarm Receiver/Display Parts List

U1	LM7805 5-V regulator IC
RX	Abacom AM-HRR3 data receiver module
C1	1-μF 35-V electrolytic capacitor
C2	4.7-μF 35-V electrolytic capacitor
DISP	ILM-216L LCD display unit (www.seetron.com)
Buz	5-V piezo buzzer
Miscellaneous	PC board, wire, header, antenna, etc.

```
'Radalarm.bs2
'all inputs tied to ground and go high +5v upon activation
'radalarm.BS2


flash                   Var     byte
io                      Var     word
timer                   Var     byte


tpin                    Con     15
LedPin                  Con     14
Siren                   Con     13
recall_delay            Con     120
bdmd                    Con     16780           '2400 baud
delay                   Con     500


init:


INA=0000
INB=0000
INC=0000
DIR13=1
DIR14=1
DIR15=1


for flash = 1 to 3
 high LedPin
 pause 1500
 low LedPin
 pause 1500
next
```

Listing 6-1 **Multichannel radio alarm program.**

```
main:
high LedPin
pause 100
low LedPin
if IN0 = 1 then gosub_chan_1
if IN1 = 1 then gosub_chan_2
if IN2 = 1 then gosub_chan_3
if IN3 = 1 then gosub_chan_4
if IN4 = 1 then gosub_chan_5


if IN5 = 1 then gosub_chan_6
if IN6 = 1 then gosub_chan_7
if IN7 = 1 then gosub_chan_8
if IN8 = 1 then gosub_chan_9
if IN9 = 1 then gosub_chan_10
if IN10 = 1 then gosub_chan_11
if IN11 = 1 then gosub_chan_12
pause 100
goto main


gosub_chan_1:
high Siren
pause 2000
serout tpin,bdmd,100,[7]
pause 100
serout tpin,bdmd,100,[7]
pause 100
serout tpin,bdmd,100,[7]
serout tpin,bdmd,100,["CHANNEL 1 ALARM"]
pause 6000
low Siren
low LedPin
serout tpin,bdmd,100,[12]
goto main


gosub_chan_2:
high Siren
pause 2000
serout tpin,bdmd,100,[7]
pause 100
serout tpin,bdmd,100,[7]
pause 100
serout tpin,bdmd,100,[7]
serout tpin,bdmd,100,["CHANNEL 2 ALARM"]
pause 6000
low Siren
low LedPin
serout tpin,bdmd,100,[12]
goto main


gosub_chan_3:
high Siren
pause 2000
serout tpin,bdmd,100,[7]
pause 100
```

Listing 6-1 Multichannel radio alarm program (*Continued*).

```
serout tpin,bdmd,100,[7]
pause 100
serout tpin,bdmd,100,[7]
serout tpin,bdmd,100,["CHANNEL 3 ALARM"]
pause 6000
low Siren
low LedPin
serout tpin,bdmd,100,[12]
goto main


gosub_chan_4:
high Siren
pause 2000
serout tpin,bdmd,100,[7]
pause 100
serout tpin,bdmd,100,[7]
pause 100
serout tpin,bdmd,100,[7]
serout tpin,bdmd,100,["CHANNEL 4 ALARM"]
pause 6000
low Siren
low LedPin
serout tpin,bdmd,100,[12]
goto main


gosub_chan_5:
high Siren
pause 2000
serout tpin,bdmd,100,[7]
pause 100
serout tpin,bdmd,100,[7]
pause 100
serout tpin,bdmd,100,[7]
serout tpin,bdmd,100,["CHANNEL 5 ALARM"]
pause 6000
low Siren
low LedPin
serout tpin,bdmd,100,[12]
goto main


gosub_chan_6:
high Siren
pause 2000
serout tpin,bdmd,100,[7]
pause 100
serout tpin,bdmd,100,[7]
pause 100
serout tpin,bdmd,100,[7]
serout tpin,bdmd,100,["CHANNEL 6 ALARM"]
pause 6000
low Siren
low LedPin
```

Listing 6-1 **Multichannel radio alarm program (*Continued*).**

```
serout tpin,bdmd,100,[12]
goto main


gosub_chan_7:
high Siren
pause 2000
serout tpin,bdmd,100,[7]
pause 100
serout tpin,bdmd,100,[7]
pause 100
serout tpin,bdmd,100,[7]
serout tpin,bdmd,100,["CHANNEL 7 ALARM"]
pause 6000
low Siren
low LedPin
serout tpin,bdmd,100,[12]
goto main


gosub_chan_8:
high Siren
pause 2000
serout tpin,bdmd,100,[7]
pause 100
serout tpin,bdmd,100,[7]
pause 100
serout tpin,bdmd,100,[7]
serout tpin,bdmd,100,["CHANNEL 8 ALARM"]
pause 6000
low Siren
low LedPin
serout tpin,bdmd,100,[12]
goto main


gosub_chan_9:
high Siren
pause 2000
serout tpin,bdmd,100,[7]
pause 100
serout tpin,bdmd,100,[7]
pause 100
serout tpin,bdmd,100,[7]
serout tpin,bdmd,100,["CHANNEL 9 ALARM"]
pause 6000
low Siren
low LedPin
serout tpin,bdmd,100,[12]
goto main


gosub_chan_10:
high Siren
pause 2000
serout tpin,bdmd,100,[7]
```

Listing 6-1 **Multichannel radio alarm program (*Continued*).**

```
pause 100
serout tpin,bdmd,100,[7]
pause 100
serout tpin,bdmd,100,[7]
serout tpin,bdmd,100,["CHANNEL 10 ALARM"]
pause 6000
low Siren
low LedPin
serout tpin,bdmd,100,[12]
goto main


gosub_chan_11:
high Siren
pause 2000
serout tpin,bdmd,100,[7]
pause 100
serout tpin,bdmd,100,[7]
pause 100
serout tpin,bdmd,100,[7]
serout tpin,bdmd,100,["CHANNEL 11 ALARM"]
pause 6000
low Siren
low LedPin
serout tpin,bdmd,100,[12]
goto main


gosub_chan_12:
high Siren
pause 2000
serout tpin,bdmd,100,[7]
pause 100
serout tpin,bdmd,100,[7]
pause 100
serout tpin,bdmd,100,[7]
serout tpin,bdmd,100,["CHANNEL 12 ALARM"]
pause 6000
low Siren
low LedPin
serout tpin,bdmd,100,[12]
goto main
```

Listing 6-1 **Multichannel radio alarm program (*Continued*).**

7

LIGHTNING ACTIVITY MONITOR

CONTENTS AT A GLANCE

Amateur radio operators, weather watchers, and boaters alike will appreciate the lightning activity monitor. Amateur radio enthusiasts of course need to know about approaching lightning so they can disconnect their antenna systems from their radio transceivers. Many hams also report storm activity via the local Skywarn system. Weather watchers monitor lightning as their hobby but also often belong to weather reporting groups. Boaters also have a keen interest in weather for advance warnings to seek shelter in the event of a bad storm. The lightning activity monitor is a unique and inexpensive way to measure intensity of lightning as it advances toward and away from you. It uses a BASIC STAMP 2 to display the number of lightning strikes per minute on a row of eight different-color LEDs. Depending on the mounting configuration, the lightning sensor is capable of detecting lightning strikes greater than 50 miles away (see Fig. 7-1).

Lightning Sensor

The BASIC STAMP lightning activity monitor consists of two parts: the lightning sensor unit (LSU) and the BASIC STAMP 2 display unit. The lightning sensor unit (LSU) performs the actual lightning detection, while the display unit indicates the storm intensity by using different-color LEDs, etc. Referring to the schematic in Fig. 7-2, the energy from a lightning strike is received by the antenna and passes through the high-gain Darlington transistor pair consisting of Q1, Q2, R2, and R3, allowing current to flow through Q2 during each lightning strike. The 9-V battery in the LSU sensing head unit provides the necessary drive to power the LED side of an optocoupler at U1, on the BASIC STAMP display board, while resistor R1 limits the current through the LED. Because current flows only during a lightning detection, the battery should last well over 2 years. Resistor R4 limits the input

Figure 7-1 Lightning activity monitor.

current received by the antenna. Diode D2 provides protection in case the battery is inadvertently connected backward.

The LSU circuit can be assembled on a small piece of perfboard (see Fig. 7-3). A suggested layout using point-to-point wiring is shown in Fig. 7-4 (also see the LSU parts list provided at the end of the chapter). When assembling the LSU board, be sure to observe the polarity of the two diodes and the orientation of the two transistors. The output of the LSU board is connected to the BASIC STAMP display board via a suitable length of coax cable; almost any good quality coax cable, such as RG-6 or RG-58, could be used. Remember the coax shield is tied to the circuit return (or minus side).

The antenna of the LSU is constructed of a 12-in length of $^3/_4$-in copper pipe. Solder a 6- to 10-in wire to the inside of the pipe using a high-wattage soldering iron or small torch. The LSU is installed in PVC pipe to weatherproof the unit. The antenna is housed in a 12-in length of $^3/_4$-in Schedule 20 pipe. Use a small amount of RTV to glue the antenna inside the PVC pipe. During assembly, splice the wire from the antenna to the antenna wire on the LSU. The battery and circuit board sit loosely inside the 1- × $^3/_4$- × 1-in tee. You can secure the battery board with some foam if desired. The coax and ground wire are routed through a hole drilled in the bottom of the 1-in cap installed on the bottom of the tee. Don't use PVC glue on the bottom cap so that the battery can be changed if necessary. An exploded view of the LSU housing is shown in Fig. 7-5. Figure 7-6 shows how the components fit inside the PVC pipe.

Since the energy received at the antenna is relative to earth, it is imperative to connect the coax shield (and LSU return) to a good earth ground. For safety reasons, this should be done close to where the LSU is mounted. A metal water pipe or a ground rod near the installation will work fine. A coax-type lightning protector/grounding block (RadioShack no. 15-909) is also advised.

To test the LSU module, connect a red LED across the unconnected end of the coax. Be sure to get the polarity correct: the anode (+) connects to the shield and the cathode (−) connects to the center conductor. Install a fresh 9-V alkaline battery. An inexpensive long-reach electric butane lighter makes a great lightning simulator. Hold the lightning

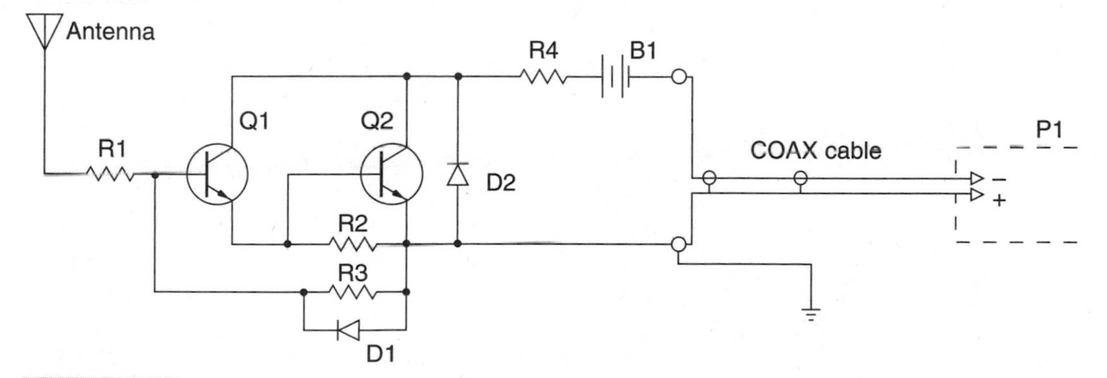

Figure 7-2 Lightning detector sensing head unit.

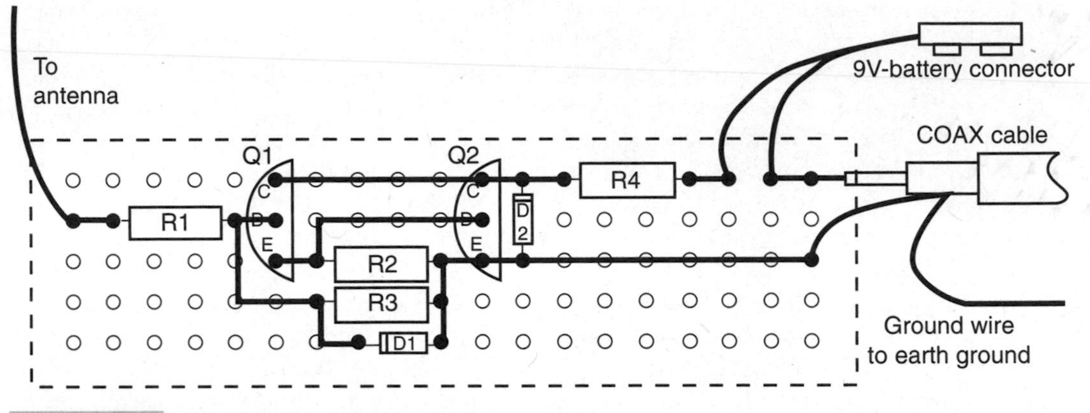

Figure 7-3 LSU circuit layout diagram.

Figure 7-4 Suggested layout using point-to-point wiring.

simulator within a couple of inches from the LSU antenna and pull the trigger. The LED should flash when the lighter sparks. If not, retrace the circuit, making sure the LED is connected correctly and the battery is good. The pole mount for the lighting activity monitor LSU board is fabricated from a 1- × $^{3}/_{4}$- ×1-in tee that has been cut in half lengthwise. Use wire ties or hose clamps to secure the LSU to the mast as shown in Fig. 7-7.

Display Board

The BASIC STAMP 2 display board is shown in Fig. 7-8. It is built from perfboard with point-to-point wiring, but a custom PC board would also work very well. The schematic for the lightning activity display board is depicted in Fig. 7-9. When assembling the light-

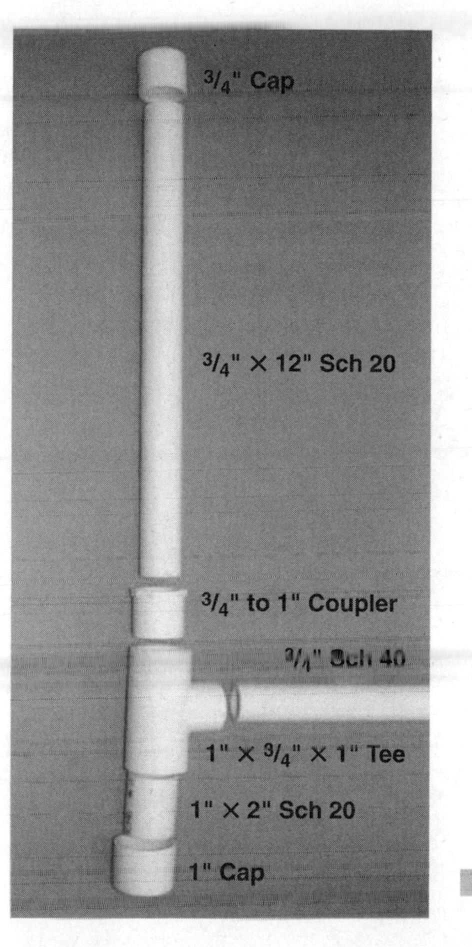

Figure 7-5 Exploded view of LSU housing.

ning monitor, be sure to use an integrated circuit socket, especially for the microprocessor. Also be careful to observe the correct orientation when installing the STAMP 2. As you construct the lightning activity display board, take an extra moment to correctly orient and install the capacitors, diodes, and LEDs to avoid damage to the board on power-up. An optoisolator is used to couple the LSU to the STAMP 2 display board and to provide ground isolation and ESD protection. The LED side of the optocoupler is connected to the LSU, and the optotransistor side is used to toggle an input pin on the BS2. Since the BS2 cannot supply the required current for all 8 LEDs, an SN74LS240 driver IC at U3 is used; this allows the LEDs to run on the V_{in} (+9 V) power. Five different-color LEDs are used for displaying the severity of lightning conditions.

Programming the Unit

Once the LSU, or sensing head, is mounted on the mast, run the coax cable inside to the BASIC STAMP 2 display board. *Note:* Connect the shield or outer conductor to the *nega-*

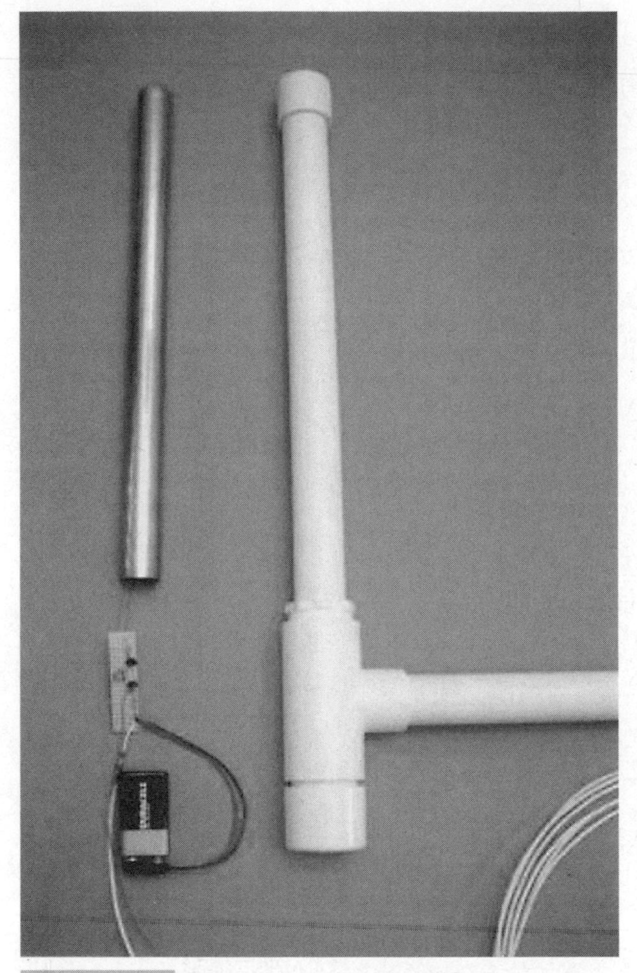

Figure 7-6 **Components inside PVC pipe.**

tive terminal on P1 and the center conductor to the *positive* terminal. Apply power to the STAMP 2 display board. Next, connect a programming cable between your personal computer and the STAMP display board on pins 1 through 4 of the STAMP 2. Once the programming cable is hooked up, you will need to go to the STAMP 2 program directory and start up the STAMP 2 Windows editor titled STAMPW.EXE. Lastly, you will need to download the BSLAM.BS2 lightning activity display program (Listing 7-1) into the BASIC STAMP 2. Once this task has been completed your lightning activity monitor is ready to go to work for you!

Once the program has been loaded to the lightning activity monitor and the program installed, all eight display LEDs should light up and then go out one by one to verify circuit operation. Next, hold the "lightning simulator" within a few inches of the LSU and spark the igniter. Within a few seconds, one or more LEDs should light up. Now wait for a thunderstorm and watch the LEDs light up!

Figure 7-7 Securing the LSU to the mast.

Figure 7-8 BASIC STAMP 2 display board.

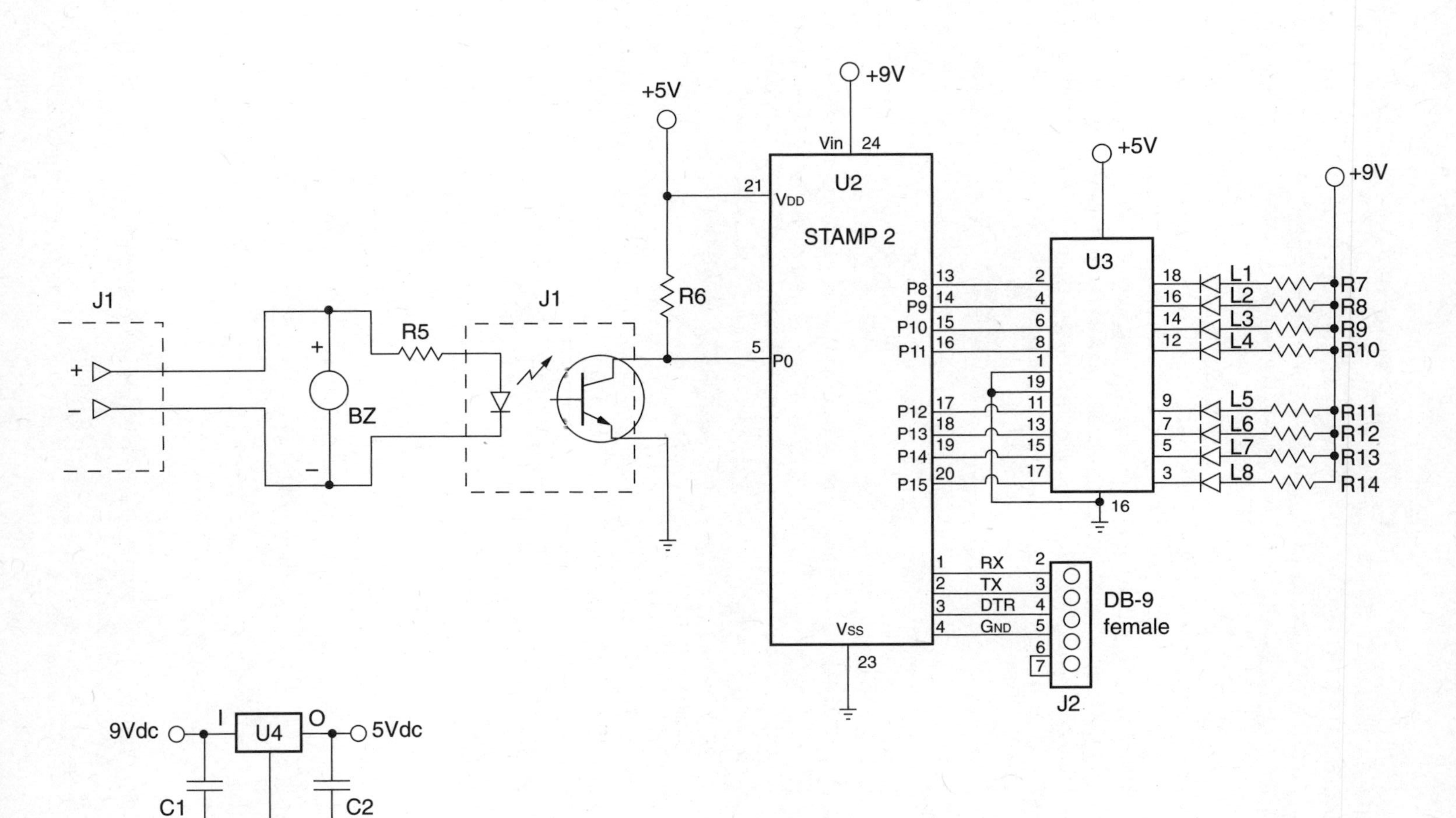

FIGURE 7-9 Lightning detector/display unit.

The software for the BASIC STAMP displays the lightning activity in strikes per minute. This allows the user to gauge the amount of lightning activity. To maximize LED activity, instead of counting lightning activity for 60 seconds, the software counts the number of lightning strikes in a 6-second period and stores the results in element 0 of a 10-element array. After each 6-second count period, all 10 array elements are summed and compared to a list of LOOKDOWN values. This returns an index value between 0 and 8. The index is then passed to a LOOKUP command to get the LED port value with 0, activating all 8 LEDs and turning them all off. Next, the contents of each array element are shifted up, discarding the tenth (oldest) element; i.e., array 9 gets copied to array 10, array 8 gets copied to array 9, and so on. This keeps a running average of the lightning counts over the 10 elements' last 6 seconds, or a 60-second period. There is also a variable to keep track of the peak counts. Each time the array is summed, the results are compared to the last recorded peak. It the new peak is larger, it replaces the old peak value.

The strikes-per-minute method allows the users to determine not only the magnitude of the lightning activity, but also whether the storm is approaching or traveling away. By monitoring the LEDs, increasing values indicate that the storm is getting closer, while decreasing values (LEDs) indicate the storm is moving away. A typical lightning strike can contain many individual flashes. This is why lightning sometimes appears to flicker. The lightning sensor design used in this project is highly sensitive and will detect each flash in a stroke. This may be noticeable when visually comparing lightning activity with the lightning activity monitor. As many as 22 detects for a single lightning flash have been monitored with intense storms. The strikes-per-minute values used in the LOOKDOWN statement can be adjusted to your location. The values are stored as strikes per minute with the largest number (blue LED 8) first, ending with 0 (all LEDs off). With smaller numbers, the LEDs light with less activity. In some areas of the country, summer thunderstorms produce over 1000 strikes per minute. The values used in the prototype LOOKDOWN are: LOOKDOWN sum, >=[1500, 1000, 500, 250, 100, 50, 10, 1, 0], index. While initially experimenting with the lightning activity monitor, you may wish to change the values until the BASIC STAMP lightning activity display works to your satisfaction!

Lightning Activity Monitor Sensing Head Assembly Parts List

R1	15-kΩ [¼]-W reslstor
R2	100-kΩ [¼]-W resistor
R3	1-MΩ [¼]-W resistor
R4	100-Ω [¼]-W resistor
Q1, Q2	2N3904 transistor (RadioShack 276-2016)
D1, D2	1N914 diode (RadioShack 276-1122)
BATT	9-V transistor radio battery

P1	Cable connector
Miscellaneous	9-V battery connector (RadioShack 270-324)
	Perfboard (RadioShack 276-1395)
	Red LED (for test) (RadioShack 276-307)
	RG58 coax (RadioShack 278-1314)

Lighting Activity Monitor Main Display Unit Parts List

U1	4N27 optoisolator (Digi-Key 4N27ND)
U2	BASIC STAMP 2 microcontroller (Parallax BS2-IC)
U3	SN74LS240N octal buffer (Digi-Key 296-1651-5-ND)
U4	LM7805 5-V regulator IC
R5	330-Ω $^1/_4$-W resistor
R6	1-kΩ $^1/_4$-W resistor
R7, R8, R9, R10	470-Ω $^1/_4$-W resistor
R11, R12, R13,	470-Ω $^1/_4$-W resistor
R14	270-Ω $^1/_4$-W resistor
C1	0.1-μF 50-V disk capacitor
C2	2.2-μF 50-V electrolytic capacitor
L1, L2	Green LED (RadioShack 276-304)
L3, L4	Yellow LED (RadioShack 276-301)
L5, L6	Orange LED (RadioShack 276-306)
L7	Red LED (RadioShack 276-307)
L8	Blue LED (RadioShack 276-311)
BZ	Piezo buzzer (RadioShack 273-066)
J1	Mating connector from P1 sensing head unit (RadioShack 276-1388)
J2	DB-9 female RS-232 connector
Miscellaneous	STAMP 2 carrier board or custom PC board

```
'bslam.bs2
' by  T. Bitson
' version... 1.1 -


' This program displays lightning activity via 8 LEDs connected
' to a BS2 via a SN74LS240N buffer. The lightning is detected by
' I/O port 0 pin 5 and is counted for 6 seconds. Ten 6 second periods
' are summed to calculate the number of lightning strikes per minute
(SPM).
' The resulting SPM is compared to a series of threshold values in a
LOOKDOWN
' statement to determine the number of LEDs to light. The peak value is
also
' stored and the highest reading is displayed on the LEDs on every other
loop.


' I/O Definitions
'device i/o    port   pin     function
'LED 1<-       8      13      output - green led
'LED 2<-       9      14      output - green led
'LED 3<-       10     15      output - yellow led
'LED 4<-       11     16      output - yellow led
'LED 5<-       12     17      output - yellow led
'LED 6<-       13     18      output - yellow led
'LED 7<-       14     19      output - red led
'LED 8<-       15     20      output - blue led


'LSU           ->     0    5        input - lightning sensor input


' Constants


LSU           con     0             ' lightning sensor input pin


' Variables
counts var    byte(10)              ' measured counts array
sum           var    word           ' sum of 10 6-second measurments
index         var    byte           ' index value for lookup/lookdown
leds          var    byte           ' led output value
oldPeak       var    byte           ' peak led value
newPeak       var    byte           ' maximum value
showPeak      var    bit            ' bit to indicate to show peak
i             var    byte           ' loop counter


'  Initialization
' do an led lamp test to start


DIRS = $FF00                 ' P0 - P7: Inputs P8 - P15: Outputs
```

Listing 7-1 Lightning activity monitor program.

```
leds = %11111111                          ' all leds on
FOR i = 0 TO 8                            ' step thru all 8 leds
      OUTH = leds                         ' turn on selected leds
      PAUSE 250                           ' wait a bit
      leds = leds/2                       ' turn off upper led
NEXT


' Main Code
loop:                                     ' main program loop
COUNT LSU, 6000, counts(0)                ' count lightning strikes
DEBUG dec ? counts
sum = 0                                   ' clear sum
FOR i = 0 TO 9                            ' add 10 6 second periods
      sum = sum + counts(i)               ' sum is total strikes for the last minute
NEXT
DEBUG dec ? sum


'  index =  0   1   2   3   4   5   6   7   8
LOOKDOWN sum,>=[400, 200, 100,  50,  25,  10,   5,   1,   0], index
LOOKUP index,  [$FF, $7F, $3F, $1F, $0F, $07, $03, $01, $00], leds
LOOKUP index,  [$80, $40, $20, $10, $08, $04, $02, $00, $00], newPeak


' check if new peak if greater than old peak
if newPeak <= oldPeak then doLeds
oldPeak = newPeak
goto doLeds
shiftValues:                                  ' loop thru array
FOR i = 8 TO 0
      counts(i+1) = counts(i)                 ' shift array values to next position
NEXT



showPeak = showPeak + 1                      ' alternate between 0 and 1


GOTO loop                                    ' loop for more


' Subroutines
doLeds:
if showPeak then doLedsWithPeak
OUTH = leds
GOTO shiftValues
doLedsWithPeak
OUTH = leds  |  oldPeak
GOTO shiftValues
```

Listing 7-1 **Lightning activity monitor program (*Continued*).**

8

TWELVE-CHANNEL RADIO REMOTE CONTROL SYSTEM

CONTENTS AT A GLANCE

The 12-channel radio remote control system is an interesting and fun project that will permit you to remotely control up to 12 devices. This unique control system utilizes a standard PS/2 keyboard with a special keyboard controller chip. The low-cost keyboard chip allows great flexibility in a STAMP 2–based circuit. With the full keyboard at your disposal, your control and communications capabilities are expanded enormously.

Transmitter

Our 12-channel remote control system revolves around two basic building blocks: a transmitter and a receiver. The transmitter section begins with an Abacom RT5 AM data transmitter. This low-cost transmitter measures about 1 in square and consumes only 4 mA. The range of the RT5 transmitter is about 70 m. This transmitter uses thick-film hybrid technology and requires no adjustments. You simply have to choose the correct matching frequency for the transmitter and receiver. This easy-to-use transmitter module sports four pinouts, which are shown in Fig. 8-1. The transmitter is powered from a 5-Vdc power supply at pin 1. Ground is applied to pin 2 and data input is coupled to pin 6. A quarter-wave vertical or half-wave dipole antenna can be connected to pin 7.

The transmitter module is driven by the keyboard encoder chip at U2, illustrated in Fig. 8-2. The keyboard encoder is a Katbrd IBM AT keyboard to RS-232 serial converter kit, available from ham radio operator K1EL, and is shown in App. 1 on the CD-ROM. The Katbrd encoder is an 8-pin specially programmed PIC. A standard PS/2 keyboard is fed to a chassis-mounted keyboard jack at J1. The keyboard jack is supplied with + 5 V at pin 5, while ground is applied to pin 2. The two active lines are the CLK, or clock, signal at pin 1 of J1 and the data signal at pin 3. The keyboard encoder can operate at two baud rates, i.e., 9600 and 1200. When pin 4 is high, 9600 baud is selected; when pin 4 is low, 1200 baud is selected. The output sensing at pin 6 is used to configure either true high marking or true low marking output. This pin allows maximum flexibility in interfacing the keyboard encoder to a transmitter or microprocessor. A reset output presents a low true reset pulse when CTL, ALT, and DELETE are pressed together. Pin 2 of the keyboard encoder was not utilized in our project.

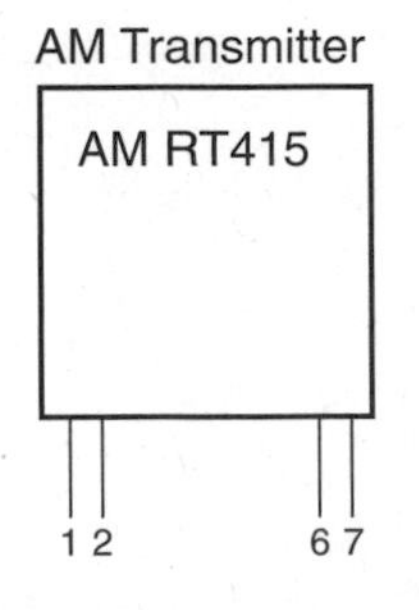

RT4	RT5	Function
1	1	Vcc (+)
2	2	GND
3	6	Data in
4	7	Ant

Figure 8-1 AM data receiver.

A 5-V regulator at U1 is used to provide a stable 5-V power source for the keyboard encoder and the RT5 transmitter module. A 9-V transistor radio battery or a 12-V rechargeable gel-cell battery could be used to power the transmitter board, which can be charged from a small "wall wart" power supply.

The transmitter board can be built on a small PC board or perfboard. The prototype was constructed on a 2- by 3-in PC board. When constructing the transmitter/encoder circuit, it is advisable to install an integrated circuit socket for the encoder chip. Pay particular attention, when installing the keyboard encoder, to orient the chip correctly in the IC socket. The encoder chip will have a notch on the top center or a small circle on the upper left corner, which denotes the top of the IC. Be careful to observe the correct polarity when installing the capacitors and the transmitter module. The keyboard encoder, transmitter module, and regulator will fit into a small metal enclosure. An aluminum chassis box is recommended, so the transmitter will not be affected by external electrical noise or hand capacitance. The entire circuit board is dwarfed by the keyboard connector soldered at one edge of the transmitter board. An RCA jack or pin jack can be installed on the enclosure in order to connect up the external antenna. An SPST slide switch or a mini toggle switch could be used to apply power to the regulator chip. Last, you will need to construct an antenna for the transmitter board. You can elect to construct either a quarter-wave vertical antenna or a half-wave horizontal dipole antenna for your transmitter. Be sure to ground the encoder and transmitter to the antenna ground. A quarter-wave antenna for 418 MHz is about 17 cm long. A quarter-wave vertical antenna is referenced against the ground plane of the transmitter and circuit board ground as shown in Fig. 8-3. If you construct a half-wave dipole antenna, you will need to reference one side of the dipole to circuit ground as depicted in Fig. 8-4. Note that if you construct a vertical antenna for the transmitter, you will need to construct a vertical antenna for the receiver; conversely, if you elect to build a dipole antenna for the transmitter, you will have to build a dipole for the receiver. It is important to have the same antenna polarization to transfer the maximum RF energy between the transmitter and receiver.

Receiver

The second half of our 12-channel radio remote control system revolves around the Abacom AM-HRR3-418 receiver and the alternative STAMP 2 microprocessor. The AM superregenerative receiver module is a low-cost, highly stable hybrid design. The receiver module pinout is shown in Fig. 8-5. The receiver module is placed on 0.1–in. PCB holes and measures 38 by 13.7 mm. The receiver has very low power consumption at 2.5 mA average. A block diagram of the AM-HRR3 receiver is shown in Fig. 8-6. The antenna lead-in feeds the RF preamp input section, which in turn is coupled to the RF oscillator. The RF oscillator is next fed to a low-pass filter section and then out to the AF or audio amplifier portion of the module. Finally, the AF amplifier is connected to a comparator, which presents the data output. Power and ground are applied to multiple pins on the

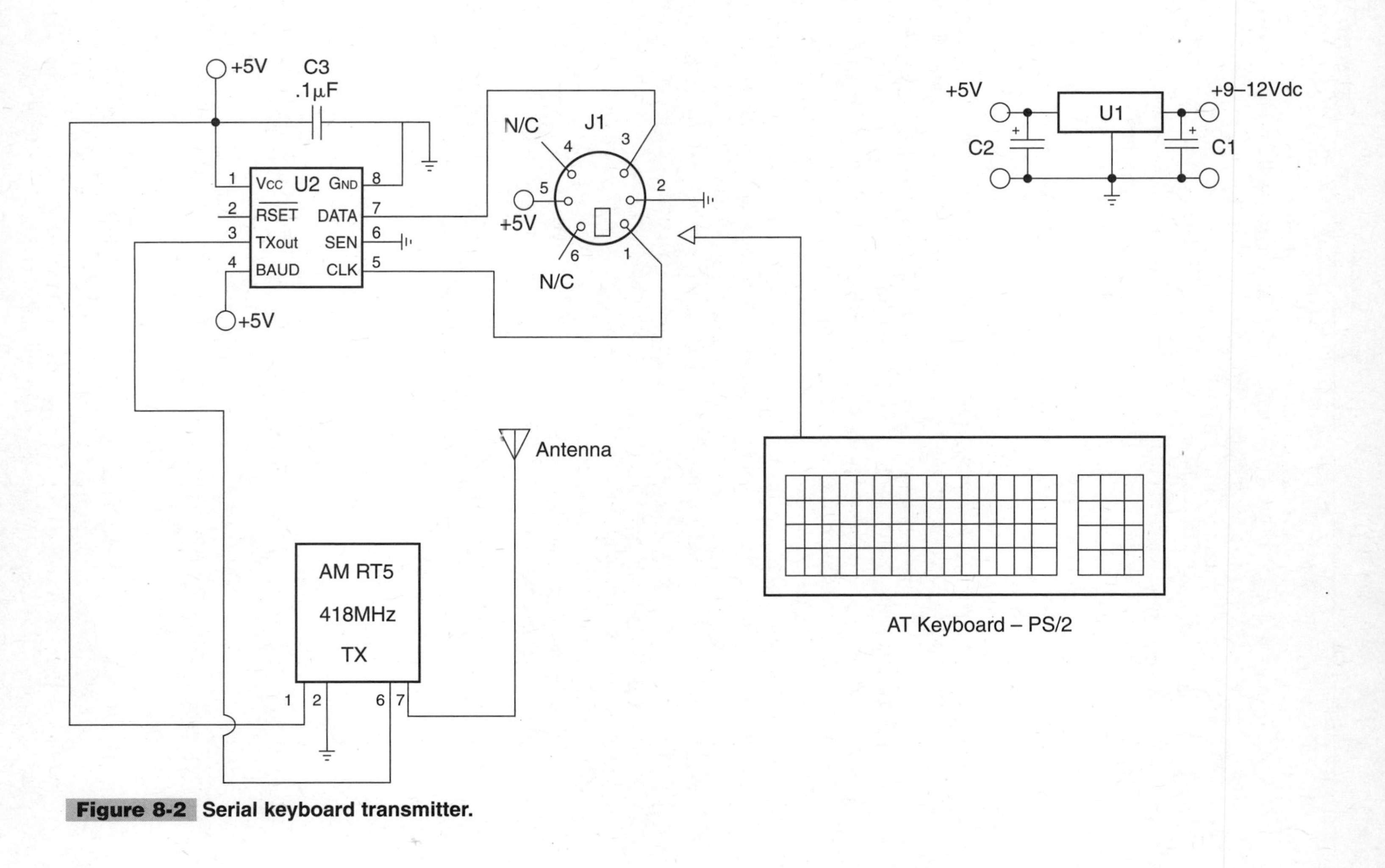

Figure 8-2 **Serial keyboard transmitter.**

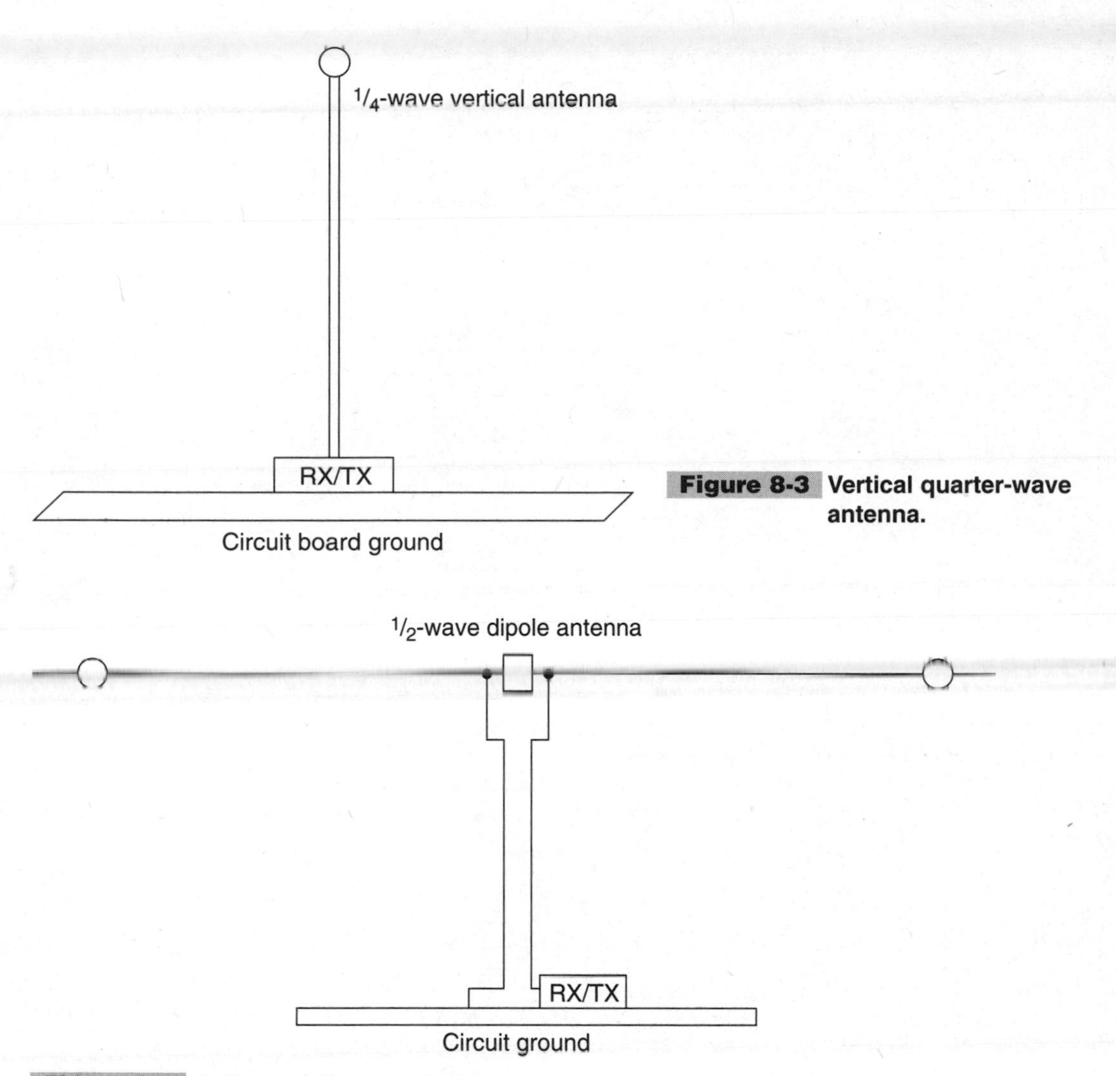

Figure 8-3 Vertical quarter-wave antenna.

Figure 8-4 Half-wave dipole antenna.

receiver. A 5-V power source is applied to pins 1, 10, 12, and 15. Ground is applied to pins 2, 7, and 11. A quarter-wave vertical or half-wave horizontal dipole antenna is connected to pin 3. Pin 13 is designated a test pin. In our 12-channel remote control system, pin 14 is used as the data output pin to the STAMP 2 controller.

The heart of the 12-channel radio remote control system is the alternative BASIC STAMP 2 microprocessor at U1 in Fig. 8-7. The receiver module is connected directly to the STAMP 2 via receiver output on pin 14, which is used to convey data to the STAMP 2. Remember the add 10 rule on alternative STAMP 2; P0 is really pin 10.

The MAX232 chip converts the STAMP 2 I/Os to true RS-232 serial level so that the STAMP 2 can communicate with your PC for programming. The MAX232 is connected to a DB-9 female to allow programming of the STAMP 2. You might consider installing a 4-pin male header on the PC board. Then you could make a programming

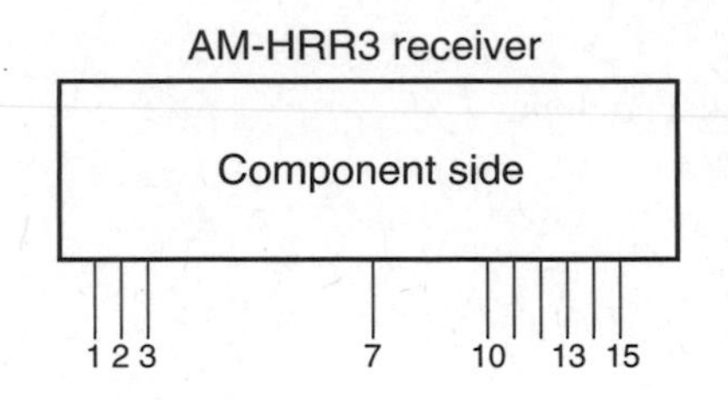

Pin #	Pin name	Pin #	Pin name
1	RF +VCC	8, 9	N/C
2	RF GND	10	AF +VCC
3	Data in	11	AF GND
4	N/C	12	AF +VCC
5	N/C	13	Test
6	N/C	14	Data out
7	RF GND	15	AF +VCC

Figure 8-5 Abacom AM-HRR3 data receiver.

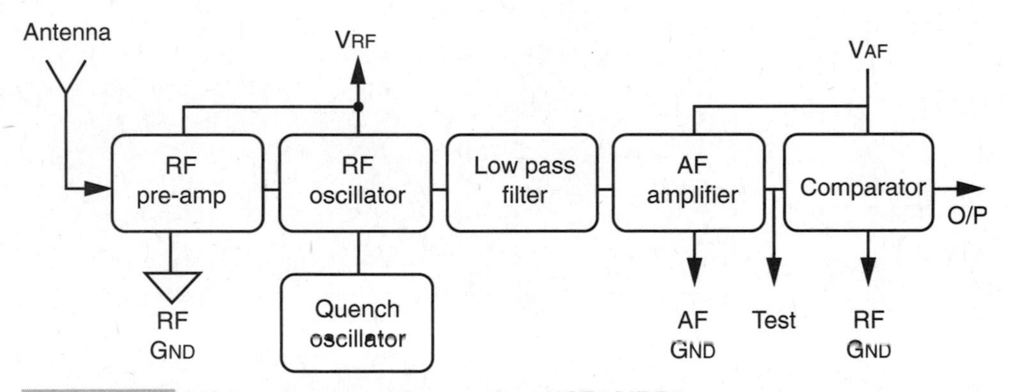

Figure 8-6 AM superregenerative receiver HRR3/HRR6.

cable with a 4-pin female header at one end and a DB-9 at the other end. The MAX232 requires only four external capacitors to operate. The DTR pin, or pin 4, of the DB-9 is connected to pin 13 of the MAX232, and the RX pin 2 is connected to pin 14 of U3. The TX pin 3 of J1 is fed to pin 14 on the U3 and ground on J1 is connected to the system ground.

Pushbutton S1 is utilized as a system reset button, which shorts pin 28 to ground momentarily. The alternative STAMP 2 processor requires a 24LC16B serial EEPROM memory chip, shown at U2. Pins 5 and 6 of the EEPROM are coupled to the STAMP 2 via pins 6 and 7. Most of the STAMP 2 I/O pins are used as outputs in this application. Beginning with P4 and continuing though P15, these outputs are all used to drive transistors, i.e., Q1 through Q12. The transistors are utilized to drive relays from RY1 through RY12. The relays are low-voltage, low-current SPST minirelays.

The twelve-channel radio remote control was built on a 4- by 6-in circuit board. The relays occupy most of the real estate on the controller's PC board. The receiver, STAMP 2,

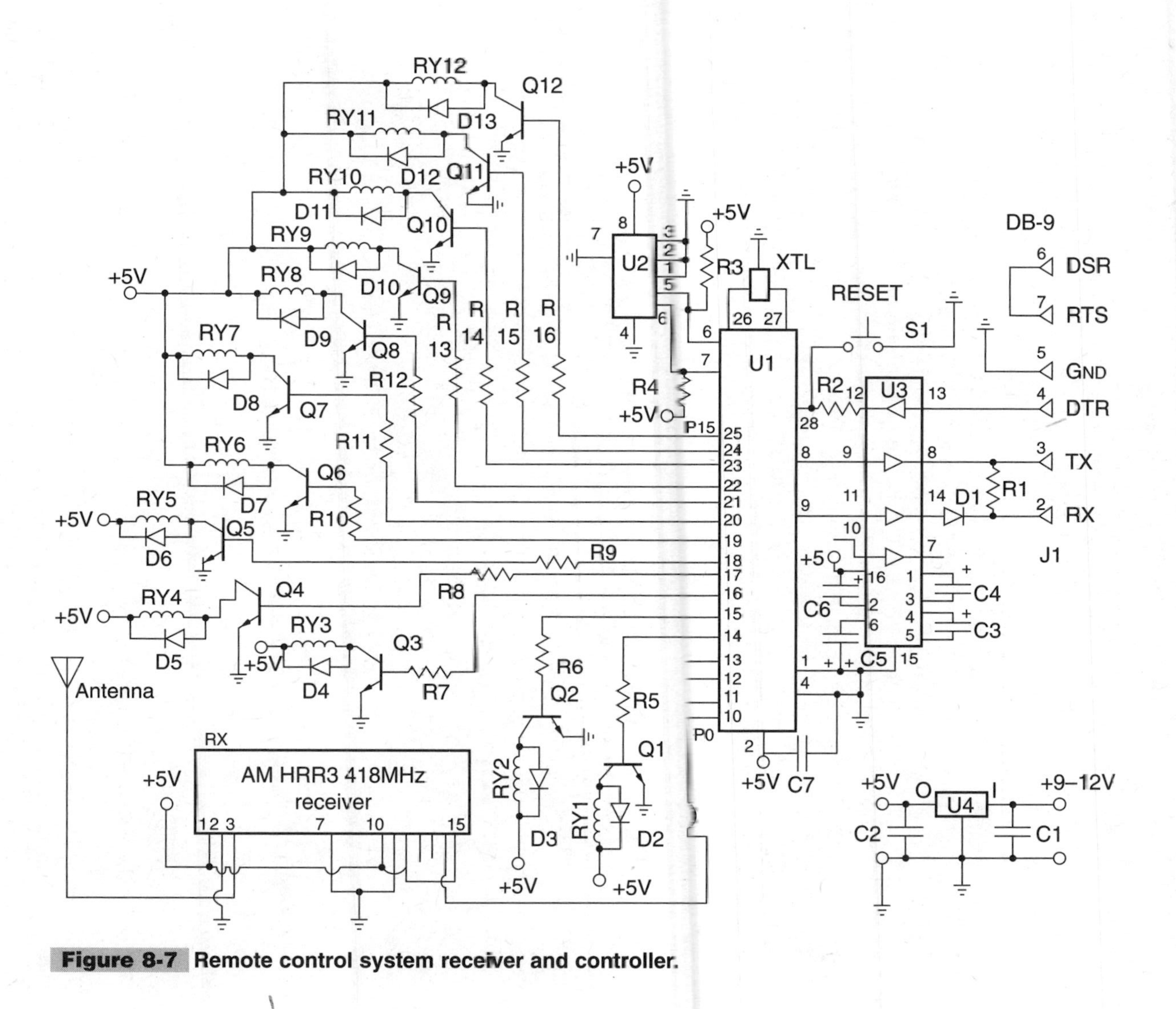

Figure 8-7 Remote control system receiver and controller.

and support chips fill one end of the board and the relays and transistors fill the rest of the PC board. It is important to keep the distances from the receiver to the STAMP 2 and from the receiver to its power source as short as possible. When constructing the receiver/controller board, be sure to use IC sockets, in case a board problem ever develops. Take care, when installing the ICs, to orient them correctly into the IC socket. Every integrated circuit contains either a notch cut into the top or a small circle cut-out on the top left of the IC to indicate the top of the IC. There are quite a few diodes on the PC board, so care should be taken to observe polarity when installing them. Also be careful when installing the electrolytic capacitors around the MAX232 as well as around the regulator chip.

The LM7805 regulator at U4 provides 5 Vdc to power the receiver and STAMP 2. A 9- to 12-V wall wart power supply could be used to power the regulator. You will need to create a quarter-wave vertical antenna or a half-wave dipole for the receiver/controller, as you did with the transmitter. Be sure to use the same polarization for both antennas. If you use a vertical antenna for the transmitter, you must also use a vertical antenna for the receiver.

Once the main circuit board is completed, be sure to carefully recheck it to ensure correct polarity for the diodes, capacitors, and integrated circuits. Once this has been completed, you can make up a programming cable, apply power to the system, and perform a "smoke test." If everything appears OK, then you can proceed to load the receiver/controller program from your PC into your BASIC STAMP 2. First, you will need to locate the STAMP 2 Windows editor program named STAMPW.EXE from the CD-ROM. Next, you will need to connect up your programming cable between your personal computer and the receiver/controller. Now, apply power to the receiver/controller board and download RADREMOTE.BS2 (Listing 8-1) to your receiver/controller board.

Using the 12-channel radio remote control is simple. With the keyboard installed and transmitter and receiver/controller switched on, you can, for example, type A0 on the keyboard to turn on the first control relay. Pressing A0 once again turns off lamp 1. Activating relay 12 is accomplished by pressing A11; turning relay 12 off is accomplished by pressing A11 once again. Your 12-channel radio remote control system is now ready to serve you!

Twelve-Channel Radio Remote Control Transmitter Parts List

U1	LM7805 5-V regulator
U2	Keyboard interface chip (Katbrd from K1EL)
C1	1-μF 35-V electrolytic capacitor
C2	4.7-μF 35-V electrolytic capacitor
C3	0.1-μF 35-V disk capacitor
TX	Abacom AM-RT5 transmitter module
J1	PS/2 keyboard connector
Miscellaneous	Wire, PC board, sockets, battery, etc.

Twelve-Channel Radio Remote Control Receiver/Controller Parts List

U1	STAMP 2 (alternative)
U2	24LC16B EEPROM memory chip
U3	MAX233 serial interface chip
U4	LM7805 5-V regulator IC
R1, R2, R3, R4	4.7-kΩ $^1/_4$-W resistor
R5, R6, R7, R8, R9	1-kΩ $^1/_4$-W resistor
R10, R11, R12, R13	1-kΩ $^1/_4$-W resistor
R14, R15, R16	1-kΩ $^1/_4$-W resistor
C1, C3, C4, C5, C6	1-μF 35-V electrolytic capacitor
C2	4.4-μF 35-V electrolytic capacitor
C7	0.1-μF 35-v disk capacitor
D1	1N914 silicon diode
D2, D3, D4, D5	1N4002 silicon diode
D6, D7, D8, D9	1N4002 silicon diode
D10, D11, D12, D13	1N4002 silicon diode
XTL	20-MHz ceramic resonator
Q1, Q2, Q3, Q4	2N2222 npn transistor
Q5, Q6, Q7, Q8	2N2222 npn transistor
Q9, Q10, Q11, Q12	2N2222 npn transistor
RY1, RY2, RY3	5-V minirelay
RY4, RY5, RY6	5-V minirelay
RY7, RY8, RY9	5-V minirelay
R10, R11, R12	5-V minirelay
S1	SPST pushbutton reset switch (normally open)
Miscellaneous	IC sockets, wire, PC board, headers, etc.

```
'radremote.bs2
stateAN0    var    bit
stateAN1    var    bit
stateAN2    var    bit
stateAN3    var    bit


stateBN0    var    bit
stateBN1    var    bit
stateBN2    var    bit
stateBN3    var    bit


stateCN0    var    bit
stateCN1    var    bit
stateCN2    var    bit
stateCN3    var    bit


letter      var    byte
number      var    byte
baud        Con    17197    '1200 baud
pin         Con    0


lamp1       Con   15
lamp2       Con   14
lamp3       Con   13
lamp4       Con   12
lamp5       Con   11
lamp6       Con   10
lamp7       Con    9
lamp8       Con    8
lamp9       Con    7
lamp10      Con    6
lamp11      Con    5
lamp12      Con    4


   letterin:
     serin pin,baud,[letter]                  ' expect ascii 65,..,75
     'branch letter-65,[LA]
     debug letter
     branch letter-97,[La,Lb,Lc]              ' letter-65 is 0,1,2,..,10
     goto letterin                            ' bad letter


   getnumber:                                 ' subroutine
     serin pin,baud,[number]
     return



   LA
     gosub getnumber                          ' expect ascii 48,50,..,57
     branch number-48  ,[AN0,AN1,AN2,AN3]     ' routine to execute 0,1,,9
     goto letterin                            ' bad number
```

Listing 8-1 Twelve-channel radio remote control system program.

```
LB:
  gosub getnumber                              ' expect ascii 48,50,..,57
  branch number-48 ,[BN0,BN1,BN2,BN3]          ' B routine to execute
  goto letterin                                ' bad number


LC:
  gosub getnumber                              ' expect ascii 48,50,..,57
  branch number-48 ,[CN0,CN1,CN2,CN3]          ' C routine to execute
  goto letterin                                ' bad number


 AN0:
   high lamp1
   branch StateAN0,[AN0on,AN0off]
 AN0on:
   StateAN0=1
   high lamp1
   goto letterin
 AN0off:
   StateAN0=0
   low lamp1
   goto letterin


 AN1:
   high lamp2
   branch StateAN1,[AN1on,AN1off]
 AN1on:
   StateAN1=1
   high lamp2
   goto letterin
 AN1off:
   StateAN1=0
   low lamp2
   goto letterin


 AN2:
   high lamp3
   branch StateAN2,[AN2on,AN2off]
 AN2on:
   StateAN2=1
   high lamp3
   goto letterin
 AN2off:
   StateAN2=0
   low lamp3
   goto letterin


 AN3:
   high lamp4
   branch StateAN3,[AN3on,AN3off]
 AN3on:
   StateAN3=1
```

Listing 8-1 Twelve-channel radio remote control system program (*Continued*).

```
  high lamp4
  goto letterin
AN3off:
  StateAN3=0
  low lamp4
  goto letterin

BN0:
  high lamp5
  branch StateBN0,[BN0on,BN0off]
BN0on:
  StateBN0=1
  high lamp5
  goto letterin
BN0off:
  StateBN0=0
  low lamp5
  goto letterin


BN1:
  high lamp6
  branch StateBN1,[BN1on,BN1off]
BN1on:
  StateBN1=1
  high lamp6
  goto letterin
BN1off:
  stateBN1=0
  low lamp6
  goto letterin


BN2:
  high lamp7
  branch StateBN2,[BN2on,BN2off]
BN2on:
  StateBN2=1
  high lamp7
  goto letterin
BN2off:
  stateBN2=0
  low lamp7
  goto letterin


BN3:
  high lamp8
  branch StateBN3,[BN3on,BN3off]
BN3on:
  StateBN3=1
  high lamp8
  goto letterin
BN3off:
  stateBN3=0
  low lamp8
  goto letterin
```

Listing 8-1 **Twelve-channel radio remote control system program (*Continued*).**

```
CN0:
  high lamp9
  branch StateCN0,[CN0on,CN0off]
CN0on:
  StateCN0=1
  high lamp9
  goto letterin
CN0off:
  StateCN0=0
  low lamp9
  goto letterin


CN1:
  high lamp10
  branch StateCN1,[CN1on,CN1off]
CN1on:
  StateCN1=1
  high lamp10
  goto letterin
CN1off:
  StateCN1=0
  low lamp10
  goto letterin


CN2:
  high lamp11
  branch StateCN2,[CN2on,CN2off]
CN2on:
  StateCN2=1
  high lamp11
  goto letterin
CN2off:
  StateCN2=0
  low lamp11
  goto letterin


CN3:
  high lamp12
  branch StateCN3,[CN3on,CN3off]
CN3on:
  StateCN3=1
  high lamp11
  goto letterin
CN3off:
  StateCN3=0
  low lamp12
  goto letterin
```

Listing 8-1 **Twelve-channel radio remote control system program (*Continued*).**

9

CALLER ID/BLOCKER PROJECT

CONTENTS AT A GLANCE

Unwanted calls are highly interruptive and very annoying; why not screen your incoming calls? Everyone knows those pesky unsolicited callers always phone you when you're indisposed or just sitting down for dinner. Screen your calls from unwanted phone callers, bill collectors, your ex-wife/husband. The Caller ID/blocker project will allow you to identify up to 10 incoming telephone calls and display the phone numbers on a liquid crystal display (LCD). The Caller ID unit will also allow you to block incoming calls, if desired. The Caller ID project is shown in Fig. 9-1.

Basic Components

The Caller ID project centers around the BASIC STAMP 2 microprocessor, a Motorola MC145447 Caller ID chip, and a handful of external components, as shown in Fig. 9-2. The Caller ID connects across your phone line with the *ring* and *tip* connections. A metal oxide varistor (MOV) is placed across the phone line to limit damaging high-voltage spikes. Capacitors C2 and C4 couple the phone line to the input of the Caller ID chip. The ring and tip phone line wires are fed to the Caller ID chip's input op-amp and then on to a bandpass filter, followed by a demodulator, and finally to the valid data detector, shown in the block diagram of Fig. 9-3. The output of the valid data detector is the actual Caller ID data information. Two additional capacitors, C1 and C2, are placed across the line ahead of a diode bridge, which is used to rectify the ringing voltage. This ringing detect signal is then input to the ring detect pins on the Motorola MC145447 Caller ID chip, pins 3 and 4. The ring detect network is formed by resistors R3, R4, R5, and R6, along with C5. The ubiquitous 3.579-MHz color burst crystal forms the basic reference oscillator. The output at pin 7 of the Caller ID chip drives an npn transistor, which in turn drives the data indicator LED at DS1. The Caller ID data is next coupled to P19 on the BASIC STAMP 2. The ring detect output of the Caller ID chip is fed to P18 of the STAMP 2. The ring detect signal determines that the circuit should pick up, or initiate. Once the ring signal is detected, the STAMP 2 pulls in the relay at K1. Indicator DS5 indicates that the relay has pulled in. The indicator at DS4 is a monitor LED; it stays on

Figure 9-1 Caller ID project, as assembled.

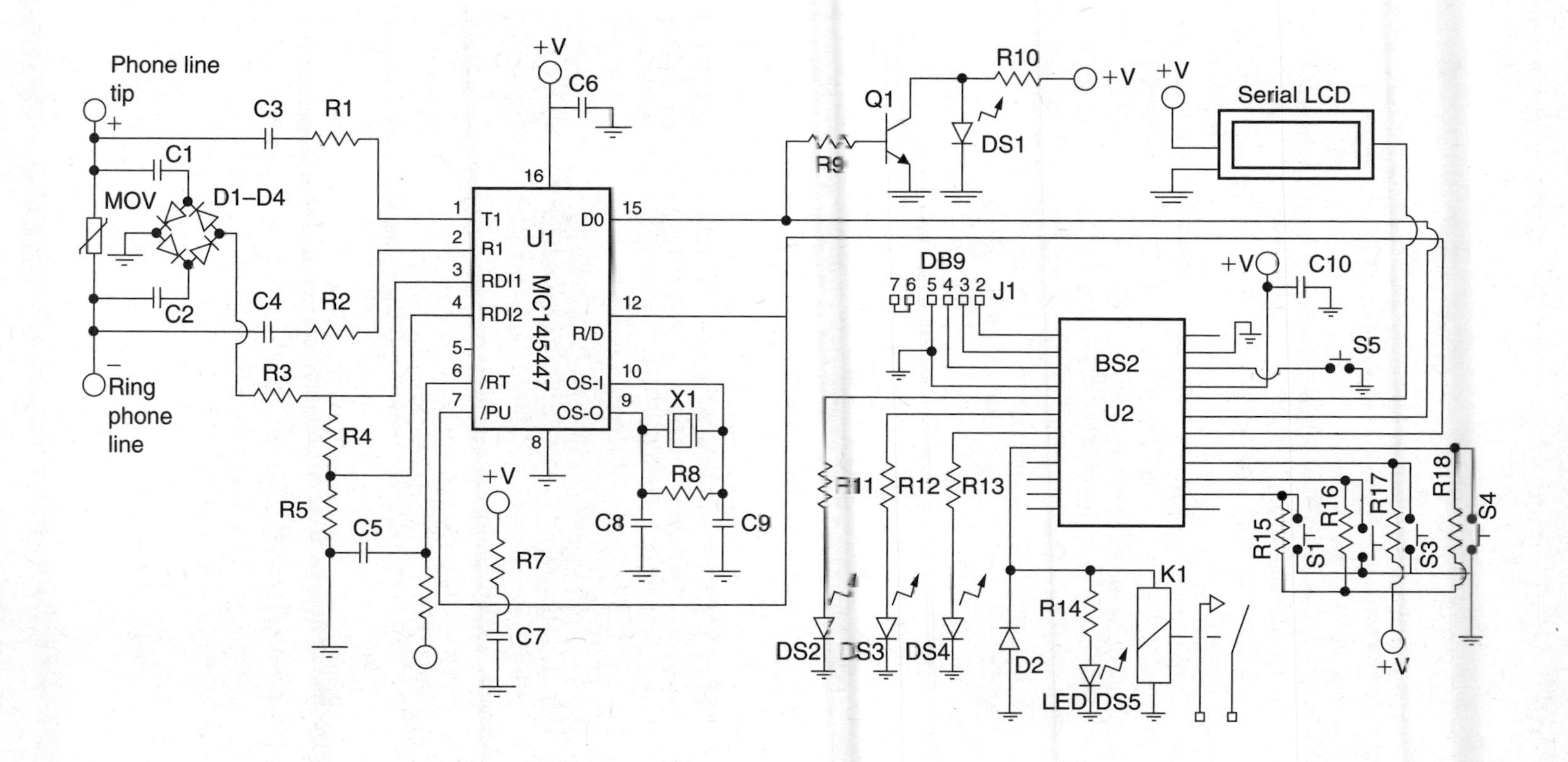

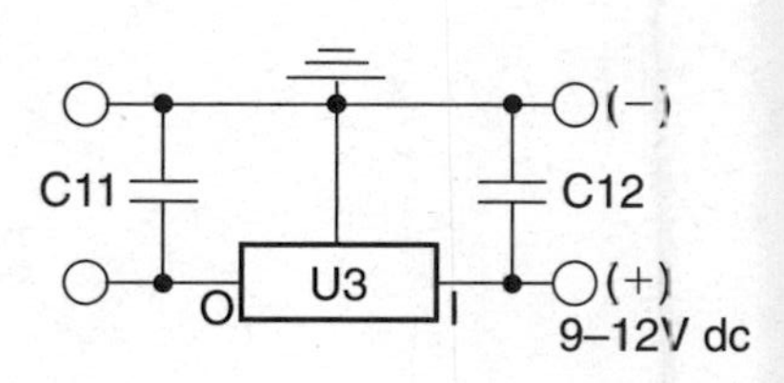

Figure 9-2 Caller ID project diagram.

normally and then toggles off when the Caller ID program is running. The indicator at DS3 is a troubleshooting LED; it goes high after the data has been read out of the LCD display unit. The last indicator at DS2 is driven by P0. This LED is also for troubleshooting; it goes high after the data has been successfully read into the MC145447.

There are four function switches on STAMP 2 pins P9 through P12; we'll discuss the functional operations later. Function A is activated by a normally open pushbutton switch at SW4. Function B is operated through SW3 at P11. Function C is implemented by SW1 at STAMP 2 pin P9. The RUN or START pushbutton is started by SW2 when the circuit is first energized. Note that all pushbutton switches are tied to high (5 V) and when pressed go to ground. The Caller ID display is connected to the STAMP 2 via pin P15. The LCD display is a serial single-wire "backpack" style from Scott Edwards Electronics at http://www.seetron.com.

Power to the Caller ID chip and the STAMP 2 is supplied by a 5-V regulator at U3. On the MC145447, power is applied to pin 16, while pin 21 is used on the STAMP 2. A system reset switch is shown at SW5, a normally open pushbutton switch. A 9- to 12-V "wall-wart" power supply can be used to power the Caller ID circuit ahead of the regulator at U3. Alternatively a 9-V transistor radio battery could be used, but a power supply is recommended.

The initial Caller ID prototype was constructed on two perfboards, as shown in Fig. 9-4. At a later date the Caller ID circuit was transferred to a glass-epoxy circuit board. You can elect to construct you own PC board or you could assemble the circuit on a STAMP 2 carrier board or perfboard if desired. When assembling the Caller ID project, try to obtain an integrated circuit socket for both the Caller ID chip and the BASIC STAMP 2 microprocessor, in the event of a chip failure in the future. It is also critical to observe the proper polarity when installing the capacitors throughout the circuit. Since there are a number of diodes in this circuit, you will need to insert the diodes in the correct direction. Pay particular attention to the correct orientation of the ICs when installing them.

The circuit board is placed in a 6- by 6-in plastic box. The pushbutton switches, power toggle switch, etc. are mounted on the front panel of the plastic enclosure, while the power cord and programming cable connector are installed on the back of the chassis box. The four programming wires from the STAMP 2 on pins 1 through 4 are brought out to a 4-pin male header (P1) on the circuit board. A programming cable is then constructed between the mating female header at J1, which is wired to a DB-9 female RS-232 connector. This connector facilitates programming the STAMP 2 when you are initially loading the Caller ID program.

Once the Caller ID circuit board has been finished and installed, you will need to apply power and connect up the programming cable between your programming PC and the STAMP 2 on the Caller ID board. Next you will have to locate the computer directory that houses the BASIC STAMP 2 Windows editor program called STAMPW.EXE and then finally load the Caller ID program titled CID23.BS2 (Listing 9-1).

Operation

Operation of the Caller ID project begins when the phone begins ringing. The ring detector tells the STAMP 2 to wake up and begin accepting data from the Caller ID chip. Upon

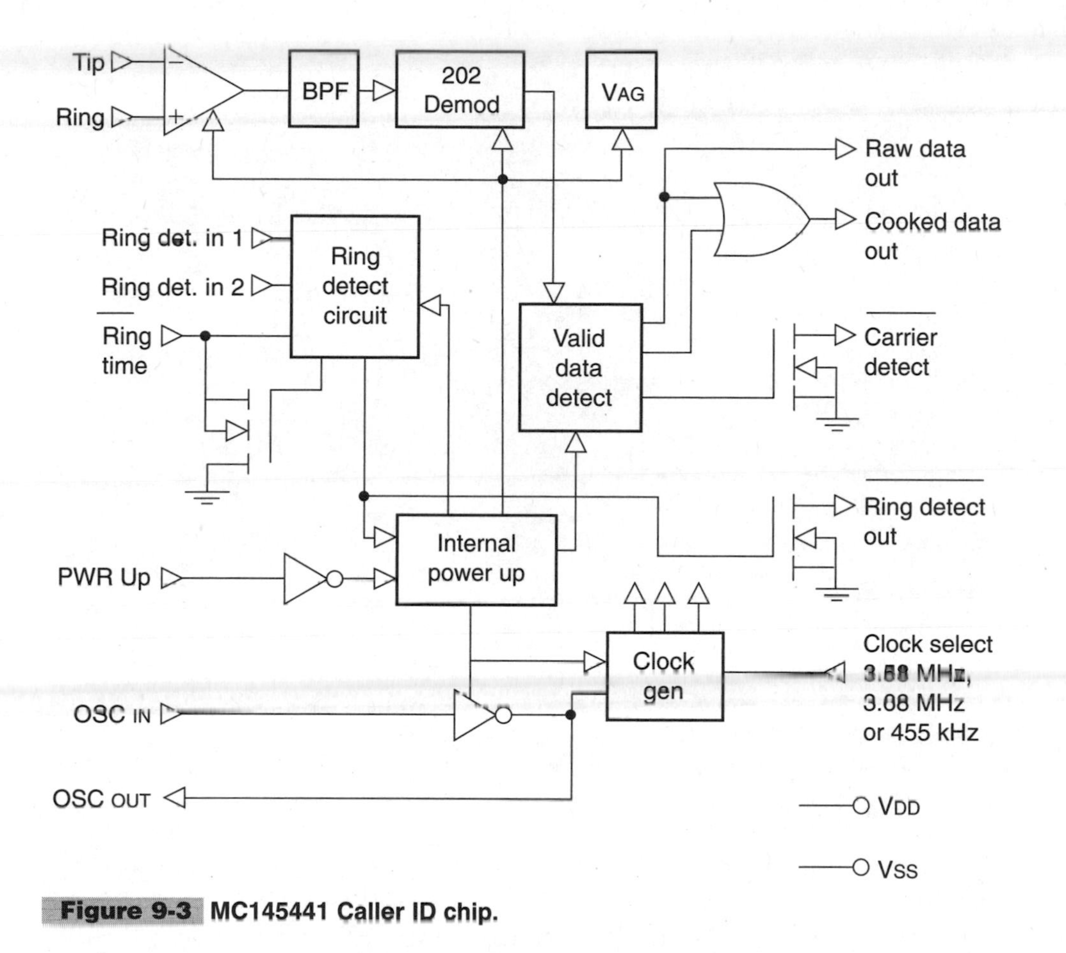

Figure 9-3 MC145441 Caller ID chip.

power-up, the LCD reads "Waiting"; then the STAMP 2 program reads in the Caller ID (CID) data, in the multiple data message format (MDMF). This format presents the number and name of the originating call. Note that the single data message format (SDMF) sends only the telephone numbers and will not work with this program in its present form. The program for this project displays the originating number on the LCD screen and stores up to 10 numbers, in the EEPROM. The system will overwrite the oldest numbers displayed as new ones are received. In order to see past numbers, you will need to press the Function A button once, or several times during a delay loop, to cycle back through the past read-ins.

The program will store up to 10 numbers in the EEPROM as numbers that are considered "blocked." To store a number in the EEPROM, simply press Function C and hold the button for about a second until the LED pin 2 is out. To see what numbers are blocked, cycle through the blocked numbers by depressing Function B as you did for Function A; note that the blocked numbers run from numbers 11 to 20. The blocked number will be stored at the memory location of the last read-out number as cycled from Function B. The

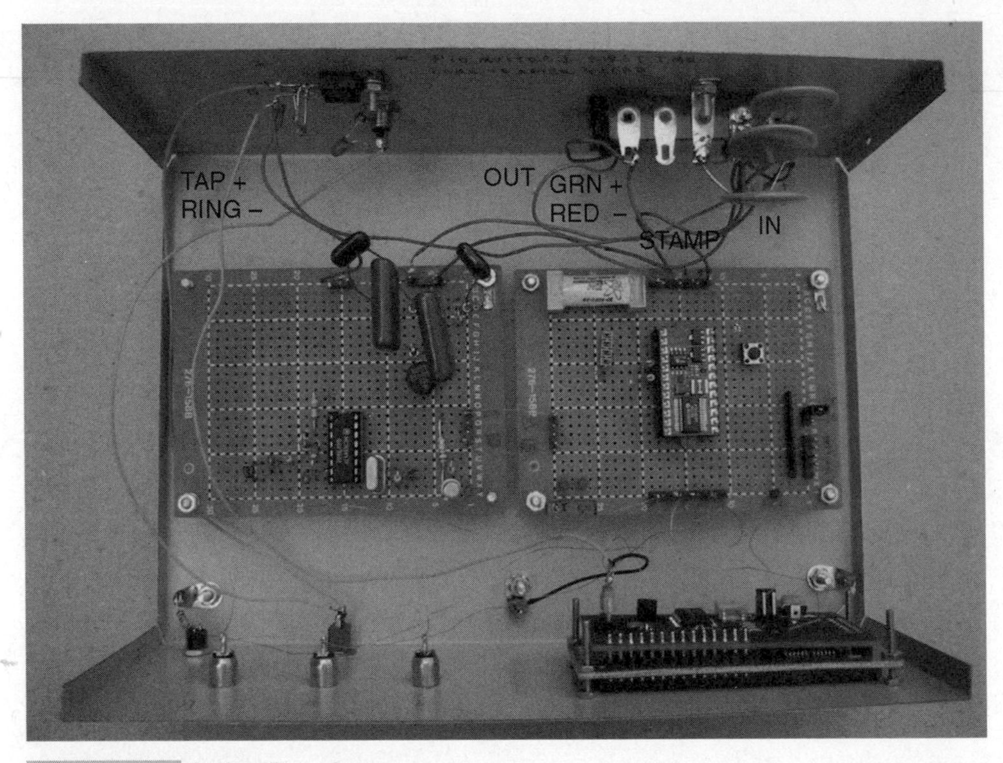

Figure 9-4 Caller ID prototype.

number that is stored in the blocked memory will be the last one read out as cycled by Function A.

In order to erase a number that has been blocked, you will need to hold down the Function C button until the LED on pin 2 goes out, and then momentarily press the Function B button. Notice that zeros will replace the blocked number (see Table 9-1).

To turn on the call blocking feature, you will need to hold down the Function C button until the LED on pin 2 goes out. Next momentarily press the Function A button. The LCD display will cycle through "Call Blocking Off," "Inhibit Blocked Calls," and "Connect Not Blocked Calls" for each push of Function A. The last two options are nearly the same except:

1 When there is an error in data read, the first will not allow the call to go through, and the second will not allow the call to go through.
2 When there is a call from a blocked number, the first will allow one ring (until the CID data has been processed), while the second will not ring at all.
3 When there is a call from a not-blocked number, the second will silence the first ring.

Orient your Caller ID unit in a location where it would be most convenient for your general use. Your Caller ID is now ready to serve you!

```
'CID23.BS2
'this program reads in and stores caller-id data from a Motorola MC145447
'Numbers which are designated blocked, reads out numbers on a serial LCD
unit
'Program Notes
'read in first world length; if #, read in #;if not # store an OTH(er)
'and fill rest of ram with ascii spaces
'tested the first (hex) word length was dec 39 for a #
'and dec 16 for out of area; I DISCARDED THIS OPTION SINCE
'read direct from MC145447
'relay is on at first
'fctc 0 (delay) -> fcta 0 -> fcta 1 -> fctc 1 =
'        cycle through blocking options
'fctc 0 (delay) store cid # in blked # register
'fctc 0 (delay) -> fctb 0 -> fctb 1  -> fctc 1 = store 00s in blked numbers
'        to erase a number
'fctc must be held down for a delay to prevent accidents
'start storing info with mem addr 12
'pin 15 is lcd output line
'pin 14 is data input line
'pin 13 is here we will come, data to be sent, line


'IF PIN 13 = 0, NRD, a ring has been detected, not ring detect
'IF PINS 12 OR 11 OR 9 = 0 = FCTA or FCTB or FCTC
'pin 12 is functiona input line, look at data, normally 1
'pin 11 is functionb input line, look at blocked #s, normally 1
'if pin 10 = 0 reset mem loc 0 (stored data), mem loc 1 (blocked #s)
'        and come up in blocking off via nb,bi
'tie pin 10 (skip) high with resistor and be able to jump to ground
'        the very first time only to set address counters to initial values
'pin 9 is functionc input line, store a blocked #, normally 1
'if you shut down or you reload the program you can save stored info
'THIS HAS FOR...NEXT EEPROM OUTPUT LOOP, SHOWS #: ########## via FUNCTION
'LIMIT 20 STORED NUMBERS
'SEE  DEFINITION BELOW: DTLMT IS FOR 10 STORED #S, NUMBERED 1-10
'BLLMT IS FOR 10 BLOCKED #S, NUMBERED 11-20
'READ OUT BLOCKED NUMBERS VIA FUNCTIONB, PIN 11
'READ OUT CID NUMBERS VIA FUNCTIONA, PIN 12
'STORE NUMBERS TO BLOCK VIA FUNCTIONC, PIN 9
'pin 0 is led for read in complete – test purposes only
'pin 1 is led for read out complete – test purposes only
'pin 2 is led for function subroutine in operation, use a led
'pin 3 is for relay
'output "WAITING" on power up, if error, after set blocking


'SET IN/OUT
input 14 : input 13 : input 12
input 11 : input 10 : input 9       'in lines
low 0 : low 1                       'out lines; set 2, 3 later
'DECLARE
'these are the ram spots that are read into
a       var     byte
b       var     byte
```

Listing 9-1 Caller ID program.

```
c        var     byte
d        var     byte
e        var     byte
f        var     byte
g        var     byte
h        var     byte
i        var     byte
j        var     byte
k        var     byte
'l       var     byte
'input lines
nrd      var     in13     'not Ring Detect from MC145447
fcta     var     in12     'function a
fctb     var     in11     'function b
skp      var     in10     'ground on very first power up to initialize

fctc     var     in9      'function c
'constants
dtst     con     12       'this is the start address for data
dtlmt    con     111      'the add. of the final data byte, 10 #s, 12-111
bllmt    con     211      'the add. of the final blk no byte, 10#s, 112-211
cidad    con     0        'this is the adr. of CID reception and storage
                          'and first read out
blnad    con     1        'this is the adr. of the blocked number
                          'read in and read out
cidro    con     2        'this is the adr. of the CID rdout when sequencing
bi       con     3        'this is the adr. of 1 inhibits blked call
nb       con     4        'this is the adr. of 1 connects line when not blked
                          'call
'more variables
x        var    byte      'scratch pad
y        var    byte      'hold addresses, etc.
z        var    byte      'scratch pad
qa       var    byte      'used in "test if # is blocked"
qb       var    byte
qc       var    byte
'THIS WILL RESET DATA COUNTERS AND BLOCKING IF PIN 10 IS LOW
      if skp = 1 then start1    'if pin 12 is high don't reset
      write cidad,dtst          'will be adr. of CID rcption
      write blnad,dtlmt+1       'will be adr. of blkd nos.
      write bi,0 : write nb,0   'come up with blocking off
'THIS IS THE SET-RESET SEGMENT FOR THE LCD
start1: pause 1000              'let lcd settle
        serout 15,396+$4000,[254,1] 'cls
        pause 1000
        serout 15,396+$4000,["WAITING"] 'at start, if error,
                                'after set blocking
'THIS IS THE MAIN HOLD LOOP
'GO TO THE BEGINNING OF SERIN-ING THE DATA IF PIN13 = 0
'OR GO TO A FUNCTION SUBROUTINE IF PINS 12 OR 11 OR 9 = 0
start2: high 2 : high 3         'set led, relay each time
start2a:if nrd = 0 then start3x 'nrd = 0, data soon
        if fcta = 0 then fnctn  'function switch pressed
        if fctb = 0 then fnctn
        if fctc = 0 then zz     'put in delay to prevents oops
        goto start2a            'loop
zz:     pause 2000              'to prevent accidents
```

Listing 9-1 Caller ID program (*Continued*).

```
        if fctc = 1 then start2a 'oops has been prevented
        goto fnctn
'COME HERE FROM ABOVE IF NRD HAS BEEN LOW
'IF "CONNECT NOT BLOCKED" IS SET TURN OFF RELAY
'NOTE HOW THIS WORKS IN CONJUNCTION WITH
'  "TEST TO SEE IF THE NUMBER IS A BLOCKED ONE AND HOW BLOCKING IS SET"
'  LATER
start3x:read nb,z          'check if "connect not blocked" is set
        if z=1 then goon   'if it is on, low 3
        goto start3        'if "connect not blocked" not set, do nil
goon: low 3
'THIS IS THE MAIN READ-IN SECTION OF THE PROGRAM
'the first CID info is hex 80 to denote MDMF, the format this program
'is written for; wait for that
'1. then read in the next byte which will be large if full data is sent
'2. or small if is "out of area" or "private"
'after waiting hex 80, read in word-length and then decide whether
'3. to wait for hex 02 = number parameter or to go to reason for absence
'4. see "notes to myself" for how I got 28, half-way between 16 and 39
'1-4 HAS BEEN CHANGED; THE STAMP APPEARS NOT TO BE FAST ENOUGH FOR
'RELIABLE OPERATION OF THIS FEATURE OF THE EARLIER PROGRAM CID23.BS2
start3: low 2              'indicator led
        serin 14,813,6000,start1,[WAIT (128)]
                           'wait hex 80
               serin 14,813,1000,nonum,[WAIT (02),a,b,c,d,e,f,g,h,i,j,k]
                           'read in word-length, then the number
                           'the word-length will be dumped
        goto anum

nonum:      b="O":c="T":d="H"     'OTH for other
      e=" ":f=" ":g=" ":h=" ":i=" ":j=" ":k=" "
      'set spaces in other variables
anum:  high 0                                              'got this far


'THIS IS THE MAIN OUTPUT SECTION OF THE PROGRAM
       serout 15,396+$4000,[254,1]                          'cls
       serout 15,396+$4000,[b,c,d," ",e,f,g," ",h,i,j,k]  'output
       high 1                                               'got this far
'HERE WE STORE DATA IN EEPROM
       read cidad,y              'define y
       gosub st      'store via subroutine
       if y = dtlmt+1 then far1 'dtlmt-dtst+1/10 #s, start with dtst again
       goto far2                 'ADDR. COUNTER ALWAYS NEXT INPUT
far1:  write cidad,dtst          'new cycle
       goto test
far2:  write cidad,y             'save how far the memory has been used+1
       goto test
'SUBROUTINE
st:    write y,b                 'STORE THE DATA IN EEPROM
       y=y+1
       write y,c
       y=y+1
```

Listing 9-1 Caller ID program (*Continued*).

```
        write y,d
        y=y+1
        write y,e
        y=y+1
        write y,f
        y=y+1
        write y,g
        y=y+1
        write y,h
        y=y+1
        write y,i
        y=y+1
        write y,j
        y=y+1
        write y,k
        y=y+1             'ready for next one
return
'TEST TO SEE IF THE NUMBER IS A BLOCKED ONE AND HOW BLOCKING IS SET
test:   qc = bllmt-dtlmt-10          'amount of blocked mem. - 10
        for qa = 0 to qc  step 10    'IS # BLOCKED?
        for qb = 1 to 10 step 1      'compare
        read dtlmt+qa+qb,y           'compare
        read cidad,x                 'compare
        read x+qb-11,z               'compare
        if y = z then here1          'compare
        goto here2                   'compare
here1:  next                         'compare
        goto test0                   'was a blocked #
here2:  next                         'compare
        goto fnctaQ                  'it wasn't a blocked #, go on
test0:  read  bi,y
        if y = 0 then fnctnQ         'INHIBIT BLOCKED off, go on
        low 3                        'if bi is 1 inhibit
        goto fnctnQ                  'go on
'NOTE HOW THESE DECISIONS WORK IN CONJUNCTION WITH EARLIER:
' "IF 'CONNECT NOT BLOCKED' IS SET TURN OFF RELAY"
'STAY AS LONG AS RINGING
fnctaQ: high 3                 'turn on relay
fnctnQ: x=0                    'scratchpad = 0
        goto stay2             'avoid a reset
stay1:  x=0
stay2:  if nrd = 0 then stay1  'do nothing as long as ringing
        pause 50               'stopped ringing
        x=x+1                  'add 1 every 50 ms
        if x < 100 then stay2  'if less than 5 sec keep testing
        low 0 : low 1          'reset 0, 1
        goto start2
'THE FUNCTION SWITCH COMES HERE
'fncta: high 3                'turn on 3 if need be
fnctn:  low 2                 'COME HERE FROM FUNCTION, blip led
        pause 250             'debounce
        if fcta = 0 then so1  'go to stored output
        if fctb = 0 then so3
        if fctc = 0 then mem1 'goto holding for function c
        pause 250             'debounce more
'       low 0                 'reset LEDs that confirmed earlier work
'       low 1
        goto start2
```

Listing 9-1 Caller ID program (*Continued*).

```
'HERE IS THE STORED OUTPUT
'STORED NUMBERS
so1:   read cidad,y                   'y is addr. of next store, in cidad
       if y <> dtst then so2          'reset cidad for highest data addr.
       y=dtlmt+1                      'if recycling
so2:   gosub socid                    'OUTPUT DATA STORE IN EEPROM
       write cidro,y                  'THIS WILL BE USED IN STORING READ
                                      'WHERE-LEFT-OFF INTO BLKED # MEM
       if y <= dtst then resty1       'reset y when first # is done
       goto hold1
resty1: y=dtlmt+1
       goto hold1
'BLOCKED NUMBERS
so3:   read blnad,y
       goto so5
so4:   y=y+10
       if y<=bllmt then so5
       y=dtlmt+1
so5:   gosub sobln
       write blnad,y
       goto hold2
       'THIS SUBROUTINE OUTPUTS THE DATA STORED IN EEPROM
socid: y=y-10
sobln: serout 15,396+$4000,[254,192]     'down a line; come here for
        z=y-2                            'consecutive stored output
        z=z/10                           'get entry # digit(s)
        serout 15,396+$4000,[dec z,": "] 'this is  entry #
        for z = y to y+9                 'OUTPUT EEPROMED DATA HERE
        read z,x
        serout 15,396+$4000,[x]
        next
        serout 15,396+$4000,["      "]   'get rid of past
return


'WHAT TO DO NEXT IN FUNCTION MODE
hold1: if fcta = 0 then hold1
hold2: if fctb = 0 then hold2
        pause 100               'debounce
        z=0
hold4: z=z+1                    'do again if another fcta=0
        pause 10                'within one second
        if fcta = 0 then so2
        if fctb = 0 then so4
        if z = 100 then fnctn   'return to function to reset leds and go
        goto hold4              'back to main wait loop
'STORE A BLOCKED # OR STORE 0S
mem1:   pause 100                   'debounce
mem2:   if fctb = 0 then do1        'store 0s if fctb = 0
        if fcta = 0 then do2
        if fctc  = 0 then mem2      'to erase
        goto on1
do1:    b=48:c=48:d=48:e=48:f=48:g=48:h=48:i=48:j=48:k=48
        goto on2                    'set zeros
on1:    read cidro,y                'from last read out
        read y,b                    'GET #S INTO b, c, d, etc.
        y=y+1
        read y,c
```

Listing 9-1 Caller ID program (*Continued*).

```
        y=y+1
        read y,d
        y=y+1
        read y,e
        y=y+1
        read y,f
        y=y+1
        read y,g
        y=y+1
        read y,h
        y=y+1
        read y,i
        y=y+1
        read y,j
        y=y+1
        read y,k
'       y=y+1          'unnecessary
on2:    read blnad,y   'to blocked data at point last looked at
        gosub st       'store it
        pause 100      'debounce
sit1:   if fctc = 0 then sit1
sit2:   if fctb = 0 then sit2
        goto fnctn
'THIS SETS WHETHER THE PROGRAM SHOULD BLOCK CALLS OR NOT
do2:   serout 15,396+$4000,[254,1] 'cls
       pause 100                   'debounce
       read bi,z                   'read setting
       if z = 0 then tgl1
       write bi,0 : write nb,1     'settings
       serout 15,396+$4000,["connect not",254,192,"blocked calls"]
       goto sit3
tgl1:  read nb,z
       if z = 0 then tgl2
       write bi,0 : write nb,0      'settings
       serout 15,396+$4000,["call blocking", 254,192,"is off"]
       goto sit3
tgl2:  write bi,1 : write nb,0      'settings
       serout 15,396+$4000,["inhibit blocked", 254,192,"calls"]
sit3:  if fcta = 0 then sit3
sit4:  if fctc = 0 then sit4
       goto start1
```

Listing 9-1 Caller ID program (*Continued*).

TABLE 9-1 MOMENTARY PUSH BUTTON SWITCH FUNCTIONS

FUNCTION A	FUNCTION B	FUNCTION C	
Read	Blocked	Store	press once
Options		X	press C then A
	Erase	X	press C then B

Caller ID/Blocker Parts List

U1	Motorola MC 145447 caller ID chip
U2	BASIC STAMP 2 (original)
U3	LM7805 5-V regulator
R1, R2, R3	10-kΩ $^1/_4$-W resistor
R4	18-kΩ $^1/_4$-W resistor
R5	15-kΩ $^1/_4$-W resistor
R6	270-kΩ $^1/_4$-W resistor
R7	4.7-MΩ $^1/_4$-W resistor
R8	10-MΩ $^1/_4$-W resistor
R9	22-kΩ $^1/_4$-W resistor
R10	470-Ω $^1/_4$-W resistor
R11, R12, R13, R14	1-kΩ $^1/_4$-W resistor
R15, R16, R17, R18	10-kΩ $^1/_4$-W resistor
C1, C2	0.2-μF 400-V Mylar capacitor
C3, C4	470-pF 400-V Mylar capacitor
C5	0.2-μF 50-V disk capacitor
C6	0.1-μF 50-V disk capacitor
C7, C11	1-μF 50-V electrolytic capacitor
C8, C9	30-pF 35-V disk capacitor
C10, C12	0.1-μF 50-V disk capacitor
MOV	300-V metal oxide varistor
D1, D2, D3, D4	1N4004 silicon diode
D2	1N4001 silicon diode
DS1, DS2, DS3	LED
DS4, DS5	LED
Q1	2N2222 npn transistor
XTL	3.579-MHz color burst crystal
K1	5-V minirelay SPST
SW1, SW2, SW3	Momentary pushbutton switch (normally open)
SW4, SW5	Momentary pushbutton switch (normally open)
Miscellaneous	Circuit board, power supply, wire, connectors, etc.

10

RADIO DTMF TONE DECODER AND DISPLAY

CONTENTS AT A GLANCE

Dual-tone multifrequency (DTMF) signaling has been around since the late 1950s. The DTMF signaling system originated from the Grand Old Dame Ma Bell. The beauty of this signal system is that two nonresonant tones are transmitted simultaneously and can be sent over land lines or wire or radio systems between and through various repeater systems around the country. The DTMF system is very reliable in that two tones must be decoded at once, so that false outputs are very unlikely. The DTMF system was and still is a very powerful method of signaling and control for both the phone system as well as for ham radio operators to control remote repeater systems or remote base transmitters over long distances. Touch-tone frequencies are listed in Table 10-1.

In the past, touch-tone frequencies had to be generated and decoded by discrete tone generators and tone decoders. Encoders and decoders were quite complex, and each would occupy large circuit boards. Today DTMF encoding and decoding can be accomplished by a single 20-pin integrated circuit chip.

TABLE 10-1 TOUCH-TONE OR DTMF TONES

	1209	1336	1477	1633
697	1	2	3	A
770	4	5	6	B
852	7	8	9	C
941	*	0	#	D

Two Projects

In this chapter, we will take a look at two different projects. The first project is a simple touch-tone generator that can be used to send DTMF signals through a radio transceiver, walkie-talkie, or wire line. The second project uses the CM8880 to decode DTMF tones on a phone line or radio circuit and display the numbers on an LCD panel. Both of our projects in this chapter will use the California Micro Devices model CM8880 touch-tone encoder/decoder chip; see App. 1 on the CD-ROM for the data sheet.

Both projects use the BASIC STAMP 2 microprocessor and the new CM8880 DTMF encoder/decoder chip, a 20-pin in-line DIP package. This encoder/decoder chip can be easily interfaced to the STAMP 2 controller and can be used to both generate and decode touch-tone tones; see App. 1 for the data sheet. The diagram in Fig. 10-1 is a block diagram of the CM8880 touch-tone IC. The chip requires only about six external components to form a complete encoder/decoder. The CM8880 chip uses the ubiquitous 3.579-MHz color burst crystal for its oscillator. The touch-tone chip produces a binary output on pins D0 through D3, which directly connects to the STAMP 2 controller; see Table 10-2. Est on pin 18 is the early steering output signal, which indicates the detection of valid tone frequencies, and StGt on pin19 is the steering input and guard time output. A time constant based on a capacitor and a resistor determines a valid signal duration time in which to accept a tone pair.

Three additional bits are used to select the modes of the CM8880: bits from the chip select (CS) pin, the read/write (RW) pin, and the resister select (RS0) pin; Table 10-3 depicts all the combinations of the three control pins. The CM8880 is active only when CS equals 0. The RW bit determines the data direction (1 = read and 0 = write), and the RS bit determines whether the transaction involves data (DTMF tones) or internal functions (i.e., 1 = instructions/status and 0 = data). Before you can use the CM8880, you have to set it up. The device has two control registers, i.e., A and B. In the beginning of the program listing you will notice the setup of the registers; also refer to Tables 10-4 and 10-5.

Touch-Tone Generator

Our first project, shown in Fig. 10-2, is a touch-tone generator that can be used to send touch-tone signals through a radio or wire circuit. This circuit can be interfaced to a micro-

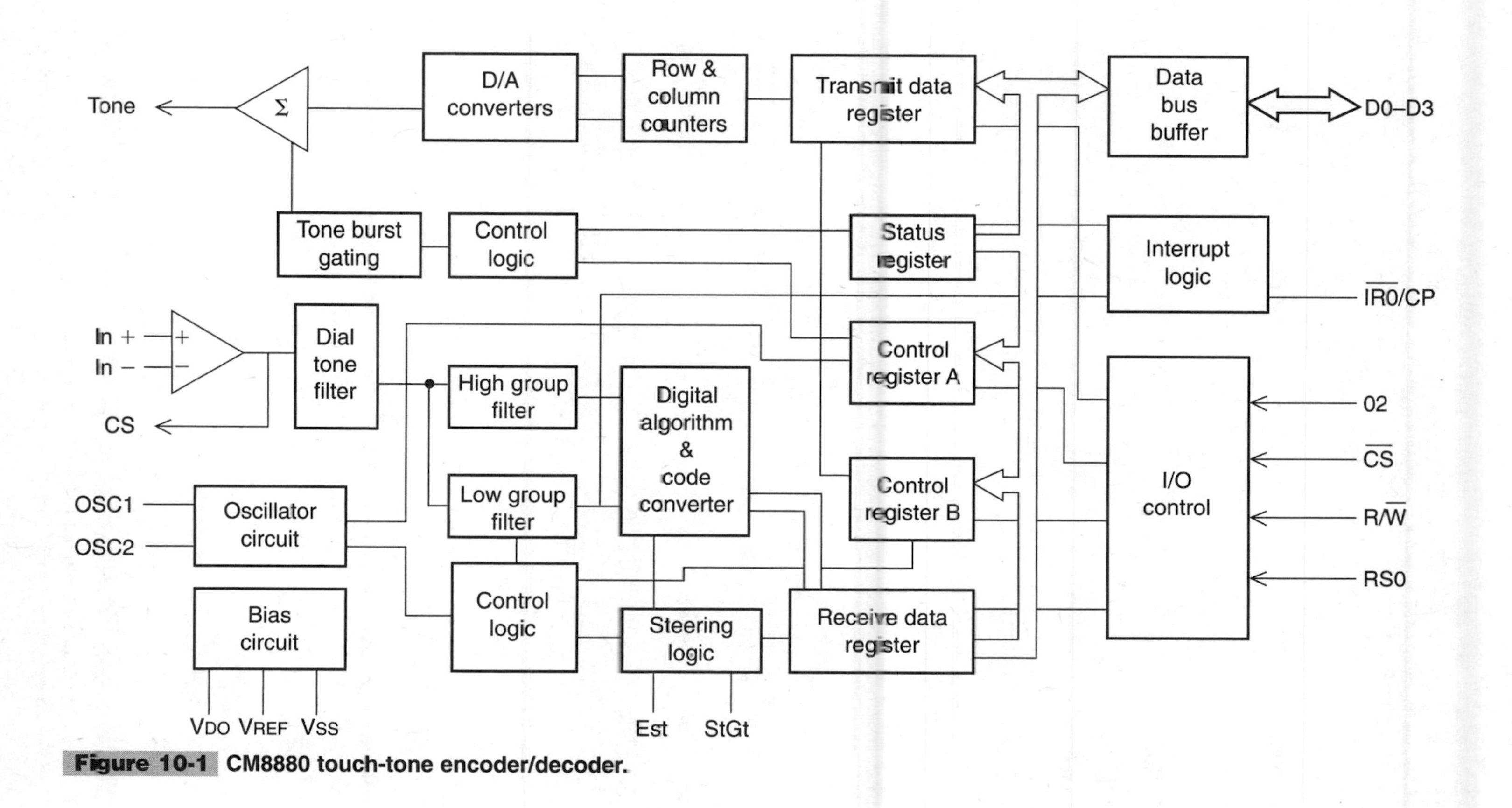

Figure 10-1 CM8880 touch-tone encoder/decoder.

TABLE 10-2 DTMF BINARY AND DECIMAL VALUES

BINARY VALUE	DECIMAL VALUE	KEYPAD SYMBOL
0000	0	D
0001	1	1
0010	2	2
0011	3	3
0100	4	4
0101	5	5
0110	6	6
0111	7	7
1000	8	8
1001	9	9
1010	10	0
1011	11	*
1100	12	#
1101	13	A
1110	14	B
1111	15	C

TABLE 10-3 CONTROL BIT CS, RW, AND RSO STATUS

CS	RW	RSO	DESCRIPTION
0	0	0	Active: write data (i.e., send DTMF)
0	0	1	Active: write instructions to CM8880
0	1	0	Active: read data (i.e., receive DTMF)
0	1	1	Active: read status from CM8880
1	0	0	Inactive
1	0	1	Inactive
1	1	0	Inactive
1	1	1	Inactive

TABLE 10-4 FUNCTIONS OF CONTROL REGISTER A

BIT	NAME	FUNCTION
0	Tone out	0 = tone generator disabled
		1 = tone generator enabled
1	Mode control	0 = send and receive DTMF
		1 = send DTMF, receive call progress tone
		(DTMF bursts lengthened to 104 ms)
2	Interrupt enable	0 = make controller check for DTMF received
		1 = interrupt controller via pin 13 when DTMF received
3	Register select	0 = next instruction write goes to CRA
		1 = next instruction write goes to CRB

TABLE 10-5 FUNCTIONS OF CONTROL REGISTER B

BIT	NAME	FUNCTION
0	Burst	0 = output DTMF bursts of 52 or 104 ms
		1 = output DTMF as long as enabled
1	Test	0 = normal operating mode
		1 = present test timing bit on pin 13
2	Single/Dual	0 = output dual (real DTMF) tones
		1= output separate row or column tones
3	Column/Row	0 = if above = 1 select row tone
		1 = if above = 1 select column tone

phone input of a radio so that you can send DTMF tones over radio to remotely control things. The touch-tone generator utilizes the STAMP 2 to animate the CM8880 encoder/decoder chip. Note that pins 9, 10, and 11 are used to enable and mode-select the CM8880 chip. Pins 14 through 17 are data output pins, which send the binary representation of the number received by the decoder portion of the chip. The touch-tone generator uses pin 8 of the CM8880 to send the analog audio signal to be amplified by U4. The output of the IC amplifier is coupled via C6 to a microphone input of a radio transceiver or wire circuit. If you have a walkie-talkie or mobile radio, you could use this circuit to send tones over radio to a remote receiver with a second CM8880 set up as a decoder to translate the remote tones into control functions; see the radio decoder project (Chap. 11). In the touch-tone generator program shown below, you will notice the phone number 459-0623. Simply substitute your own phone number or tone sequence for your particular

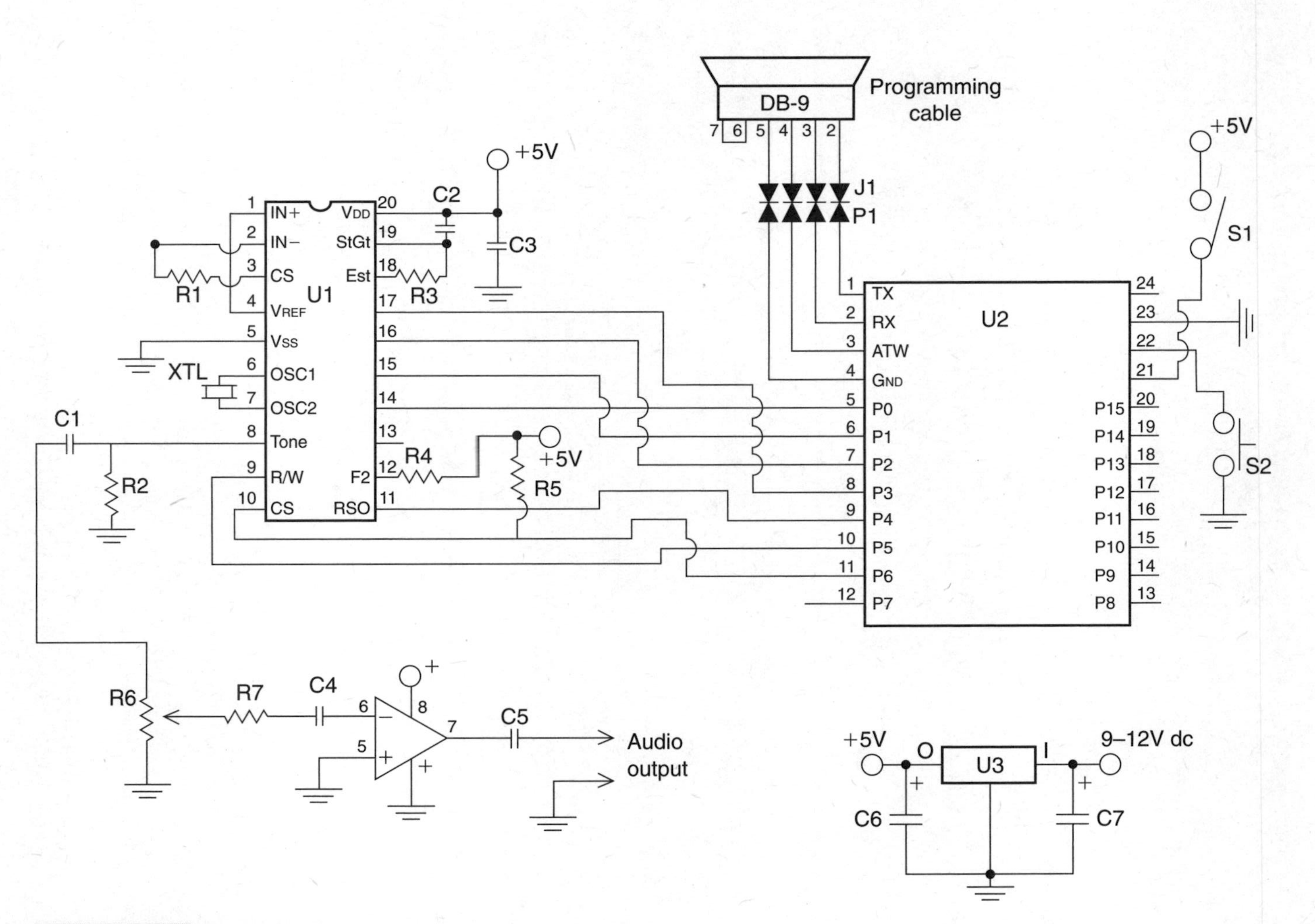

Figure 10-2 Touch-tone generator.

application. The touch-tone generator circuit is powered by 5 Vdc, which is supplied via regulator U3. The input of the regulator can be fed by a 9- to 12-Vdc "wall wart" power supply or can be powered by batteries if desired.

The touch-tone generator project can be built on a STAMP 2 carrier board or on a dedicated printed circuit board. When possible, use integrated circuit sockets for the ICs; this will make life much easier if a component fails later. Also take care to observe the proper orientation of the ICs before inserting them into the sockets. Be careful to observe the polarity of the capacitors. A 4-pin male header is placed on the circuit board at P1. This provides a simple connector for programming the STAMP 2. An external programming cable can then be constructed between the female header at J1 and the DB-9 female RS-232 connector. The completed touch-tone generator circuit can be next mounted in a plastic or aluminum chassis box. Once the touch-tone generator circuit has been completed, connect your programming cable and power-up the circuit. Next locate the DIAL.BS2 program shown in Listing 10-1, start up the STAMP 2 editor program, and download and run the DIAL.BS2 program. You are now ready to generate touch tones.

Decoder Display

The second project utilizes the CM8880 DTMF encoder/decoder and the STAMP 2 to form a touch-tone decoder display project, shown in Fig. 10-3. This decoder/display unit will allow you to monitor touch-tone signals, decode the phone numbers dialed, and display the number on a serial LCD display. This project can be placed across the phone line to capture phone numbers, or it could be used with a radio transceiver to monitor access tones sent over the air and display the captured numbers on the display. The touch-tone decoder project uses the CM8880 in the decoder mode. The tone signals are received via C1 and C2, using the balanced input network shown in Fig. 10-4. The STAMP 2 animates the control pins 9, 10, and 11, and accepts the binary output of the CM8880 via pins 14 through 17. The LCD panel is driven by the BASIC STAMP 2 on pin 7. The program accepts numbers and displays them sequentially, and if the delay between characters is too longs then the display will show a <space> until another stream is decoded. Power switch

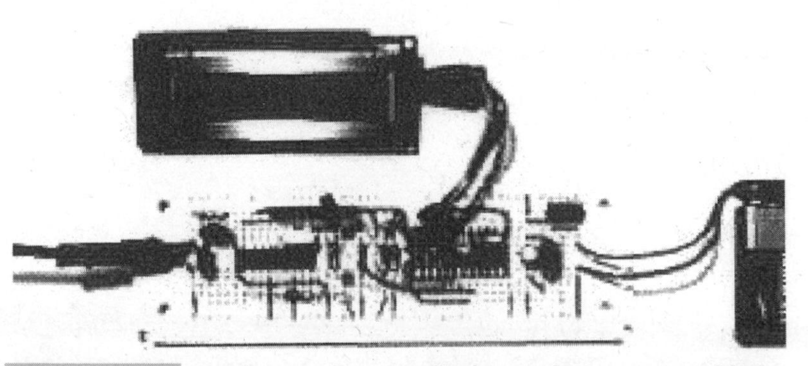

Figure 10-3 Touch-tone decoder display project, as assembled.

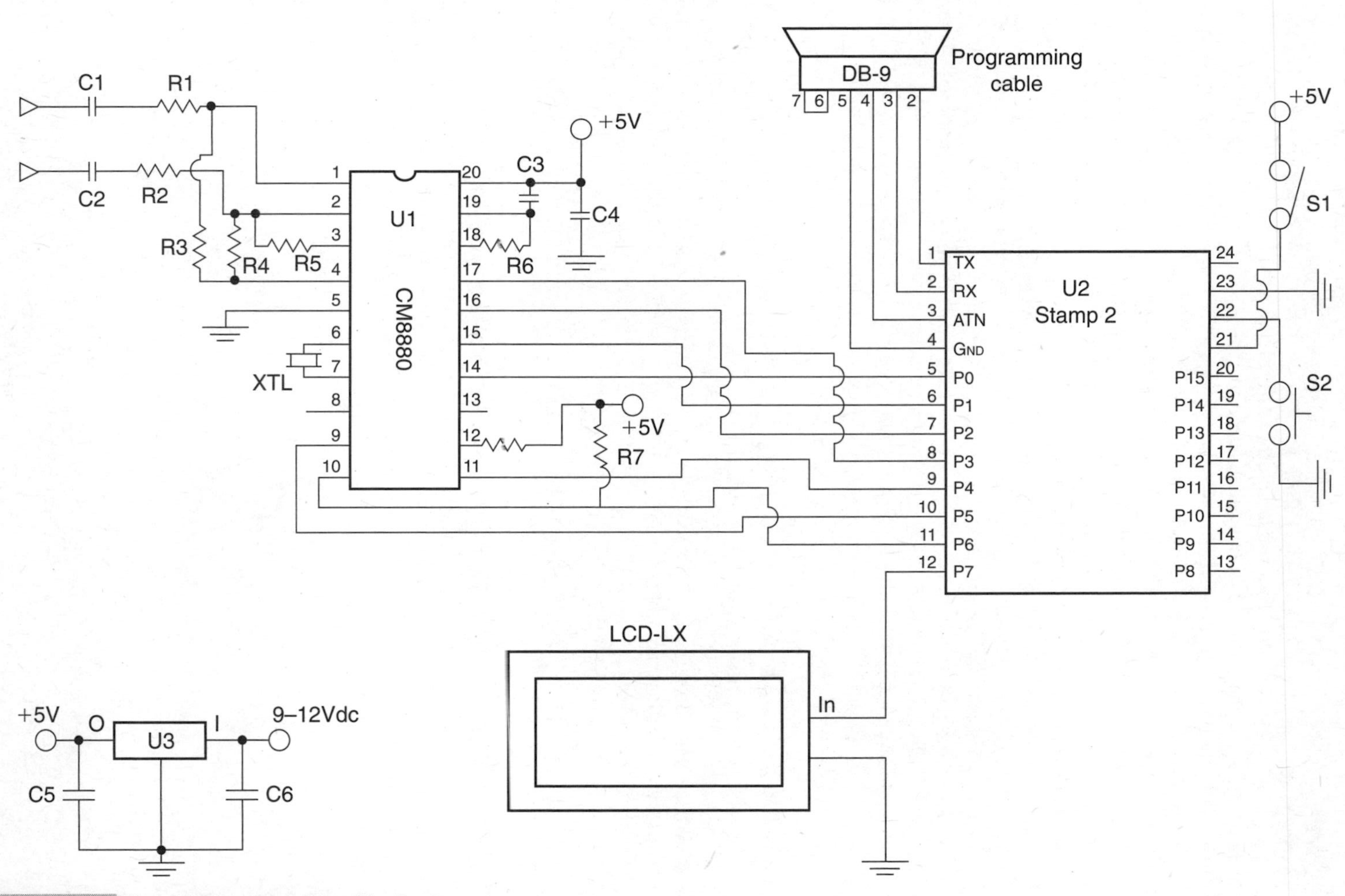

Figure 10-4 Touch-tone decoder with display.

S1 applies 5 V to the STAMP 2 via the 5-V regulator at U3. Momentary pushbutton S2 is the system reset button, which is used in the event of a system lockup.

You will need to locate or build a 9- to 2-Vdc power supply in order to power the radio DTMF controller. You can elect to build a power supply or try to locate a wall wart power supply from a surplus electronic supplier.

The circuit can be assembled on a custom circuit board, a perfboard, or a BASIC STAMP 2 carrier board, and then mounted in a small plastic enclosure box to protect the circuit board. The power switch S1 and the reset switch S2 are placed on the front of a plastic enclosure along with the serial LCD display unit. A two-circuit jack for the audio input is placed at the rear of the plastic case along with the hole and grommet for the serial programming connector.

Once the circuit has been mounted in an enclosure, you are now ready for the "smoke test." Connect your programming computer to the DB-9 connector and apply power to the touch-tone decoder/display unit and you're ready to begin. Next locate the STAMP 2 Windows editor titled STAMPW.EXE. Now locate the DTMF_RCV.BS2 (Listing 10-2) program and download it into your touch-tone decoder/display circuit, remove the programming cable, and you're ready to "sniff" the phone line or radio waves, looking for touch-tone numbers.

The radio DTMF controller can be used for a number different applications. We've only illustrated two; your imagination may devise more. Our project utilizes a VHF/UHF ham radio transceiver, but a scanner could have been used for the receiver. This circuit could also be adapted to Family Radio Service (FRS) or the long-range General Mobile Radio Service (GMRS) radios if desired.

Touch-Tone Generator/Dialer Parts List

U1	CM8880 touch-tone encoder/decoder IC
U2	BASIC STAMP 2 (original)
U3	LM 7805 5-V regulator IC
U4	LM358 op-amp
R1	100-kΩ $^1/_4$-W resistor
R2, R5	10-kΩ $^1/_4$-W resistor
R3	390-kΩ $^1/_4$-W resistor
R4	3.3k-Ω $^1/_4$-W resistor
R6	10-kΩ potentiometer
R7	22-kΩ $^1/_4$-W resistor
C1	0.1-μF 250-V polyester capacitor
C2, C3, C7	0.1-μF 35-V disk capacitor
C4, C6	2.2-μF 35-V electrolytic capacitor

C5	10-μF 35-V electrolytic capacitor
XTL	3.579-MHz color burst crystal
S1	SPST toggle switch
S2	SPST pushbutton switch (normally open)
P1	DB-9, nine-pin female RS-232 connector
Miscellaneous	Circuit board, headers, wire, jack, plugs

Touch-Tone Decoder/Display Parts List

U1	CM8880 touch-tone encoder/decoder IC
U2	BASIC STAMP 2 (original)
U3	LM7805 5-V regulator
R1, R2, R5	100-kΩ $^1/_4$-W resistor
R3	37.5-kΩ $^1/_4$-W resistor
R4	60-kΩ $^1/_4$-W resistor
R6	390-kΩ $^1/_4$-W resistor
R7	10-kΩ $^1/_4$-W resistor
C1, C2	0.1-μF 250-V polyester capacitor
C3, C4	0.1-μF 35-V disk capacitor
C5	2.2-μF 35-V electrolytic capacitor
XTL	3.579-MHz color burst crystal
S1	SPST toggle switch
S2	SPST pushbutton switch (normally open)
LX	LCD module Scott Edwards ILM-216 (http://www.seetron.com)
P1	DB-9 nine-pin female RS-232 connector
Miscellaneous	Circuit board, headers, wire

```
'DIAL.BS2
' This program sends DTMF tones via the 8880
' This program demonstrates how to use the CM8880 as a DTMF tone
' generator. All that's required is to initialize the 8880 properly,
' then write the number of the desired DTMF tone to the 8880's
' 4-bit bus.
' The symbols below are the pin numbers to which the 8880's
' control inputs are connected, and one variable used to read
' digits out of a lookup table.
RS          con 4              ' Register-select pin (0=data).
RW          con 5              ' Read/Write pin (0=write).
CS          con 6              ' Chip-select pin (0=active).
digit       var     nib        ' Index of digits to dial, 1-15.
' This code initializes the 8880 for dialing by writing to its
' internal control registers CRA and CRB. The write occurs when
' CS (pin 6) is taken low, then returned high. See the accompanying
' article for an explanation of the 8880's registers.
OUTL = 127                              ' Pins 0-6 high to deselect 8880.
DIRL = 127                              ' Set up to write to 8880 (pins 0-6 outputs).
OUTL = %00011011                        ' Set up register A, next write to register B.
high CS


OUTL = %00010000                        ' Clear register B; ready to send DTMF.
high CS
' This for/next loop dials the seven digits of my fax number. For
' simplicity, it writes the digit to be dialed directly to the output
' pins. Since valid digits are between 0 and 15, this also takes RS,
' RW, and CS low—perfect for writing data to the 8880. To complete
' the write, the CS line is returned high. The initialization above
' sets the 8880 for tone bursts of 200 ms duration, so we pause
' 250 ms between digits. Note: in the DTMF code as used by the phone
' system, zero is represented by ten (1010 binary) not 0. That's why
' the phone number 459-0623 is coded 4,5,9,10,6,2,3.
for digit = 0 to 6
lookup digit,[4,5,9,10,6,2,3],OUTL          ' Get digit from table.
high CS                                     ' Done with write.
pause 250                                   ' Wait to dial next digit.
next
end
```

Listing 10-1 Touch-tone generator/dialer program.

```
'DTMF_RCV.BS2
' This program: (Receives/displays DTMF using CM8880)
' This program demonstrates how to use the 8880 as a DTMF decoder. As
' each new DTMF digit is received, it is displayed on an LCD Serial
' Backpack screen. If no tones are received within a period of time
' set by sp_time, the program prints a space (or other selected character)
' to the LCD to record the delay. When the display reaches the right-hand
' edge of the screen, it clears the LCD and starts over at the left edge.
RS              con 4             ' Register-select pin (0=data).
RW              con 5             ' Read/Write pin (0=write).
CS              con 6             ' Chip-select pin (0=active).
dtmf            var     byte      ' Received DTMF digit.
dt_Flag         var     bit       ' DTMF-received flag.
dt_det          var     INL.bit2  ' DTMF detected status bit.
home_Flag       var     bit       ' Flag: 0 = cursor at left edge of LCD.
polls           var     word      ' Number of unsuccessful polls of DTMF.
LCDw            con 16            ' Width of LCD screen.
LCDcol          var     byte      ' Current column of LCD screen for wrap.
LCDcls          con 1             ' LCD clear-screen command.
I               con 254           ' LCD instruction toggle.
sp_time         con 1500          ' Print space this # of polls w/o DTMF.
n24n            con $418D         ' Serout constant: 2400 baud inverted.
' This code initializes the 8880 for receiving by writing to its
' internal control registers CRA and CRB. The write occurs when
' CS (pin 6) is taken low, then returned high.
OUTL = %01111111                        ' Pin 7 (LCD) low, pins 0 through 6 high.
DIRL = %11111111                        ' Set up to write to 8880 (all outputs).
OUTL = %00011000                        ' Set up register A, next write to register B.
high CS
OUTL = %00010000                        ' Clear register B; ready to send DTMF.
high CS
DIRL = %11110000                        ' Now set the 4-bit bus to input.
high RW                                 ' And set RW to "read."
serout 7,n24n,[I,LCDcls,I]              ' Clear the LCD screen.
' In the loop below, the program checks the 8880's status register
' to determine whether a DTMF tone has been received (indicated by
' a '1' in bit 2). If no tone, the program loops back and checks
' again. If a tone is present, the program switches from status to
' data (RS low) and gets the value (0-15) of the tone. This
' automatically resets the 8880's status flag.
again:
high RS                                 ' Read status register.
low CS                                  ' Activate the 8880.
dt_flag = dt_det ' Store DTMF-detected bit into flag.


high CS                                 ' End the read.
if dt_Flag = 1 then skip1               ' If tone detected, continue.
polls = polls+1                         ' Another poll without DTMF tone.

if polls < sp_time then again          ' If not time to print a space, poll.
if LCDcol = LCDw then skip2             ' Don't erase the screen to print spaces.
dtmf = 16                               ' Tell display routine to print a space.
gosub Display                           ' Print space to LCD.
```

Listing 10-2 Touch-tone decoder/display program.

```
polls = 0                                   ' Clear the counter.
goto again                                  ' Poll some more.
skip1:                                      ' Tone detected:
polls = 0                                   ' Clear the poll counter.
low RS                                      ' Get the DTMF data.
low CS                                      ' Activate 8880.
dtmf = INL & %00001111                      ' Strip off upper 4 bits using AND.
high CS                                     ' Deactivate 8880.
gosub display                               ' Display the data.
goto again                                  ' Do it all again.
Display:
if LCDcol < LCDw then skip3                ' If not at end of LCD, don't clear screen.
serout 7,N24N,[I,LCDcls,I]                  ' Clear the LCD screen.
LCDcol = 0                                  ' And reset the column counter.
skip3: ' Look up the symbol for the digit.
if LCDcol=0 AND dtmf=16 then ret            ' No spaces at first column.
lookup dtmf,["D1234567890*#ABC-"] ,dtmf
serout 7,N24N,[dtmf]                        ' Write it to the Backpack display.
LCDcol = LCDcol + 1                         ' Increment the column counter.
ret:
return
```

Listing 10-2 Touch-tone decoder/display program (*Continued*).

11

RADIO DTMF TONE DECODER/ CONTROL

CONTENTS AT A GLANCE

Dual-tone multifrequency (DTMF) signaling has been around for over 40 years now and continues to be the preferred tone signaling method. As mentioned previously the beauty of this signaling system is that two nonresonant tones out of eight are transmitted simultaneously and can be sent over land lines or wire or radio systems between and through various repeater systems around the country. The DTMF system is very reliable in that two nonharmonically related tones must be decoded at once, so that false outputs are very unlikely. The DTMF system was and still is a very powerful method of signaling and control for the phone system as well as for ham radio operators in controlling remote repeater systems or remote base transmitters over long distances. The 2-of-8 touch-tone frequencies versus numerical digits are illustrated in Table 11-1.

TABLE 11-1 TOUCH-TONE OR DMTF TONES

	1209	1336	1477	1633
697	1	2	3	A
770	4	5	6	B
852	7	8	9	C
941	*	0	#	D

In the past, touch-tone frequencies had to be generated and decoded by discrete tone generators and tone decoders. Encoders and decoders were quite complex, and each could fill up an entire circuit board. Today, DTMF encoding and decoding can be accomplished by the same single 20-pin integrated circuit chip.

In this chapter, we will take a look at a remote control touch-tone radio decoder that can be used to remotely control your garage door or other home appliance via an amateur radio transceiver, scanner, or FRS radio. The remote control touch-tone decoder can also be used to mute the radio speaker until a specific message needs to be sent over radio. If, for example, you are sharing a frequency and you want to call a family member, when you decode the proper tone sequence your radio will allow the speaker to operate and let your specific message through.

Basic Components

The remote control radio decoder uses the BASIC STAMP 2 microprocessor and the new CM8880 DTMF encoder/decoder chip, in a 20-pin in-line DIP package, from California Micro; see App. 1 on the CD-ROM for data sheets. This encoder/decoder chip can be easily interfaced to the STAMP 2 controller and can be used to both generate and decode touch-tone tones. A block diagram of the CM8880 is shown in Fig. 11-1. The chip requires only about six external components to form a complete encoder/decoder. The CM8880 chip uses the ubiquitous 3.579-MHz color burst crystal for its oscillator. The chip produces a binary output on pins D0 through D3, which directly connects to the STAMP 2 controller; see Table 11-2. Est on pin 18 is the early steering output signal, which indicates the detection of valid tone frequencies, and StGt on pin 19 is the steering input and guard time output. A time constant based on a capacitor and a resistor determines a valid signal duration time in which to accept a tone pair. Three additional bits are used to select the modes of the CM8880: from the chip select (CS) pin, the read/write (RW) pin, and the resister select (RSO) pin; Table 11-3 depicts all the combinations of the three control pins. The CM8880 is active only when CS equals 0. The RW bit determines the data direction (1 = read and 0 = write), and the RS bit determines whether the transaction involves data (DTMF tones) or internal functions (i.e., 1 = instructions/status and 0 = data). Before you can use the CM8880, you have to set it up. The device has two control registers, A and B.

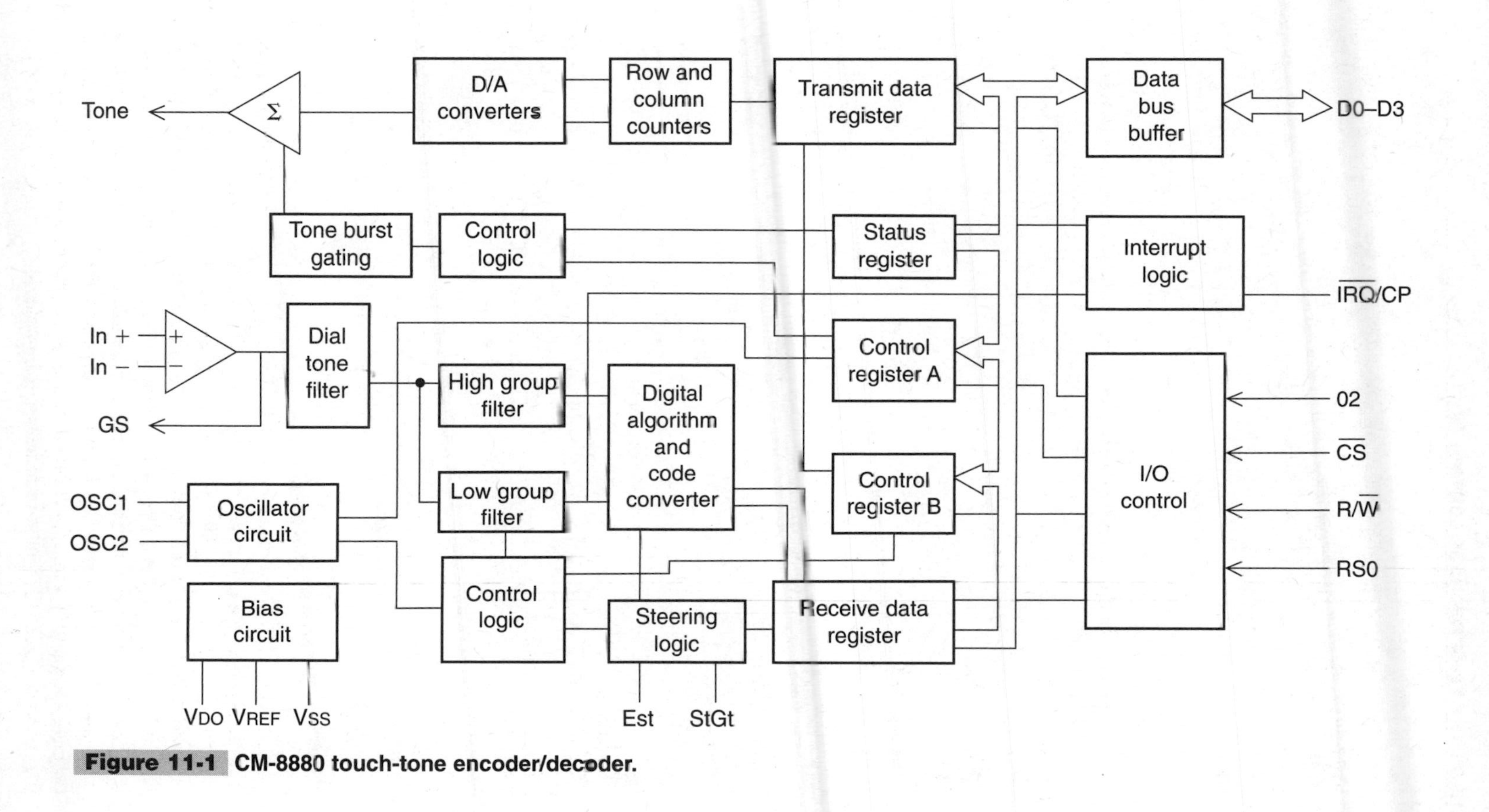

Figure 11-1 CM-8880 touch-tone encoder/decoder.

In the beginning of the program listing you will notice the setup of the registers; also refer to Tables 11-4 and 11-5.

The radio DTMF tone control circuit is shown in Fig. 11-2. The circuit begins with the input/output section of the CM8880 encoder/decoder chip. The audio input signal enters the CM8880 decoder section via pins 2 and 3, while the audio output section of the encoder enters the chip on pin 8. Note, the tone output on pin 8 is shown connected to the microphone input jack of the radio transceiver; this feature was not implemented in this project, but could be easily set up. Operationally you would use the STAMP 2 frequency output, or Freqout command, to send DTMF tones through your receiver The color burst crystal (XTL) at pins 6 and 7 sets up the reference oscillator section of the tone decoder chip. The decoded binary digital outputs from the CM8880 are presented on pins 14 through 17, i.e., D0 to D3, respectively. The R/W control bit is at pin 9, while the CS bit is on pin 10. The RS0 bit is input at pin 11. Data outputs of the CM8880 are connected to the STAMP 2 on pins 5 through 8. Setup or control data bits are presented to the BS2 controller on pins 9 through 11. The control relay at RLY1 is activated from STAMP 2 pin P7, which drives an npn transistor at Q1. The relay used was a low-current DPDT relay. Power is applied to the DTMF chip on pin 20 and ground is connected to pin 5. Reset pin 22 is coupled through a normally open pushbutton switch to ground in the event of a system lockup problem.

Power to the BASIC STAMP 2 is applied to the 5-V pin of the STAMP 2 at pin 21, and is fed from the 5-V regulator at U3. System ground is applied to pin 23 of the BS2. You will need to locate or build a 9- to 12-Vdc power supply in order to power the radio DTMF controller. You can elect to build a power supply or try to locate a "wall wart" power supply from a surplus electronic supplier.

Programming the STAMP 2 controller is achieved using the TX, RX, ATN, and ground pins 1 though 4. These pins are brought out through a header at P1, a 4-pin male 0.010-in header strip. A cable is then made between a female header at J1 and a 9-pin DB-9 female connector.

The touch-tone radio decoder circuit can be assembled on a custom circuit board, a perfboard, or a BASIC STAMP 2 carrier board, and then mounted in a small plastic enclosure box to protect the circuit board. When constructing the circuit board, be sure to observe the correct polarity while installing the capacitors and diodes, in order to avoid problems upon power-up. It is advisable to install IC sockets for the integrated circuits, in the event of a faulty component at a later date. Remember to install header strips for the programming cable as well as for the audio input and audio output connections. Once the circuit board has been built and checked, you can begin to look for a suitable plastic or aluminum chassis box.

DPDT Relay

The radio DTMF controller circuit can be used in many ways, but we are going to describe two possible applications. The first application, shown in Fig. 11-2, features a DPDT relay that can be utilized to activate a garage door opener, an electric dead-bolt "strike" or lock, or an air conditioner, fan, or some other home appliance. A second set of relay contacts can be wired in parallel to the first set of contacts for high-power-device control. The second set of contacts could also be used to control a second or simultaneous appliance.

TABLE 11-2 DTMF BINARY AND DECIMAL VALUES

BINARY VALUE	DECIMAL VALUE	KEYPAD SYMBOL
0000	0	D
0001	1	1
0010	2	2
0011	3	3
0100	4	4
0101	5	5
0110	6	6
0111	7	7
1000	8	8
1001	9	9
1010	10	0
1011	11	*
1100	12	#
1101	13	A
1110	14	B
1111	15	C

TABLE 11-3 CONTROL BIT CS, RW, AND RSO STATUS

CS	RW	RS0	DESCRIPTION
0	0	0	Active: write data (i.e., send DTMF)
0	0	1	Active: write instructions to CM8880
0	1	0	Active: read data (i.e., receive DTMF)
0	1	1	Active: read status from CM8880
1	0	0	Inactive
1	0	1	Inactive
1	1	0	Inactive
1	1	1	Inactive

TABLE 11-4 FUNCTIONS OF CONTROL REGISTER A

BIT	NAME	FUNCTION
0	Tone out	0 = tone generator disabled
		1 = tone generator enabled
1	Mode control	0 = send and receive DTMF
		1 = send DTMF, receive call progress tone
		(DTMF bursts lengthened to 104 ms)
2	Interrupt enable	0 = Make controller check for DTMF received
		1 = Interrupt controller via pin 13 when DTMF received
3	Register select	0 = Next instruction write goes to CRA
		1 = Next instruction write goes to CRB

TABLE 11-5 FUNCTIONS OF CONTROL REGISTER B

BIT	NAME	FUNCTION
0	Burst	0 = output DTMF bursts of 52 or 104 ms
		1 = 1 output DTMF as long as enabled
1	Test	0 = normal operating mode
		1 = present test timing bit on pin 13
2	Single/Dual	0 = output dual (real DTMF) tones
		1 = output separate row or column tones
3	Column/Row	0 = if above = 1 select row tone
		1 = if above = 1 select column tone

Silent Radio

Our second application for the remote touch-tone radio decoder, shown in Fig. 11-3, depicts a "silent" radio scheme. In this application, the radio receiver is kept quiet until the tone sequence is received, and on activation the remote speaker is connected to the radio. The audio from the radio receiver is fed both to the touch-tone decoder and to the input gain control ahead of the audio amplifier at U4. When the proper tone sequence is received, the relay at RLY1 is activated and the audio path from the receiver is connected to audio amplifier at U4, which is used to drive the 8-Ω speaker, and when an important or desired message is received, the speaker comes to life! In this application, the radio or scanner could be placed in the kitchen if desired, and you could use the setup to announce

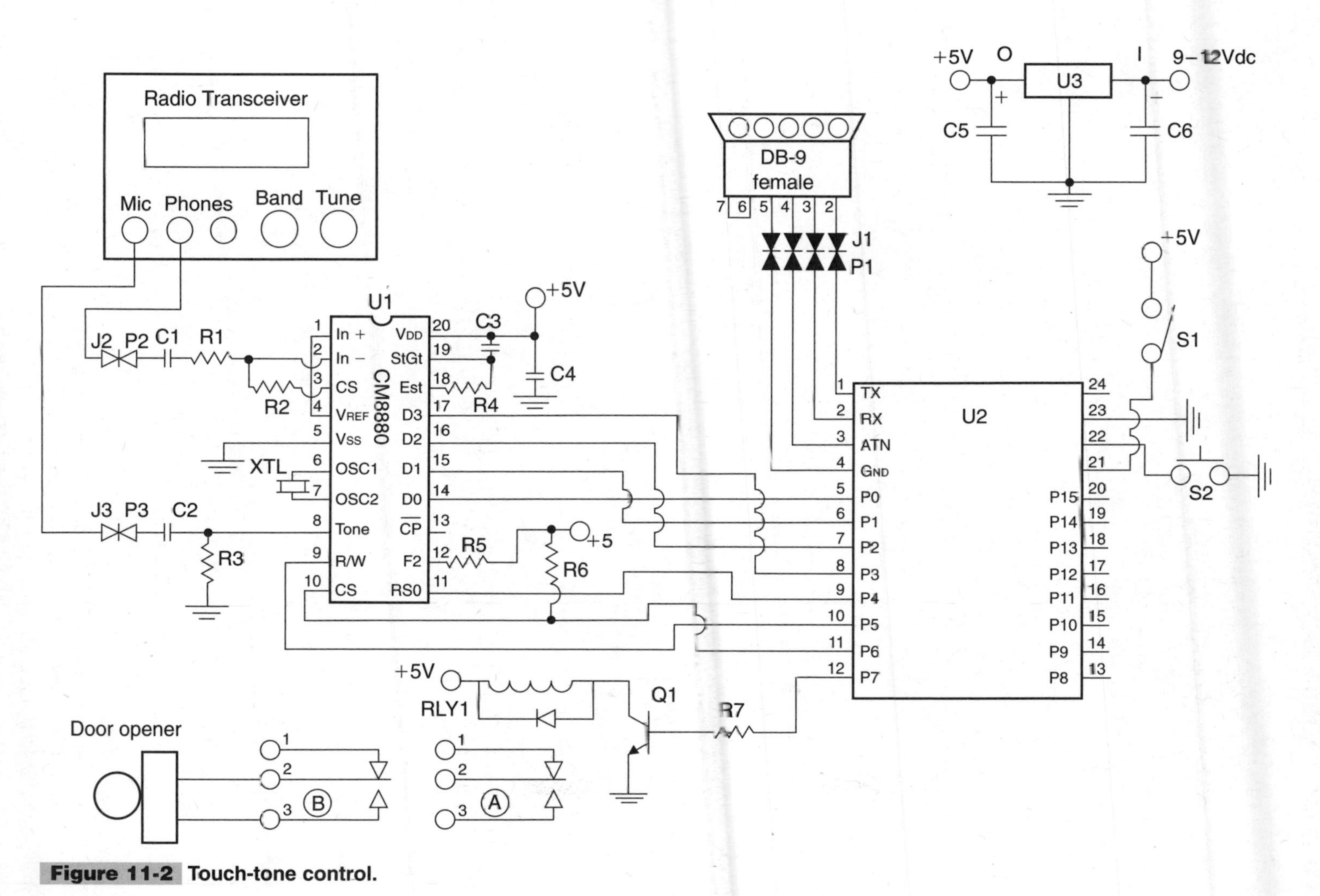

Figure 11-2 Touch-tone control.

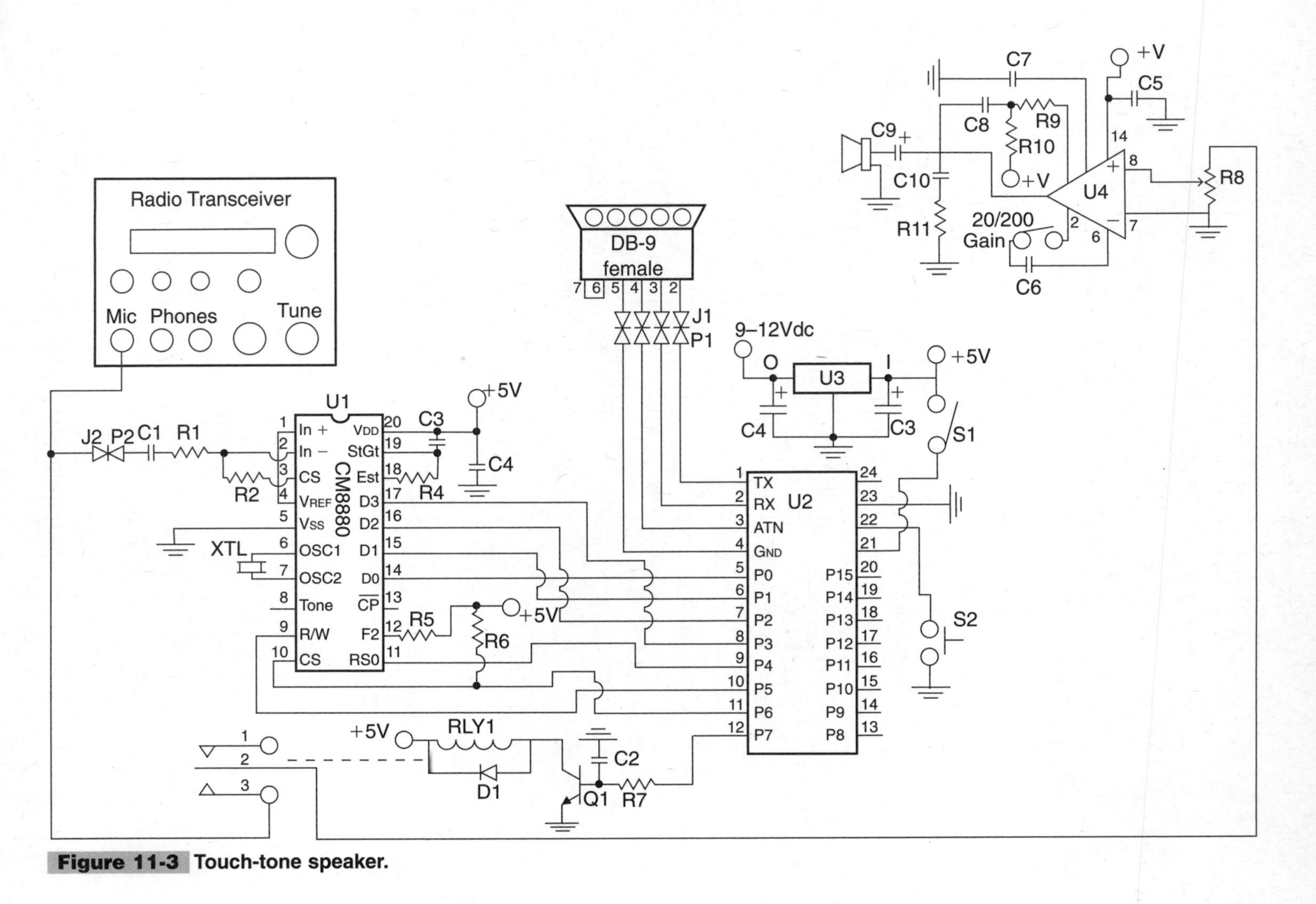

Figure 11-3 Touch-tone speaker.

your arrival remotely, thus silencing the radio most of the time to eliminate other conversations. This is great, since most family members are not particularly interested in a lot of "conversational noise."

The Radio DTMF controller can be used for many different applications; we illustrated only two but your imagination can think up many more. You can expand the program to decode other tone sequences and more control relays if desired. The BASIC STAMP 2 program can be modified to suit your particular future needs. Our project utilizes a VHF/UHF ham radio transceiver, but a scanner could have been used for the receiver. This circuit could also be adapted to FRS or the long range GMRS radios if desired.

Programming

With your remote radio decoder built and ready to go, you will next need to connect your remote radio tone decoder to your personal computer in order to program the system. Once the programming link cable has been connected and power applied to the circuit, you are ready to roll. Go to the STAMP 2 Windows Editor STAMPW.EXE, download the DECODE2.BS2 (Listing 11-1) program, and then run it. Remember that the remote radio DTMF controller program uses the sequence *1234567# as the code for the lock portion of the program. Note the code is placed in the lock lookup string, 1 to 9 digits in length, which has 16 corresponding string positions: D, 1, 2, 3, 4, 5, 6, 7, 8, 9, 0, *, #, A, B, C. The value at each position of the string is the key sequence of that digit. A string value of 0 forces a reset (wrong digit) and is used to turn off the relay-latched output device. A zero is used for all remaining digits *not* used in the combination code. With incorrect code attempts, one needs to input an additional reset DTMF tone ahead of the code string (any unused digit) for proper reset before use. When the proper sequence is received, a relay routine is activated as an output device. Pressing the digits #9 will reset the decoder when a wrong digit is pressed and you wish to begin again.

The remote DTMF radio decoder project is fun to build and will lend itself to many different applications around your home or ham shack!

Radio DTMF Controller (Figure 11-1) Parts List

U1	CM8880 touch-tone encoder/decoder IC
U2	BASIC STAMP 2 (original)
U3	LM7805 5-V regulator
R1, R3, R6	10-kΩ $^{1}/_{4}$-W resistor

R2	100-kΩ 1/4-W resistor
R4	390-kΩ 1/4-W resistor
R5	3.3-kΩ 1/4-W resistor
R7	1-kΩ 1/4-W resistor
C1, C2, C3, C4	0.1-μF 50-V disk capacitor
C5	4.7-μF 50-V electrolytic capacitor
C6	0.2-μF 50-V electrolytic capacitor
Q1	2N2222 npn transistor
D1	1N4001 silicon diode
S1	SPST power switch
S2	Momentary pushbutton switch (normally open)
RLY1	5–6-V DPDT relay (RadioShack)
Miscellaneous	Male and female headers, PC board, speaker, 9-pin RS-232 connector

```
'decode2.bs2
'  " DECODE.BS2 " by Joe Altieri
'  The "Listen" routine is based on Scott
'  Edwards'  DTMF_RCV.BS2  program. Pin 4 is the "register-select"
'  pin ( 0=data ). Pin 5 is the R/W pin ( 0=write ). Pin 6 is the
'  "chip-select" pin ( 0=active ). Contact Scott Edwards Electronics
'  for CM8880 chips, DTMF programs, and circuit diagrams
T    var  byte              ' Received DTMF digit
df   var  bit              ' DTMF-received flag
ds   var  INL.bit2            ' DTMF-detected status bit
C    var  word              ' Tones rcvd counter variable
G    var  byte             ' Sequence counter variable
J    var  byte             ' Table counter variable
  OUTL = %01111111                 ' pin 7 low, pins 0-6 high
  DIRL = %11111111               ' set up write to 8880 (out)
  OUTL = %00011000                ' set up CRA; write to CRB
  high 6                   ' low 6 to high 6 = "write"
  OUTL = %00010000                ' clear CRB; rdy/snd DTMF
  high 6                   ' low 6 to high 6 = "write"
  DIRL = %11110000                ' set 4-bit bus to input
  high 5                   ' set R/W to "read"
Off:                                      ' Off routine
  low 7                                   ' turn off output device
Begin:                                    ' Reset routine
  C=0 : G=0 : J=                          ' clear all variables to zero
Listen:                                   ' Listen for DTMF
  high 4                                  ' read status register
  low 6                                   ' activate chip-select pin
  df = ds                                 ' store DTMF-det bit in flag
  high 6                                  ' de-activate chip-select pin
  if df = 1 then Lock                     ' if tone, then continue
```

Listing 11-1 Radio touch-tone decoder program.

```
  goto Listen                              ' listen again
Lock:                                      ' Tone detected, test code
  low 4                                    ' get DTMF data (low rs pin)
  low 6                                    ' activate chip-select pin
  T = INL & %00001111                      ' remove upper bits using AND
  high 6                                   ' de-activate chip-select pin
  C=C+1                                    ' increment tone counter
  if C>2000 then Begin                     ' resets >2000 wrong entries
  lookup T,["0234567800019000"],J          ' convert tone to string
  J=J-48                                   ' convert string to value
  if J=0 then Off                          ' wrong digit "reset"
  if C=1 and J=1 then Key                  '  IF all digits are in the
  if C=2 and J=2 then Key                  '  correct order, AND after
  if C=3 and J=3 then Key                  '  only one attempt, THEN
  if C=4 and J=4 then Key                  '  increment the sequence
  if C=5 and J=5 then Key                  '  counter variable : "G" ...
  if C=6 and J=6 then Key                  '
  if C=7 and J=7 then Key                  '
  if C=8 and J=8 then Key                  '
  if C=9 and J=9 then Key                  '
  goto Listen                              ' listen again
Key:                                       ' Sequence counter
  G=G+1                                    ' increment
  if C=9 and J=9 and G=9 then Relay        ' activate output device
  goto Listen                              ' listen again
Relay:                                     ' Relay output routine
  pause 800                                ' wait time delay
  high 7                                   ' turn on output device
  pause 200                                ' wait time delay
  goto Begin                               ' begin again
```

Listing 11-1 Radio touch-tone decoder program (*Continued*)

Radio DTMF Controller (Figure 11-2) Parts List

U1	CM8880 touch-tone encoder/decoder IC
U2	BASIC STAMP 2 (original)
U3	LM7805 5-V regulator
R1, R3, R6	10-kΩ 1/4-W resistor
R2	100-kΩ 1/4-W resistor
R4	390-kΩ 1/4-W resistor
R5	3.3-kΩ 1/4-W resistor
R7	1-kΩ 1/4-W resistor
R8	10-kΩ 10-turn potentiometer

R9, R10	510-Ω 1/4-W resistor
R11	2.7-Ω 1/4-W resistor
C1, C2, C3, C5, C6	0.1-μF 50-V disk capacitor
C4	2.2-μF 50-V electrolytic capacitor
C7	10-μF 50-V electrolytic capacitor
C8	4.7-μF 50-V electrolytic capacitor
C9	100-μF 50-V electrolytic capacitor
C10	0.05-μF 35-V disk capacitor
Q1	2N2222 npn transistor
D1	1N4001 silicon diode
S1	SPST power switch
S2	Momentary pushbutton switch (normally open)
RLY1	5–6-V DPDT relay (RadioShack)
Miscellaneous	Male and female headers, PC board, speaker, 9-pin RS-232 connector

12

MORSE CODE RADIO KEYER PROJECT

CONTENTS AT A GLANCE

Morse code is probably the oldest form of serial communication protocol still in use today. Despite the age of Morse code, it has some virtues that make it viable as a means of communication even today. Morse code offers inherent compression; the letter E is transmitted in one-thirteenth the time it takes to send the letter Q. Morse code requires much less transmitted power compared with other modulation methods. Morse code also requires much less bandwidth and can often get through when other methods of communication fall short. Morse code can be sent and received by humans or by automated equipment.

About Morse Code

Although Morse code has fallen out of favor as a means of sending large messages, it is still legal and often a preferred method to communicate when operating QRP or low power, or when conditions are marginal. As the old saying goes, "When nothing else can get through,

CW can!" Those in the ever-growing QRP movement build low-power portable transceivers often run by batteries. Continuous-wave (CW) transceivers require far fewer parts and are simpler to build. CW is also a preferred way to identify automated repeater and beacon stations. The Morse keyer is great for local VHF/UHF code practice sessions. The BASIC STAMP 2, with its ease of programming and low power requirement, make it ideal for the Morse code keyer project.

The Morse code keyer project can be utilized to key your favorite amateur radio transceiver and send Morse code automatically for you. The Morse keyer can be programmed for up to four preset messages. Canned messages are popular with amateur radio operators for sending basic parameters of exchange. One message can be used for calling CQ, i.e., looking to contact others. One message can give your location, and another message could be used to describe your station operating conditions.

The characters of the Morse code are represented by sequences of long and short beeps known as dots and dashes (or dits and dahs). There are one to six beeps or elements in the characters of the standard Morse code. The first step in writing a program to send Morse code is to devise a compact way to represent sequences of elements, and an efficient way to play them back.

The chart shown in Table 12-1 illustrates the encoding scheme used in this program. A single byte represents a Morse character. The highest 5 bits of the byte represent the actual dots (0s) and dashes (1s), while the lower 3 bits represent the number of elements in the character. For example, the letter F is a dot, dash, dot, dot, so it is encoded as 0010X100, where X is a don't care bit. Since Morse characters can contain up to six elements, we have to handle the exceptions. Fortunately, there is some excess capacity in the number-of-elements portion of the byte, which can represent numbers up to 7. Therefore, we can assign a six-element character ending in a dot the number 6, while a six-element character ending in a dash gets the number 7. The program shows how these bytes can be played back to produce Morse code. The table of symbols at the beginning of the program listing contains the timing data for the dots and dashes themselves. If you need or want to change the program sending speed, you will just enter new values for the dit_length, dah_length, etc. Remember, you will need to keep the timing relationships roughly the same; a dash should be about 3 times as long as a dot.

The Morse keyer program uses the BASIC STAMP's *lookup* function to play sequences of Morse characters. *Lookup* is a particularly modern feature of STAMP BASIC in that it is an object-oriented data structure. It not only contains the data, but it also knows how to retrieve it.

Building the Keyer

Building the Morse code keyer is very straightforward. You can use either the original BASIC STAMP, as shown in Fig. 12-1, or the alternative STAMP to build this project. Simply translate the correct input/output pins on the alternative STAMP pins. The Morse keyer can be built on a perfboard, a point-to-point wiring board, or the STAMP 2 carrier board, or you could build your own custom circuit board if desired. Using an integrated

TABLE 12-1 MORSE CODE CHARACTERS AND THEIR ENCODED EQUIVALENTS

CHARACTER	MORSE	BINARY	DECIMAL	CHARACTER	MORSE	BINARY	DECIMAL
A	* –	01000010	66	S	* * *	00000011	3
B	– * * *	10000100	132	T	–	10000001	129
C	– *–*	10100100	164	U	** –	00100011	35
D	– * *	10000011	131	V	* * * –	00010100	20
E	0	00000001	1	W	* – –	01100011	99
F	* * – *	00100100	36	X	– * * –	10010100	148
G	– – *	11000011	195	Y	– * – –	10110100	180
H	* * * *	00000100	4	Z	– – * *	11000100	196
I	* *	00000010	2	0	– – – – –	11111101	253
J	* – – –	01110100	116	1	* – – – –	01111101	125
K	– * –	10100011	163	2	* * – – –	00111101	61
L	* – * *	01000100	68	3	* * * _	00011101	29
M	– –	11000010	194	4	* * * * –	00001101	13
N	– *	10000010	130	5	* * * * *	00000101	5
O	– – –	11100011	227	6	– * * * *	10000101	133
P	*– – *	01100100	100	7	– – * * *	11000101	197
Q	– – * –	11010100	212	8	– – – * *	11100101	229
R	* – *	01000011	67	9	– – – – *	11110101	245

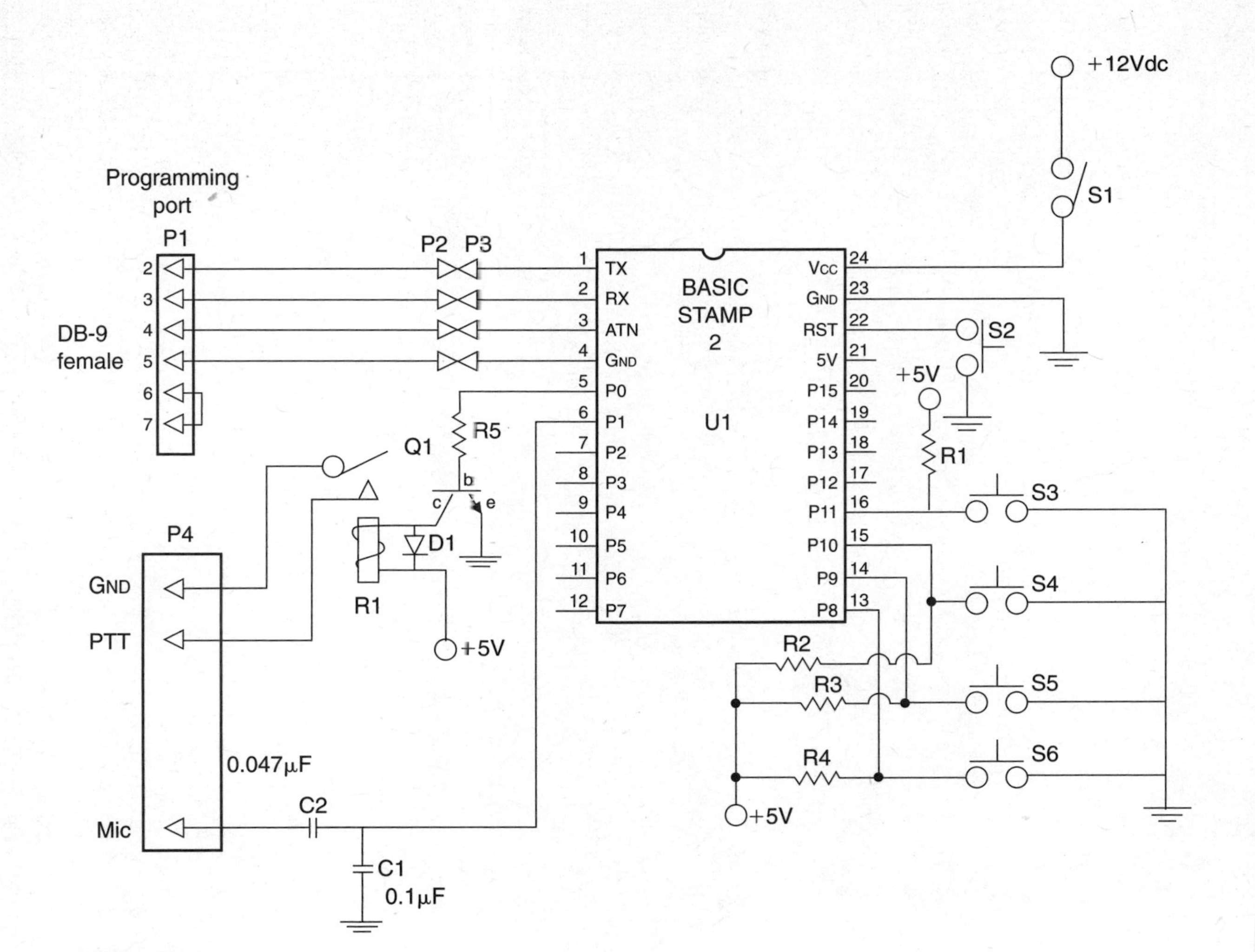

Figure 12-1 Morse keyer.

circuit socket for the BASIC STAMP 2 microprocessor is highly recommended. When assembling your Morse code keyer be sure to observe the correct polarity when installing capacitors, diodes, and transistors, to ensure your circuit will operate correctly when power is first applied.

The Morse keyer circuit is mounted in a small plastic enclosure with the on/off switch and reset switch and the four custom message buttons on the front or top of the case. The power switch S1 is an SPST toggle switch, while the system reset button S2 is a momentary pushbutton switch. Switches S3, S4, S5, S6 are the custom message momentary push buttons. Note the 10-kΩ pull-up resistors across the input pushbutton switches. The two outputs from the Morse keyer are at pins 5 (P0) and 6 (P1). The audio output signal is provided at P1 through two capacitors C1 and C2, which are interfaced to your transceiver. The output at P0 is the push to talk (PTT) line. The output signal at P0 is fed to a 1-kΩ resistor coupled to an npn transistor. The transistor Q1 is used to drive the PTT relay which is used to turn on your ham radio transceiver, thus keying your radio. A low-current reed relay is used to send Morse code. Reed relays are very fast, have a long life, and consume low power, which is ideal for this application. The PTT and audio output lines are brought out to the rear panel of the Morse code keyer and connected to a $^1/_8$-in stereo jack. You may wish install an external coaxial power jack on the rear panel as well. You could build a cable with a $^1/_8$-in stereo plug at one end and a suitable microphone connector at the opposite end of the cable. The programming cable is connected to pins 1 through 4 of the BASIC STAMP 2. A 4-pin male header could be mounted on the circuit board for the programming cable. A programming cable could be fabricated with a 4-pin female header and a length of flexible 4-conductor telephone cable connected to a 9-pin serial RS232 connector. The Morse keyer can be powered by a 9-V transistor radio battery if desired; this would facilitate building a compact or portable unit. Or the keyer could be powered from a 12-V system power bus.

A variation on the Morse code keyer project would be to design a multipurpose keyer for home station use, with the addition of a voice message chip such as an ISD2590 chip, used in the radio mailbox project (Chap. 14). The ISD voice recorder chip could be used for voice message keying in addition to Morse code keying if desired. The voice message chip could be implemented using some additional parts if desired, as shown in Fig. 12-2.

Programming the Keyer

Connect your Morse keyer to the programming cable and a PC, and apply power to the Morse code keyer. Note that the Morse keyer program must be customized for your particular application, i.e., call letters, location, and radio gear. First locate the STAMP 2 Windows editor program named STAMPW.EXE. Now locate the Morse keyer program, MORSE2.BS2 (Listing 12-1), and download it to the microprocessor in the Morse keyer. Finally, take a look at the program, and substitute your station information in the message data section of the program for your preferences. Once the program has been saved with your own particular parameters, you are now ready to download the finished program to your Morse code keyer. This is a fun project which you will find very useful in your ham radio shack!

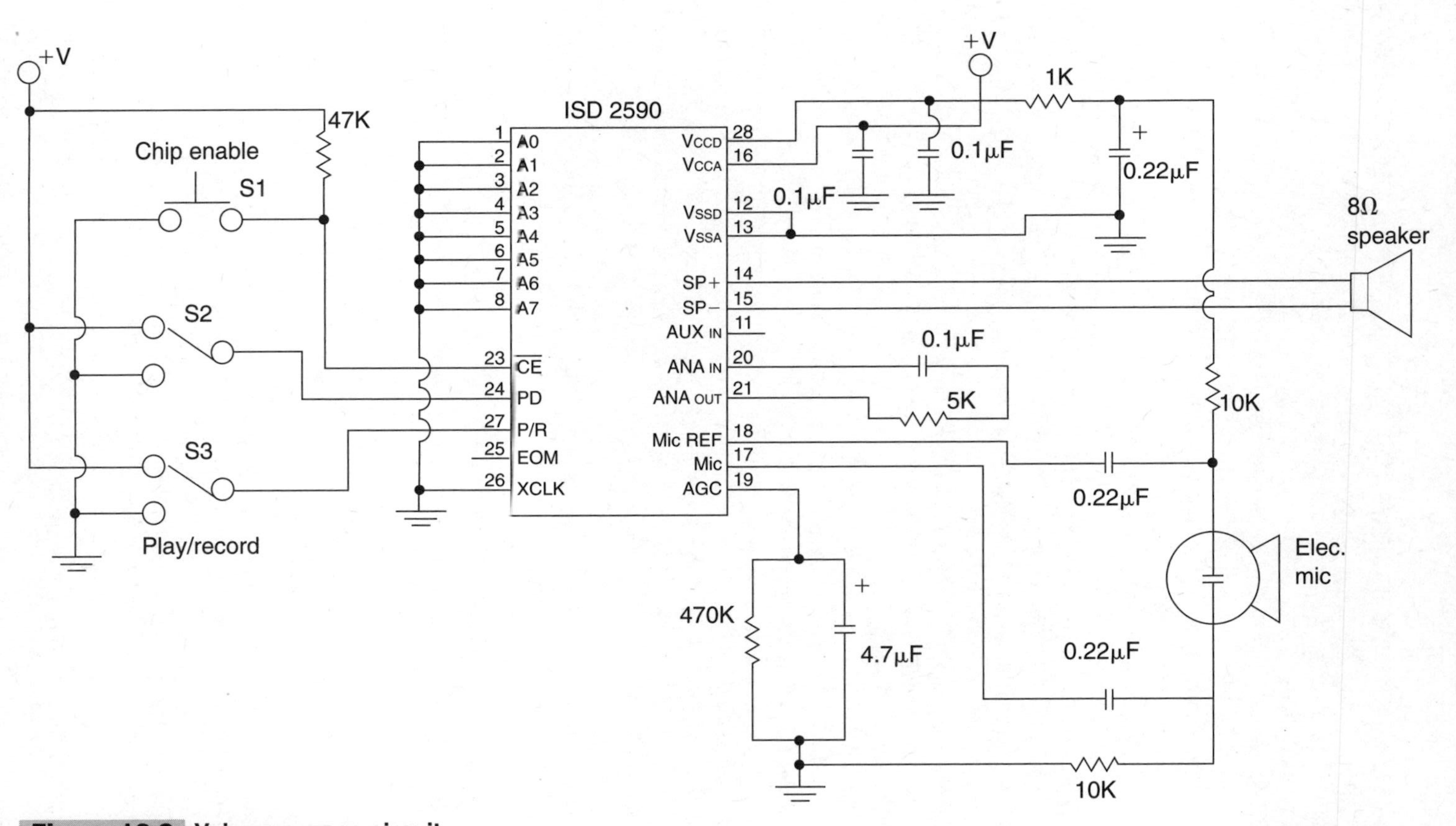

Figure 12-2 Voice message circuit.

```
'MORSE2.BS2
' four message keyer for Amateur Radio.
'Richard Clemens, KB8AOB
'The user can input four different messages in msg1-msg4,
'their character counts in msc1-msc4, and use four
'buttons to display the output and/or key a transmitter.
'NB: the messages are limited to the upper and lower case
'letters, digits 0-9, space, period, comma, ? and /
'load the EEPROM with data
'switch, special characters, A-M, N-Z, and 0-9
'(switch is so we only convert data after an initial load)
swtc     data    0
spec     data    0,87,207,54,149
leta     data    66,132,164,131,1,36,195,4,2,116,163,68,194
letn     data    130,227,100,212,67,3,129,35,20,99,148,180,196
nums     data    253,125,61,29,13,5,133,197,229,245
'user can change the messages below
msg1     data    "CQ CQ CQ de KB8AOB KB8AOB KB8AOB K"
msg2     data    "QTH QTH is Buckhannon, WV Buckhannon, WV"
msg3     data    "UR RST RST is 599 599  NAME NAME HR IS RICH RICH"
msg4     data    "TNX for QSO and 73 de KB8AOB"
'character counts must match messages above
msc1     con       34
msc2     con       40
msc3     con       48
msc4     con       28
tabl     con       41
tone     con     1200
spc      con      380
dit      con       70
dah      con      210
chr      var       b0
ele      var       b1
x1       var       b2
x2       var       b3
x3       var       b4
x4       var       b5
'initialize pins 0 and 1 for output
high  0
low   1
'clear the button variables
b8  = 0
b9  = 0
b10 = 0
b11 = 0
'test to see if we just downloaded, if yes, set the
'switch so we don't try this again and then convert all
'the messages into their morse code bytes
read 0,chr
if chr <> 0 then loop
write 0,255
'using the ASCII codes to shorten the lookdown
'replace each character with its morse code byte
```

Listing 12-1 Morse code keyer program.

```
'tabl is the number of elements in the table of codes
cvt_msgs:
x2 = tabl + 1
x3 = tabl + msc1 + msc2 + msc3 + msc4
for x1 = x2 to x3
read x1,chr
lookdown chr,[32,46,44,63,47],chr
if chr > 4 then nspec
chr = chr + 1
goto find
'if not a space, period, comma, ? or / then
'sort out the numbers, small letters, and capitals
'and compute a pointer for the table of codes


nspec:
  if chr > 90 then sml
  if chr > 57 then cap
  chr = chr - 16
  goto find
  sml:
  chr = chr - 91
  goto find
  cap:
  chr = chr - 59
  find:
  read chr,chr
  write x1,chr
next
'Jump here if not a new program load and then
'wait for a button to be pressed. (All ready to go
'for the Stamp Experiment Board!)
'input on pins 8-11 for messages 1-4
loop:
button 8,0,255,0,b8,1,msg1_o
button 9,0,255,0,b9,1,msg2_o
button 10,0,255,0,b10,1,msg3_o
button 11,0,255,0,b11,1,msg4_o


goto loop
'output the desire message
msg1_o:
x3 = tabl + 1
x4 = msc1 + tabl
goto msg_o
msg2_o:
x3 = msc1 + tabl + 1
x4 = msc1 + msc2 + tabl
goto msg_o
msg3_o:
x3 = msc1 + msc2 + tabl + 1
x4 = msc1 + msc2 + msc3 + tabl
goto msg_o
msg4_o:
x3 = msc1 + msc2 + msc3 + tabl + 1
x4 = msc1 + msc2 + msc3 + msc4 + tabl
```

Listing 12-1 Morse code keyer program (*Continued*).

```
msg_o:
for x1 = x3 to x4
gosub code
next
goto loop


'routines to send the 0's and 1's of the morse
'code bytes as dits and dahs (See Note 8)
code:
read x1,chr
  if chr <> 0 then sendit
     pause spc
  return
sendit:
  gosub morse
  return
morse:
ele = chr & %00000111
if ele = 7 then Adj1
if ele = 6 then Adj2
key:
for x2 = 1 to ele
  if chr >= 128 then dah_o
  goto dit_o
shift:
  chr = chr * 2
next
  pause dah
return

Adj1:
  ele = 6
  goto key
Adj2:
  chr = chr & %11111011
  goto key
dit_o:
  low 0
  freqout 1,dit,tone
  high 0
  pause dit
goto shift
dah_o:
  low 0
  freqout 1,dah,tone
  high 0
  pause dit
goto shift
```

Listing 12-1 Morse code keyer program (*Continued*).

Morse Keyer Parts List

U1	BASIC STAMP 2 (original)
S1	SPST toggle switch
S2, S3, S4, S5, S6	Normally open momentary pushbutton switches
R1, R2, R3, R4	10-kΩ $^1/_4$-W resistors
R5	1-kΩ $^1/_4$-W resistor
Q1	2N2222 npn transistor
D1	1N4001 silicon diode
R1	5-V reed relay
P1	9-pin RS-232 serial female connector
P2	4-pin female header
P3	4-pin male header
P4	3-pin connector (radio keyer)

13

AMATEUR RADIO FOX/BEACON PROJECT

CONTENTS AT A GLANCE

An amateur radio "fox" is a hidden radio transmitter. Many ham radio clubs all around the world sponsor hidden transmitter hunts to introduce new club members to the subhobby of radio direction finding. This fun event encourages new hams to meet in the great outdoors and to work with and meet new club members. Amateur radio "fox hunts" are great ice breakers for new ham radio operators; often the event is followed by a cookout or get-together. The fox hunt had its origins in British outdoor sporting events where hunt club members rode horses and trained their dogs to hunt foxes, hence the name fox hunt. Radio direction finding (RDF) can also be used to find or track sources of interference from wireless electronic communications, including broadcast and two-way radio, television, and telephones. It is also used to track missing or stolen cars and other property. Search and rescue workers use it to find persons in distress. Emergency locator transmitters in downed aircraft are tracked with RDF techniques.

Hams use RDF to track jamming stations and stolen equipment, but more often they use it just for fun. Hidden transmitter hunting has been done by hams for about 50 years, and it is a growing activity. *T hunting* refers specifically to hunts involving hams driving in RDF-equipped vehicles. A mobile T hunt is best described as hide-and-seek for all ages with radio gear. When you set out on a T hunt, you never know where you'll end up, and you have no idea what you're going to find. No form of ham radio contesting is more fun! Mobile T hunting is done in cities and towns all over the United States and elsewhere. Mobile T hunting is called fox hunting in some parts of the United States, but everywhere else in the world, the terms *fox hunting* and *amateur radio direction finder* (ARDF) refer to another kind of RDF contest, done completely on foot in large woods and parks. It's a map-and-compass sport similar to orienteering, with about a half-dozen fox transmitters to find in a period of 2 hours or so. Someday this sport, which is also called foxtailing, fox-teering, and radio-orienteering, may become an Olympic event. Meanwhile, it's a fun-filled activity for your hamfests and Scout Jamborees.

An amateur radio fox is a low-power VHF/UHF transmitter running off batteries, hidden in a box, under a tree, or under a vehicle. The object of the game, or hunt, is to be the first person or team to locate the hidden transmitter. Hams use radio direction finding techniques and directional antennas in some cases. Some amateurs use Doppler shift/multi-antenna receivers or receivers with a signal strength meter, and often hand-held radios. Usually, numbered cards are left beside the fox. The first person who locates the fox will pick up the first card, the second person to locate the fox would pick up the second card, etc. Often prizes are given for the first and second persons finding the hidden transmitter. The fox hunt is fun and a way to sharpen skills and meet new club members. The fox/beacon project can also be used by people who are not amateur radio hams, but who seek the fox by using FRS-type hand-held radios.

Use as a Beacon

The fox/beacon project can also be used as a beacon transmitter. A beacon is an automated transmitter that sends its call letters and Morse code sequence of identification. Data sent by a beacon station could be location, purpose, call letters, and frequency. The transmitter turns on and off at selected intervals of time. The purpose of a beacon transmitter is to test frequency propagation conditions, or to test a particular band condition, in order to see if a band is open or usable at various times of the day and night. Beacon transmitters are set up for VHF/UHF bands as well for HF or shortwave bands.

The study of radio propagation is an interesting and very useful aspect of amateur radio. There are numerous free and commercial software packages on the Internet for beacon clocks and timing programs as well as propagation prediction software; see the list of Web sites at the end of this chapter. Beacon clocks help in identifying which international high-frequency beacon is on the air at a particular time; see Fig. 13-1, which illustrates the location of the international HF Beacon stations around the world. Propagation software will assist you in predicting a possible usable path between you and a particular country or beacon location.

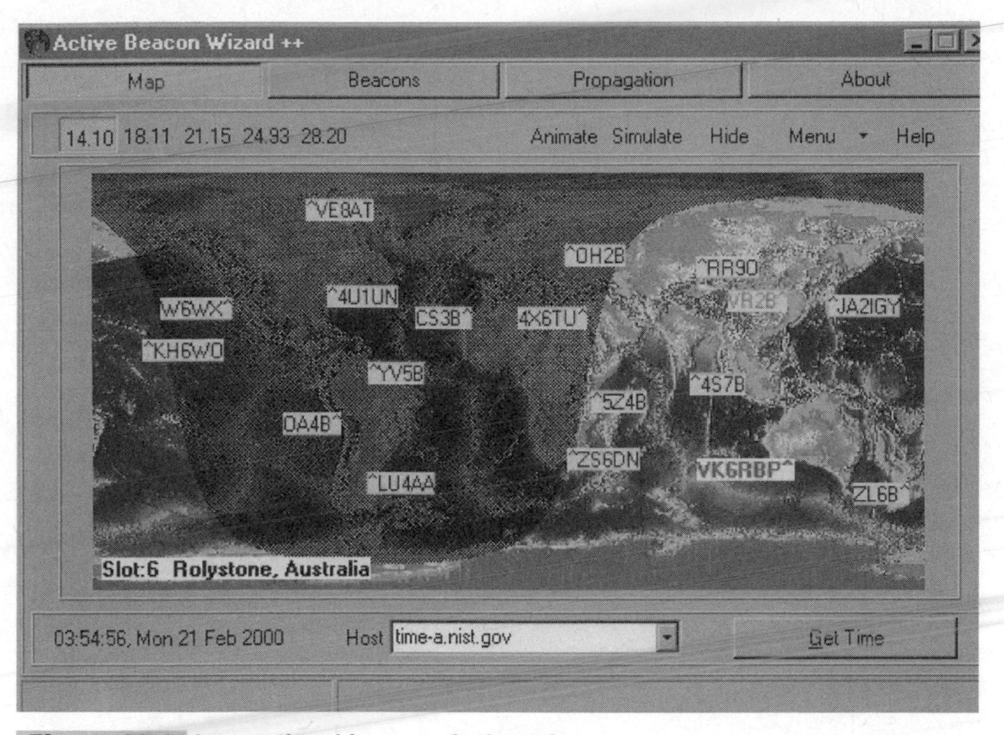

Figure 13-1 International beacon clock stations.

The Northern California DX Foundation (NCDXF), in cooperation with the International Amateur Radio Union (IARU), constructed and operates a worldwide network of high-frequency radio beacons on 14.100, 18.110, 21.150, 24.930, and 28.200 MHz. These beacons help both amateur and commercial high-frequency radio users assess the current condition of the ionosphere. The entire system is designed, built, and operated by volunteers at no cost except for the actual price of hardware components; see the beacon timing chart in Table 13-1.

The NCDXF/IARU International Beacon Network provides reliable signals on the air, around the clock, from fixed locations worldwide. With 3 minutes of listening for the beacons, one can find out either where a particular band is open or which band has the best propagation to a particular part of the world.

In principle, one can simply listen on the beacon frequencies and copy the CW call signs of the various beacons to figure out where the band is open, but in practice, not every ham operator can copy calls at 22 words per minute and some beacons may be heard at too low a signal strength to catch the call. Because the beacons transmit at known times, it is easy to know which beacon one is hearing without actually copying the CW call sign. Since the beacons are running 100 W to a vertical antenna, even a weak beacon signal may indicate a path with excellent propagation for stations using higher power and directive antennas.

In order to know which beacon is transmitting at any particular time, you can either refer to the beacon transmission schedule or use your computer and one of the beacon clock programs. If you want to know where to point your antenna or decide which beacons are the most interesting to you, you can refer to the beacon locations. If you have a computer and

TABLE 13-1 HF BEACON SCHEDULE*

			FREQUENCY, MEGAHERTZ				
SLOT	COUNTRY	CALL SIGN	14.100	18.110	21.150	24.930	28.200
1	UN	4U1UN	00:00	00:10	00:20	00:30	00:40
2	Canada	VE8AT	00:10	00:20	00:30	00:40	00:50
3	U.S.A.	W6WX	00:20	00:30	00:40	00:50	01:00
4	Hawaii	KH6WO	00:30	00:40	00:50	01:00	01:10
5	New Zealand	ZL6B	00:40	00:50	01:00	01:10	01:20
6	Australia	VK6RBP	00:50	01:00	01:10	01:20	01:30
7	Japan	JA2IGY	01:00	01:10	01:20	01:30	01:40
8	Russia	open	01:10	01:20	01:30	01:40	01:50
9	China	open	01:20	01:30	01:40	01:50	02:00
10	Sri Lanka	4S7B	01:30	01:40	01:50	02:00	02:10
11	South Africa	ZS6DN	01:40	01:50	02:00	02:10	02:20
12	Kenya	5Z4B	01:50	02:00	02:10	02:20	02:30
13	Israel	4X6TU	02:00	02:10	02:20	02:30	02:40
14	Finland	OH2B	02:10	02:20	02:30	02:40	02:50
15	Madeira	CS3B	02:20	02:30	02:40	02:50	00:00
16	Argentina	LU4AA	02:30	02:40	02:50	00:00	00:10
17	Peru	OA4B	02:40	02:50	00:00	00:10	00:20
18	Venezuela	YV5B	02:50	00:00	00:10	00:20	00:30

*Transmission cycle, minutes:seconds.

a computer-compatible radio and would like a record of when various beacons can be heard at your QTH, you will want to learn about automated beacon monitoring.

The beacons come on the air with 100 W, sign the beacon call sign, and then step down to each of four power levels, 100, 10, 1, and 0.1 watt, and finally back to 100 W for the sign-off call. Each power level would last about 10 seconds before automatically switching to the next level. The total number of beacons in the network is presently 18. The entire beacon network allows a listener to hear beacons from all parts of the world on a given frequency in a 3-minute period. Alternatively, the listener can follow a single beacon through the bands and determine the best band open to that area.

Radio beacons are very popular with VHF/UHF amateur radio operators around the world. Since VHF/UHF frequencies are usually limited to shorter ranges than HF frequencies, there are many more regional VHF/UHF beacons set up by private individuals around the world. The radio fox/beacon is ideal for VHF/UHF for determining band open-

```
'FOX10.BS2
'by G Crenshaw
'WD4BIS
'Constants
tone con 1200            '1200 hz tone for Morse ID
dit_con 100              'Time in ms for Dit subroutine
dah_con 300              'Time in ms for Dah subroutine
'Program
start:
gosub key PTT            'Key the transmitter
gosub dit                'Send W
gosub dah
gosub dah
pause dah
gosub dah                 'Send D
gosub dit
gosub dit
pause dah
gosub dit                 'Send 4
gosub dit
gosub dit
gosub dit
gosub dah
pause dah
gosub dah                 'Send B
gosub dit
gosub dit
gosub dit
pause dah
gosub dit                 'Send I
gosub dit
pause dah
gosub dit                 'Send S
gosub dit
gosub dit
pause dah
gosub bubbleup            'Bubble up tone
freqout 1,60000,100000    'Send a 1000hz tone for 60 seconds out I/O pin 1
gosub bubbledown tone     'Bubble down tone
gosub unkey PTT           'Unkeys transmitter
pause 60000               'Wait 60 seconds
goto start                'Returns to beginning of program
'Subroutines
dah:                      'Dah subroutine
  freqout 1,dah_,tone
  Pause dit_
return
dit:                       'Dit subroutine
  Freqout 1,dit_,tone
  Pause dit_
return
key PTT:                    'Key xmtr,takes I/O pin 0 high,takes I/O pin 3 low
   High 0                   'Take P0 high
   Low 3                    'Takes P3 low
```

Listing 13-1 Fox/beacon program.

```
   pause 30000 Pause .3 second to give xmtr time to key
return
unkey PTT:                    'Drop xmtr returns I/O pin 1 and 3 to original state
   low 0                      'Take P0 low
   high 3                     'Take P3 high
return
Bubbleup:                     'Bubbleup tone
  Freqout 1,dit_,500
  Freqout 1,dit_,600
  Freqout 1,dit_,700
  Freqout 1,dit_,800


return
Bubbledown:                   'Bubbledown tone
  Freqout 1,dah_,800
  Freqout 1,dah_,700
  Freqout 1,dah_,600
  Freqout 1,dah_,500
return
```

Listing 13-1 Fox/beacon program (*Continued*).

ings and propagation studies. Space does not permit listing all of the VHF/UHF beacons, but Table 13-2 lists the new band plans for beacon frequencies on the VHF/UHF bands. The new Beacon-Net system allows VHF/UHF ham radio operators to study propagation between two specific stations and will alert the hams at each when a path is available and working.

Description of the Project

The radio fox is built around the STAMP 2 microprocessor. In this project we use the original STAMP 2, but you could easily substitute the alternative STAMP 2 by translating the input/output pin numbers. The radio fox first keys the amateur. The STAMP 2 then sends out an audio alert tone and then your ham radio call letters. Nonhams can have STAMP 2 send a tone sequence and then Morse code representation of the name of a person, club, etc. The fox sends a 1000-Hz tone for about a minute and then unkeys the radio and rests for 1 minute. It then repeats the cycle over and over. The dit and dah routines in the program (see Listing 13-1) will of course have to be edited to send your own call sign.

The radio fox is built around the original BASIC STAMP 2 as shown in Fig. 13-2. You can build your fox on a perfboard with point-to-point wiring, use the BASIC STAMP 2 carrier board, or build a small circuit board for the circuit. It is advisable to use an integrated circuit socket for the BASIC STAMP 2 rather than solder the microprocessor directly to the circuit board, in the rare event the microprocessor fails. When constructing the fox, be sure to observe the correct polarity of the capacitors when installing them. The

TABLE 13-2 VHF/UHF BEACON BAND PLAN

BEACON FREQUENCY, MEGAHERTZ	FREQUENCY BAND, MEGAHERTZ
50	50.060–50.080
144	144.275–144.300
432	432.300–432.400
903	903.070–903.080
1296.07	1296.070–1296.080

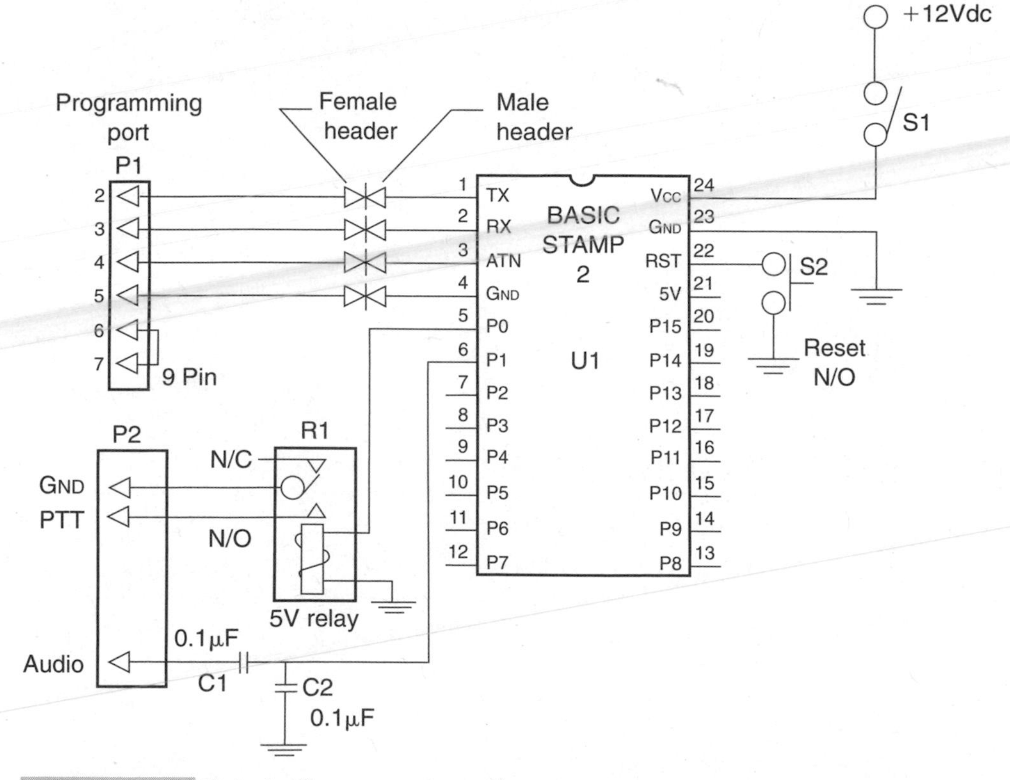

Figure 13-2 **Main fox/beacon schematic.**

audio output pin at pin 6 or (P1) is used to feed audio to your transmitter's microphone input connector. A capacitor is fed in series to the audio line of your transmitter, while a 0.1-μF capacitor is connected across the microphone pins. A relay is used to key the transmitter and is used to isolate ground between the fox and your transmitter. The PTT control line from pin 5 (or P0) is used to key the 5-V minirelay, which is used to key your radio on and off. The minirelay is a RadioShack 30-mA coil relay. The normally open contacts of the relay are used for this project. A reset switch is connected between RST pin 22 and ground. The reset switch could be a normally open pushbutton switch PC-board-mounted switch if desired. The reset switch is used in the event the processor locks up and can be used to restart the STAMP 2.

Power for the fox is supplied to pin 24 (Vcc). A 12-V gel-cell rechargeable battery can be used to power the circuit. The ampere-hour capacity of the battery used depends on the power output of the transmitter used. A 2- to 7-A-h battery is recommended. A power charge connector can be installed on the fox enclosure, in order to recharge the gel-cell. The first four pins of the STAMP 2 are used for programming. Pin 1 is the TX pin, while pin 2 is the RX pin. Pin 3 is the ATN pin, and ground is shown at pin 4. The four pins can be connected to a 4-pin male header, or to a connector on the side of the plastic enclosure. The programming connector is then wired via 4-conductor telephone wire. A 9-pin female RS-232 connector is used to interface to a PC for programming.

The diagram in Fig. 13-3 illustrates the fox connected to a battery and a suitable transmitter or transceiver via the microphone connector. The fox can be installed in an ammo can, which is sealed from moisture. A BNC connector can be installed on the side of the ammo can, and a “rubber duck” antenna can then be connected to the fox. A locking hasp and padlock can be added to keep the equipment from being stolen. You can add a piece of chain and chain the fox to a tree. You can also add a small clipboard with 3- by 5-in cards, each with a printed number. As fox hunters find the fox, they take a numbered card from the clipboard so they can claim their place in the sequence of fox finders. It is also a good idea to make up a sheet with your name, phone number, and call sign, and a brief description of what the box does. Then laminate the sheet and affix it to the side of the ammo box. In the event it is found by an unsuspecting person, the local bomb squad will not be called in to investigate.

If you choose to build the fox/beacon project as a beacon transmitter, you may wish to house the project differently. You may want to build the fox/beacon circuit and interface in an extruded aluminum chassis box that can be placed near an HF or VHF/UHF transceiver for table-top operation in a separate corner of your radio room.

The fox is now almost completed. Next you will have to program the STAMP 2. First locate the Fox STAMP 2 Windows editor program titled STAMPW.BS2. Next you will need to connect up your programming cable, apply power to the circuit, find the fox program labeled FOX10.BS2, simply change the text in the program to reflect your own call, save the program, and you’re ready to go. Your radio fox is now ready—go have some fun!

Fox/Beacon Project Parts List

U1	STAMP 2 (original)
C1, C2	0.01-μF 35-V capacitor
S1	Power switch, SPST toggle
S2	Normally open pushbutton switch

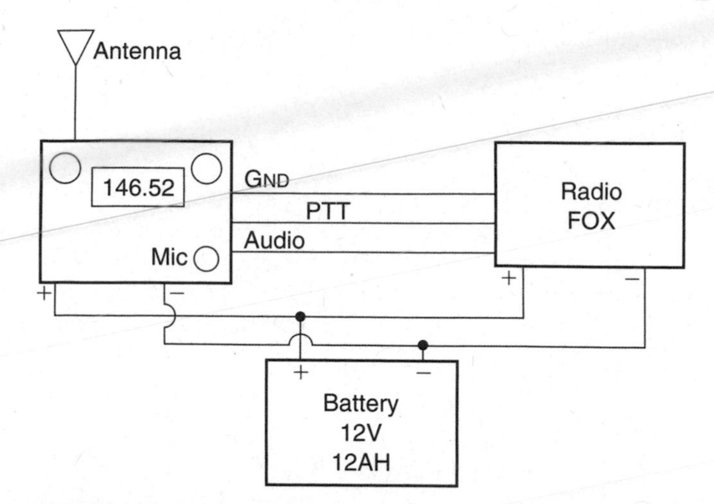

Figure 13-3 Fox/beacon hookup.

RY1	5-V 30-mA minirelay (RadioShack)
P1	9-pin RS-232 female connector
P2	Radio microphone connector
Miscellaneous	STAMP 2 carrier board, PC board, headers, wire, chassis box, etc.

Fox Hunting Clubs

http://members.aol.com/homingin/

http://www.pe.net/~dopplers/

http://home.att.net/~wb8wfk/

http://www.west.net/~jay/hunt/

http://www.frontiernet.net/~n2ki/

http://members.aol.com/joek0ov/nfw.html

Beacon Software Web Sites

http://www.taborsoft.com/abw/index.shtml

http://www.huntting.com/beaconclock/

http://www.dxzone.com/catalog/Software/Propagation/

http://www.ncdxf.org/beacon.htm

http://www.iaru.org/articles/

http://www.mutadv.com/kawin/pages/bcnclk.htm

http://www.wa7x.com/beacon.shtml

14

RADIO MAILBOX

CONTENTS AT A GLANCE

The radio mailbox is the radio amateur's dream project. This fun and useful project is one you'll want to build. The radio mailbox is much like a cell phone with a voice messaging feature. As an amateur radio enthusiast, you will immediately see the benefit of this project. As with your phone, you can't always be standing by, waiting for a call. The radio mailbox connects to your VHF/UHF ham radio transceiver's speaker or headphone jack and acts as a voice mailbox for incoming voice messages while you are away from your radio; see Fig. 14-1. This personal radio mailbox is simple to operate and will work with any radio transceiver or receiver. The radio mailbox sits in the background and looks for a touch-tone sequence to come across your radio's speaker. When the proper touch-tone sequence is generated, the radio mailbox immediately begins recording a voice message. After a message has been left, the radio operator leaving the message simply presses the digits #9 (pound-nine) to reset the radio mailbox for the next message. The radio mailbox will take a number of messages up to 90 seconds long. The radio mailbox is ideal for leaving short messages to inform your friends of an upcoming meeting or event.

Basic Components

The radio mailbox centers around three major components: the touch-tone decoder chip, the ISD2590 voice message chip, and of course the BASIC STAMP 2 microprocessor (see Fig. 14-2). We use the original STAMP 2, but you can easily substitute the alternative STAMP 2. The first component we see in the project is the touch-tone decoder. The decoder chip is connected to your amateur radio transceiver's speaker or headphone jack, via capacitor C1 and resistors R1 and R2. The CM8880 touch-tone encoder/decoder is shown in the block diagram in Fig. 14-3. The CM8880 touch-tone decoder looks for a combination of tones for each numerical digit, as seen in Table 14-1. The touch-tone decoder receives and decodes the tone pairs and produces a binary value based on the valid digit received; the binary value is then fed to the STAMP 2. The CM8880 touch-tone IC decodes the incoming touch tones and produces a binary code, as shown in Table 14-2. If the number 4 is pressed on a remote touch-tone keypad, a binary value of 0100 will show up at the outputs D0 through D3 on pins 14, 15, 16, and 17 on the CM8880. This binary value is next sent to the BASIC STAMP 2 at U2.

The touch-tone decoder IC is clocked by a 3.579-MHz color burst crystal at pins 6 and 7. The resistor and capacitor at C2 and R3 form a guard/time network that controls the incoming tone duration time limits. The CM8880 must be powered from a 5-Vdc power

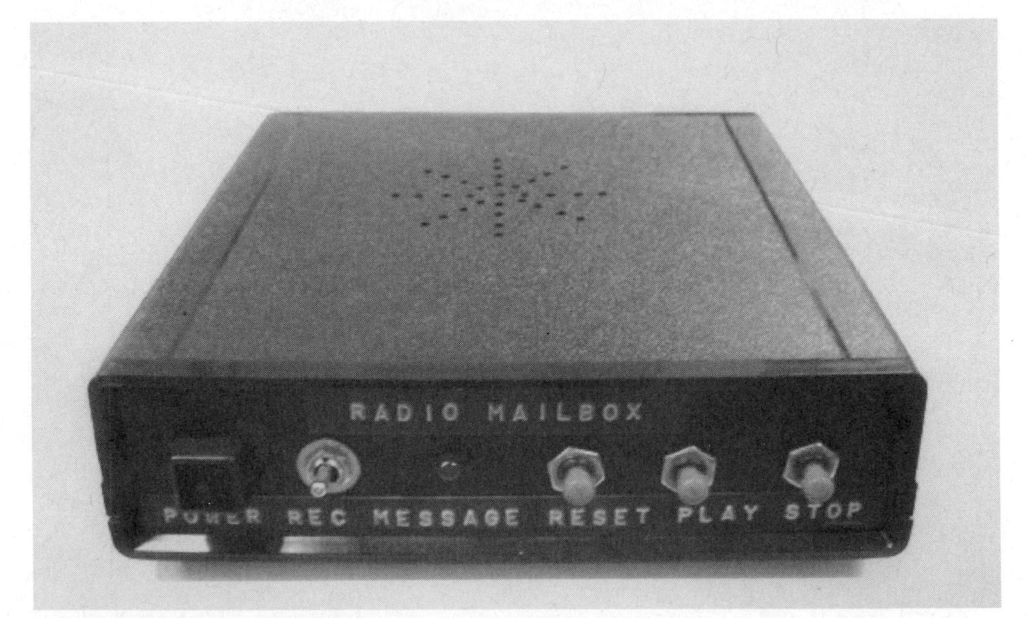

Figure 14-1 Radio mailbox.

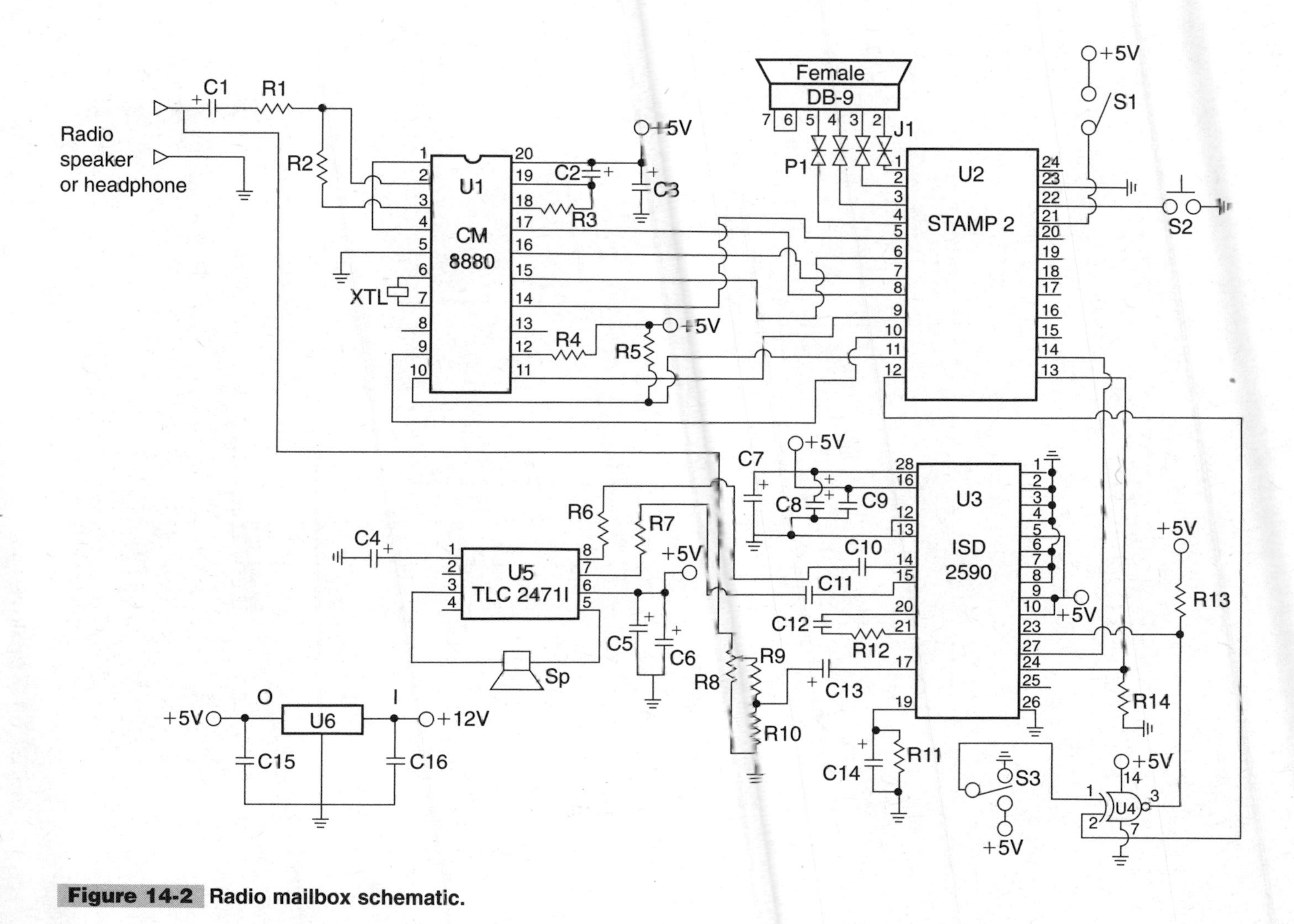

Figure 14-2 Radio mailbox schematic.

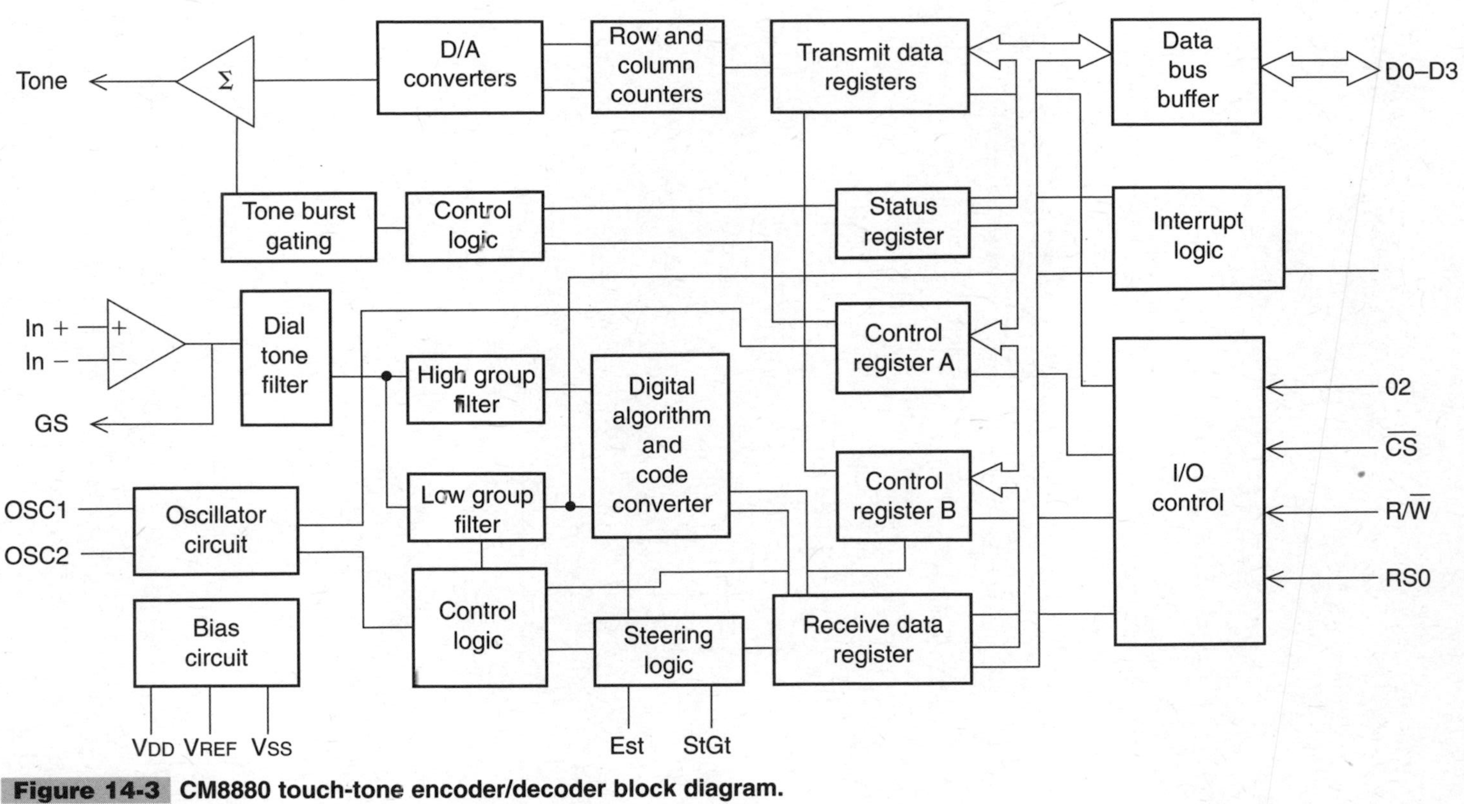

Figure 14-3 **CM8880 touch-tone encoder/decoder block diagram.**

TABLE 14-1 TOUCH TONE (DTMF) FREQUENCY

	1209	1336	1477	1633
697	1	2	3	A
770	4	5	6	B
852	7	8	9	C
941	‘	0	#	D

TABLE 14-2 DTMF BINARY AND DECIMAL VALUES

BINARY VALUE	DECIMAL VALUE	KEYPAD SYMBOL
0000	0	D
0001	1	1
0010	2	2
0011	3	3
0100	4	4
0101	5	5
0110	6	6
0111	7	7
1000	8	8
1001	9	9
1010	10	0
1011	11	*
1100	12	#
1101	13	A
1110	14	B
1111	15	C

source. The CM8880 requires input/output control lines to animate the touch-tone decoder via the microprocessor. These 3 data-bit pins are used to select the modes of the CM8880: the chip select (CS) pin, the read/write (R/W) pin, and the resistor select (RS0) pin. Table 14-3 depicts all the combinations of the three control pins. The CM8880 is active only when CS equals 0. The R/W bit determines the data direction: 1 = read and 0 = write. The RS bit determines whether the transaction involves data (DTMF tones) or internal functions, i.e., 1 = instructions/status and 0 = data. Before you can use the CM8880, you have

TABLE 14-3 CONTROL BIT CS, RW, AND RSO STATUS

CS	RW	RSO	DESCRIPTION
0	0	0	Active: write data (i.e., send DTMF)
0	0	1	Active: write instructions to CM8880
0	1	0	Active: read data (i.e., receive DTMF)
0	1	1	Active: read status from CM8880
1	0	0	Inactive
1	0	1	Inactive
1	1	0	Inactive
1	1	1	Inactive

to set it up. The device has two control registers, i.e., A and B. In the beginning of the program listing you will notice the set-up of the registers; also refer to Tables 14-4 and 14-5.

The STAMP 2 I/O lines P0 through P6 are used for controlling the touch-tone decoder chip, while P7, P8, and P9 are used to control the ISD voice message IC. The ISD2590 is a 28-pin CMOS voice message chip containing a 480-kbyte EEPROM memory, as shown in Fig. 14-4. The ISD2590 contains preamplifiers, filters, address controller, and memory. The most complex aspect of the ISD2590 is the address selection and mode selection. You can use a microprocessor to address each of the address pins for precise message duration recording and playback. We chose a more simplified approach to operating the ISD2590. Connecting address pins A8/pin 9 and A9/pin 10 to 5 V allows the chip to go into the mode selection mode; see Table 14-6. The ISD2590 can operate in six different modes of operation. Table 14-6 lists jointly compatible modes that can be used together for more complex mode selection. Note that address pin 5 is connected to 5 V, while all other address pins are connected to ground.

The voice message chip is controlled via three lines on the voice recorder chip, beginning with the power down (PD) line, which is controlled by the STAMP 2 on pin 13. The second control line is the record/play button on the ISD2590 pin 27, which is controlled by pin 14 on the STAMP 2 pin 14. The third control line is the CE or Start/Pause input at pin 23. The automatic gain control (AGC) on the voice message chip is controlled by the network of resistor R12 and capacitor C14. When the microphone input is used, the ANA-IN and ANA-OUT are wired together using R12 and C12. You can also have the ISD2590 accept "line-in" audio, by connecting the audio source directly to pin 20 and removing R6 and C12. The audio output lines on the ISD2590 at pin 14 and pin 15 are coupled through capacitors C10 and C11 to resistors R6 and R7, which connect the voice chip with an external audio amplifier at U5. An 8-Ω speaker is connected to pins 3 and 5 of the TLC2471I audio amplifier chip. The audio amplifier is powered via a 5-V source, as is the entire circuit. Bypass capacitors C5 and C6 are connected across the 5-V supply to ground.

Audio input signals from the radio transceiver are fed to the touch-tone decoder as well as to the ISD2590 voice recorder chip via the voltage divider network formed by resistors

TABLE 14-4 FUNCTIONS OF CONTROL REGISTER A

BIT	NAME	FUNCTION
0	Tone out	0 = tone generator disabled
		1 = tone generator enabled
1	Mode control	0 = send and receive DTMF
		1 = send DTMF, receive call progress tone
		(DTMF bursts lengthened to 104 ms)
2	Interrupt enable	0 = make controller check for DTMF received
		1= interrupt controller via pin 13 when DTMF received
3	Register select	0 = next instruction write goes to CRA
		1 = next instruction write goes to CRB

TABLE 14-5 FUNCTIONS OF CONTROL REGISTER B

BIT	NAME	FUNCTION
0	BURST	0 = output DTMF bursts of 52 or 104 ms
		1 = output DTMF as long as enabled
1	Test	0 = normal operating mode
		1 = present test timing bit on pin 13
2	Single/dual	0 = output dual (real DTMF) tones
		1= output separate row or column tones
3	Column/row	0 = if above = 1 select row tone
		1 = if above = 1 select column tone

R8, R9, and R10. The volume control at R8 is set to ensure minimum distortion levels to the ISD2590 chip. Bypass capacitors C7, C8, and C9 are all placed across the power supply as close to the pins as possible.

PROGRAMMING CONNECTIONS

The STAMP 2 I/O pins 1 through 4 are used to program the radio mailbox system. These four pins from the microprocessor are brought out to a 4-pin male header (P1) on the circuit board. Next you will need to make up a programming cable. A 4-pin female header (J1) is connected to a length of four wire telephone cable and a DB-9 female connector is soldered to the other end of the cable.

POWER SUPPLY

The entire radio mailbox circuit is powered from the 5-V regulator at U6. The regulator accepts a 9- to 12-V input and produces a regulated 5-Vdc supply for the circuit. A 9- to

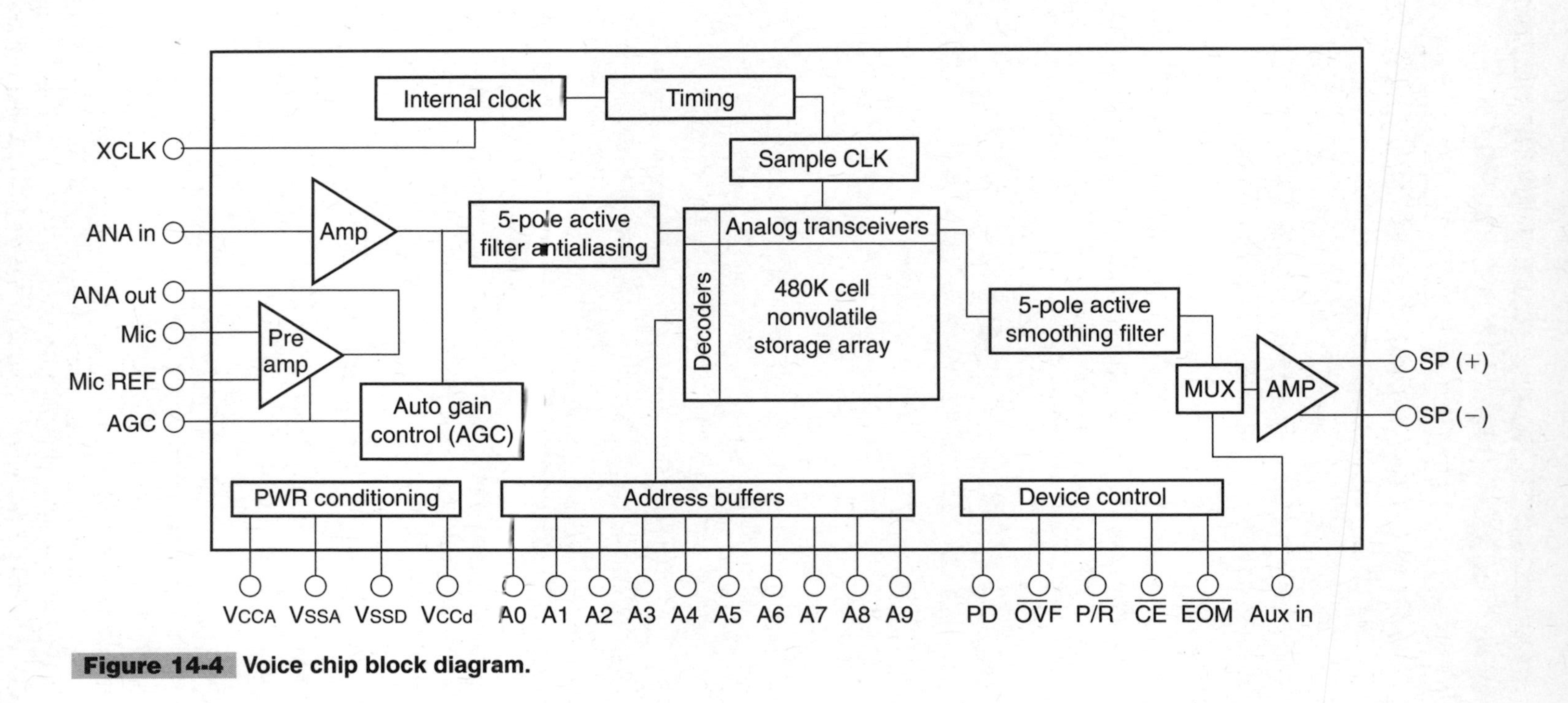

Figure 14-4 Voice chip block diagram.

TABLE 14-6 ISD2590 OPERATIONAL MODE SELECTION

MODE CONTROL	FUNCTION	TYPICAL USE	JOINTLY COMPATIBLE
M0	Message queing	Fast forward through messages	M4, M5, M6
M1	Delete EOM* marks	Position EOM marker at end	M3, M4, M5, M6
M2	Not applicable	Reserved	Not applicable
M3	Looping	Continuous play from address 0	M1, M5, M6
M4	Consecutive address	Record/play consecutive messages	M0, M1, M5
M5	CE level activate	Allows message pausing	M0, M1, M3, M4
M6	Pushbutton operations	Simplified device interface	M0, M1, M3

*EOM = end of message.

12-V "wall wart" power supply could be used to power the circuit at home or on the bench, while the radio mailbox could be powered via the 12-V automotive battery if operated in your car, truck, or motor home.

Radio Mailbox Construction

Construction of the radio mailbox is straightforward. It can be assembled on a 4- by 6-in circuit board and placed in a plastic or aluminum extruded enclosure. Integrated circuit sockets are used for each of the ICs, in the event of a rare microprocessor failure. There are a number of capacitors in this project, and particular attention should be paid to polarity when you are installing them, to ensure the circuit will correctly power up. The programming pins of the STAMP 2 are connected to a 4-pin 0.10-in male header to facilitate connecting a programming cable to the outside world. The power switch at S1 is a toggle type that is connected to a 2-pin male header for supplying power to the radio mailbox circuit. Switch S2 is provided as a reset device, in the event that the processor locks up. Pressing S2 connects pin 22 to ground momentarily. Switch S3 is message playback button. This button is used to play back the recorded messages once you return to the radio mailbox. Switch S3 must be left high for the entire listen cycle and then returned to the standby position, which is low.

Operating the Radio Mailbox

The radio mailbox is quite simple to operate. Connect up the programming cable from the radio mailbox to your programming PC. Next, locate the STAMP 2 Windows editor program titled STAMPW.BS2. Now, connect up your power supply and download the RADIOMAIL.BS2 program (Listing 14-1) into the STAMP 2. Connect up the radio mailbox across your radio transceiver's speaker or headphone jack and you are ready to let the

```
'RADIOMAIL.BS2
T    var   byte                         ' Received DTMF digit
df   var   bit                          ' DTMF-received flag
ds   var   INL.bit2                     ' DTMF-detected status bit
C    var   word                         ' Tones rcvd counter variable
G    var   byte                         ' Sequence counter variable
J    var   byte                         ' Table counter variable
PD   con     8
P_R  con     9
 OUTL = %01111111                       ' pin 7 low, pins 0-6 high
 DIRL = %11111111                       ' set up write to 8880 (out)
 OUTL = %00011000                       ' set up CRA; write to CRB
 high 6                                 ' low 6 to high 6 = "write"
 OUTL = %00010000                       ' clear CRB; rdy/snd DTMF
 high 6                                 ' low 6 to high 6 = "write"
 DIRL = %11110000                       ' set 4-bit bus to input
 high 5                                 ' set R/W to "read"
 OUTH = %00000011
 DIRH = %00000011
 Init:                                  ' init routine
 high 7
 pause 50
 low PD
 high P_R                               ' turn off output device
Begin:                                  ' Reset routine
 C=0 : G=0 : J=0                        ' clear all variables to zero
 Listen:                                ' Listen for DTMF
 high 4                                 ' read status register
 low 6                                  ' activate chip-select pin
 df = ds                                ' store DTMF-det bit in flag
 high 6                                 ' de-activate chip-select pin
 if df = 1 then Lock                    ' if tone, then continue
 goto Listen                            ' listen again
Lock:                                   ' Tone detected, test code
 low 4                                  ' get DTMF data (low rs pin)
 low 6                                  ' activate chip-select pin
 T = INL & %00001111                    ' remove upper bits using AND
 high 6                                 ' de-activate chip-select pin
 C=C+1                                  ' increment tone counter
 if C>2000 then Begin                   ' resets >2000 wrong entries
 lookup T,["0234567800019000"],J          ' convert tone to string
  J=J-48                                  ' convert string to value
  if J=0 then Init                        ' wrong digit "reset"
  if C=1 and J=1 then Key                 '  IF all digits are in the
  if C=2 and J=2 then Key                 '  correct order, AND after
  if C=3 and J=3 then Key                 '  only one attempt, THEN
  if C=4 and J=4 then Key                 '  increment the sequence
  if C=5 and J=5 then Key                 '  counter variable : "G" ...
  if C=6 and J=6 then Key
  if C=7 and J=7 then Key
  if C=8 and J=8 then Key
  if C=9 and J=9 then Key

  goto Listen                             ' listen again
Key:                                      ' Sequence counter
```

Listing 14-1 Radio mailbox program.

```
  G=G+1                                 ' increment
  if C=9 and J=9 and G=9 then Relay     ' activate output device
  goto Listen                           ' listen again
Relay:
  pause 100
  low P_R                               ' Relay output routine
  pause 800                             ' wait time delay
  low 7                                 ' turn on output device
  pause 200                             ' wait time delay
  goto Begin                            ' begin again
```

Listing 14-1 **Radio mailbox program (*Continued*).**

radio mailbox do its magic. You may have to adjust the proper audio level to prevent distorted audio from reaching the radio mailbox; this might take a few tries to get right.

Operation begins with your radio buddies sending the proper touch-tone sequence to your radio. Once the proper tone sequence is received and decoded, the STAMP 2 takes the output at pin 12 to a low state, which propagates a low through the exclusive OR gate at U4, thus triggering the ISD2590 to begin recording the first message. After the voice message has been recorded, the radio operator sends the digits #9 (pound-nine) to reset the tone decoder. Once the decoder is reset, it waits for the proper tone sequence. A number of recorded messages can be recorded by different radio friends, and, on returning to your radio, you can elect to listen to all the recorded messages by simply toggling S3 to the play position. When you have listened to all the recorded messages, just return S3 to the standby position. This project has many possibilities, and it can be expanded with many more features if desired. Just enter the magic code *1234567# and the radio mailbox is ready to serve you!

Radio Mailbox Parts List

U1	CM8880 touch-tone decoder chip
U2	BASIC STAMP 2 (original)
U3	ISD2590 voice recorder chip
U4	LM7486 exclusive OR gate
U5	TLC2471I audio amplifier chip
U6	LM7805 5-V regulator IC
R1, R5	10-kΩ $^1/_4$-W resistor
R2, R13, R14	100-kΩ $^1/_4$-W resistor
R3	390-kΩ $^1/_4$-W resistor
R4	3.3-kΩ $^1/_4$-W resistor
R6, R7	47-kΩ $^1/_4$-W resistor

R8	200-Ω potentiometer (20 turns)
R9	220-kΩ $^1/_4$-W resistor
R10	1-kΩ $^1/_4$-W resistor
R11	470-kΩ $^1/_4$-W resistor
R12	5.1-kΩ $^1/_4$-W resistor
C1	0.1-μF 200-V polyester capacitor
C2, C3, C5	0.1-μF 35-V ceramic capacitor
C7, C8, C9, C10	0.1-μF 35-V ceramic capacitor
C11, C12, C16	0.1-μF 35-V ceramic capacitor
C4	3-μF 35-V electrolytic capacitor
C6, C15	10-μF 35-V electrolytic capacitor
C13	0.22-μF 35-V ceramic disk capacitor
C14	4.7-μF 35-V electrolytic capacitor
XTL	3.579-MHz color burst crystal
S1, S3	SPST toggle switch
S2	Momentary pushbutton switch (normally open)
Miscellaneous	IC sockets, PC board, wire, male and female headers, DB-9 connectors

15

ORB-TRACKER: AUTOMATIC SATELLITE AZ-EL ROTOR CONTROL

CONTENTS AT A GLANCE

The Orb-Tracker project is designed to remotely control the azimuth and elevation of a satellite antenna system. Many amateur radio operators and weather satellite enthusiasts need a method to control their high-gain satellite antenna systems. The Orb-Tracker will control two single low-cost antenna rotators such as the Gemini Orbit 360 or the dual-rotator Yeasu 5500 azimuth/elevation (AZ-EL) rotator using a BASIC STAMP 2 microcontroller. In order to track low-earth-orbit satellites, amateur radio and other satellite enthusiasts use satellite tracking software, which is readily available. NASA originates data sets that are used to track the many orbiting satellites that cross the heavens every day and night. These satellite data sets are readily available and can be downloaded every month with updated parameters for many satellites now in the sky.

A few satellite tracking programs, such as WISP, NOVA, and The Station Program, send out serial data from the standard serial port of the PC. The EASYCOMM data format from these programs can be easily utilized to control the Orb-Tracker's STAMP 2, which in turn controls the AZ-EL rotators. Polar satellites cross the sky in ascending or descending orbits around the globe, crossing over the north and south poles during each revolution around the earth. In order to track these moving satellites and maintain the maximum signal emitted by satellites, amateur radio operators must move their antennas both in elevation and azimuth, meaning one rotator must be able to rotate 360° in the horizontal plane and the second rotator must move up and down from the horizontal plane through at least 90° to pole or vertical.

The Orb-Tracker can simultaneously control two rotators, using the data from the serial port, once the proper satellite data set is plugged into the tracking program. The Orb-Tracker is designed to be used with two inexpensive TV rotators, which use positive and negative drive voltages and variable resistors to indicate position, such as Orbit 360 rotators. The Orb-Tracker, as mentioned earlier, will also drive the dual-rotator Yaesu 5500 AZ-EL rotator.

Basic Components for Orbit

The heart of the Orb-Tracker is the BASIC STAMP 2 processor, an A/D converter, a power supply, and a handful of discrete components. The diagram in Fig. 15-1 illustrates the Orb-Tracker for use with the Orbit 360–type low-cost TV rotators. The serial data output from your PC tracking program (NOVA/WISP) is fed to the DB-9 serial connector at the top left of the diagram. The serial port connector also does double duty, since it is also used to program the BASIC STAMP 2 microprocessor. The BASIC STAMP 2 serial port connections are on pins 1 through 4, respectively (the TX, RX, ATN, and ground pins).

The Orb-Tracker's A/D converter samples the position voltages from both the azimuth and elevation rotators. These analog voltages are converted into digital numbers by the LTC1298 converter. The position voltage from the rotators varies between 0 and 5 Vdc, depending on the position of the rotator. This value is converted into a digital number that varies from 0 to 4095. The resolution of the Orb-Tracker is theoretically 360/4095, or 0.087°. This digital number is sent serially to the BASIC STAMP 2. Simultaneously, the STAMP 2 is also getting new position data from the tracking program, which it uses to compare the last known position of the rotators to the new positions coming from the track program. The BASIC STAMP 2 compares these two voltages in order to decide whether the rotator needs to be rotated clockwise or counterclockwise in order to make the rotator position value equal to the satellite's new position value. The Orbit 360 rotators are dc-controlled motors. This means that reversing the power leads reverses the direction of rotation. This makes direction control fairly easy, using two relays per rotator. One relay serves as a switch to turn +12 V on or off. The other relay switches −12 V. The outputs from these two relays are tied together to one power lead of the rotator. The other lead is connected to ground. Since both relays are never turned on at the same time, there are no problems with two relay outputs tied together.

In order to protect the Orb-Tracker system in the event of a microprocessor crash, two 50-Ω 10-W current-limiting resistors are placed in both the +12-V and the −12-V power leads. The relays chosen for the Orb-Tracker are Potter & Brumfield JWD-107-5 units. The

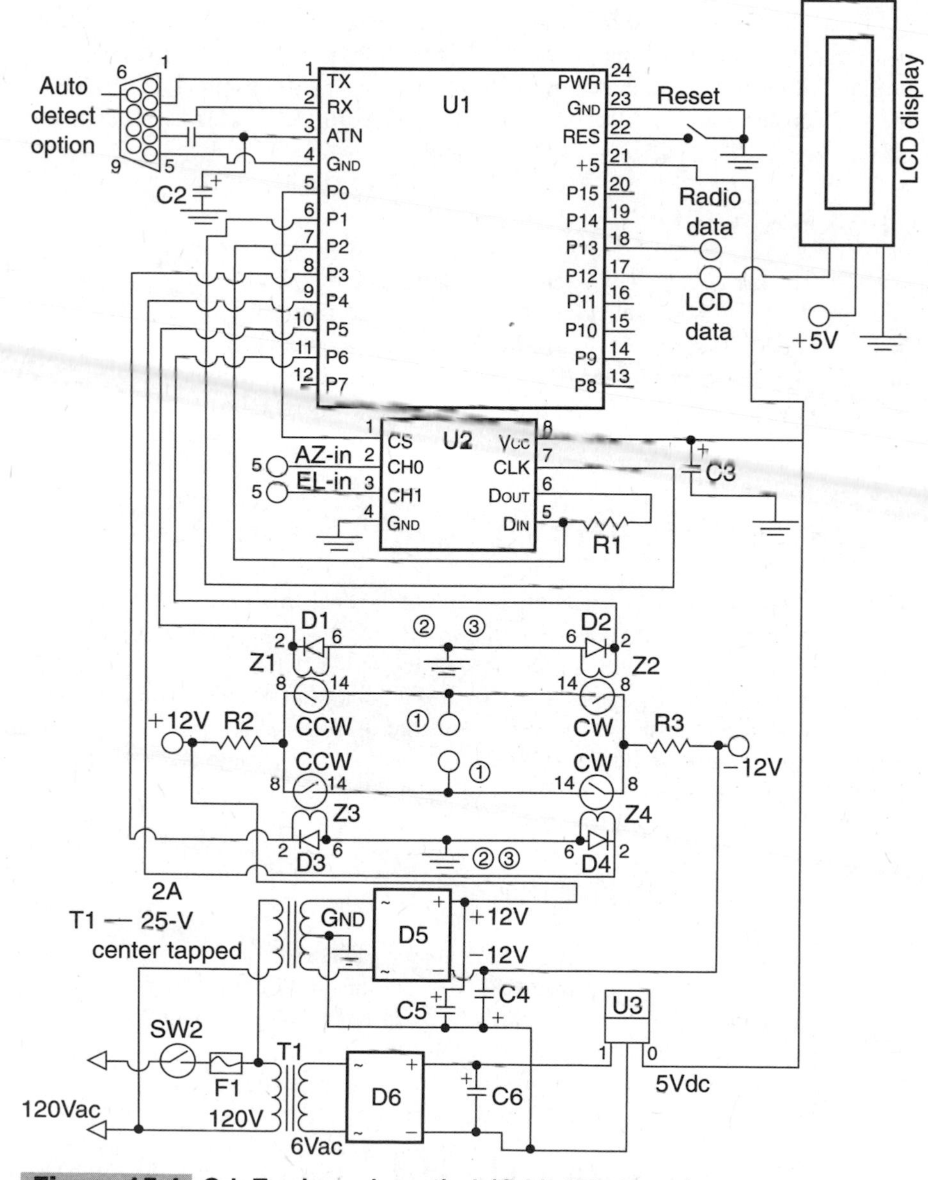

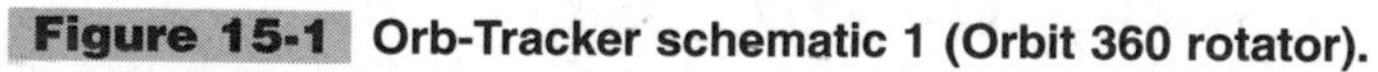
Figure 15-1 Orb-Tracker schematic 1 (Orbit 360 rotator).

STAMP 2 I/O pins can source 20 mA and sink 25 mA; fortunately, 5-V relays draw only 15 mA. If other relays are to be used, then driver transistors would be needed to drive them. Note that silicon diodes are placed across the relays for circuit protection. The two tagged circles marked 1 in the center of the schematic are used to drive the rotators.

The azimuth and elevation feedback information from the rotator's variable position resistors are coupled to input pins 2 and 3 of the LTC1298 A/D converter. The chip select (CS) pin of the A/D is animated by the STAMP 2 at pin P0. The clock data (CLK) input to the A/D converter is controlled by pin P1 of the STAMP 2. A 1-kΩ resistor is placed across pins 5 and 6 of the LTC1298.

POWER SUPPLY

The Orb-Tracker circuitry is powered from a triple-output dc power supply, which produces both +12 and −12 V as well as 5 V. A 25-V center tap, 2-A transformer is used to supply ac to two diode bridges, one for the 12-V supply and the other for the 5-V supply. The output from the 5-V supply is coupled to an LM7805 5-V three-lead regulator IC. The 5-V supply is used to power the BASIC STAMP 2 and the LM1298 A/D chip.

Basic Components for Yeasu 5500

The diagram in Fig. 15-2 depicts the circuit for the dual-rotator Yeasu 5500 series AZ-EL rotator. The circuit is quite similar to that in Fig. 15-1, except the Yaesu 5500 rotator supplies its own 13.8-V power. This version of the Orb-Tracker requires only a 5-V power source. The wiring configuration to the rotator is somewhat different in that the tag marked 3 is the EL-down control, tag 5 is the EL-up control, and the AZ right and left controls are marked 2 and 4, respectively.

Positional Readout

The Orb-Tracker also provides an LCD positional readout on pin P12. A single serial line LCD display such as the BPI-216 or the ILM-216L from Scott Edwards, http://www.seetron.com, can be readily used for displaying position data of the controller in action.

Construction of the Orb-Tracker

Either Orb-Tracker circuit can be constructed on a 4- by 6-in circuit board, and then mounted in an aluminum chassis box; see Fig. 15-3. When you are constructing the Orb-Tracker it is advisable to utilize IC sockets for the integrated circuits and the microprocessor. When assembling the circuit board, pay particular attention to the polarity of the electrolytic capacitors and diodes to avoid problems upon initial power-up of the circuit.

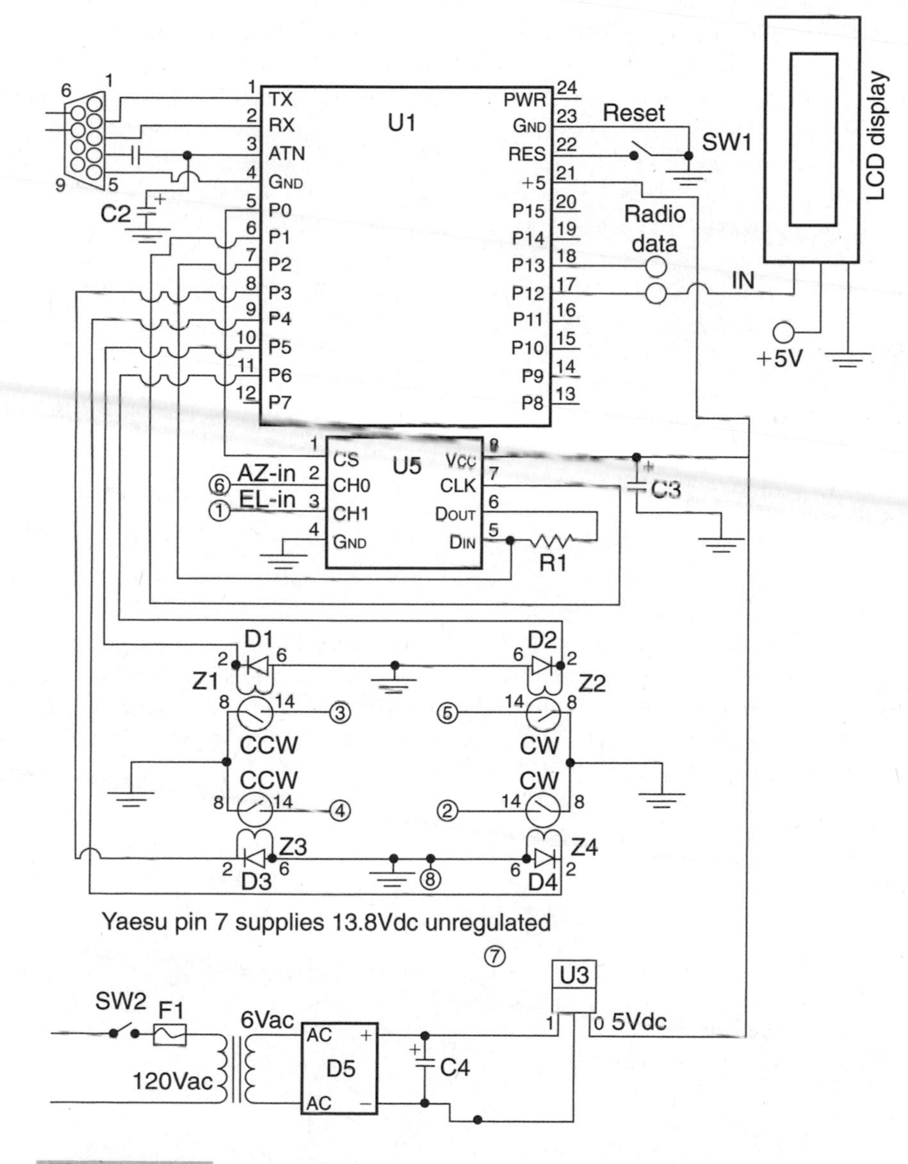

Figure 15-2 Orb-Tracker schematic 2 (Yaesu 5500 rotator).

As you are constructing the Orb-Tracker, you should install a 4-pin 0.010-in male header for the programming pins, 1 through 4 on the STAMP 2. Note pin 1 on the STAMP 2 is the TX pin, which connects to pin 2 of a 9-pin female RS-232 connector, while pin 2 of the STAMP 2 is the receive pin on pin 3 of the serial connector. The ATN line on pin 3 of the STAMP 2 is connected to pin 4 on the serial connector, while the ground on the microprocessor is connected to pin 5 of the RS-232 connector.

Remember one version of the Orb-Tracker uses a dual power supply while the other uses a triple power supply. These supplies can be easily built with center-tapped transformers and discrete diode bridges, a couple of 3-pin regulator ICs, and a few electrolytic capacitors.

Finally, you will need to connect up your programing cable and download the Orb-Tracker program into your STAMP 2 processor. First you will need to locate the Windows STAMP 2 editor program titled STAMPW.EXE. Next connect up your programming cable and apply power to the Orb-Tracker circuit. Finally, locate the appropriate Orb-Tracker program (see Listings 15-1 and 15-2) and download it into the STAMP 2. Note that the serial port connector on both system diagrams are used for double duty. The BASIC STAMP 2 is initially programmed via the serial port of a PC, using the STAMP 2 Windows editor program. Once the microprocessor has been loaded with either the SST.BS2 or Yaesu.BS2 programs, the Orb-Tracker will begin to look for tracking information from a tracking program, such as WISP or NOVA.

The software for the Orb-Tracker uses a "software brake," i.e., a 100-ms burst of opposite voltage to quickly stop "coasting" antennas. Coasting is due to the mass of the antennas on the rotator. On small antennas arrays, this system works rather well, but with large antenna systems, you may want to not use the "software brake."

The Orb-Tracker was not intended to be used with very large antenna arrays, but it can handle the standard VHF/UHF uplink and downlink satellite antenna combinations. The Orb-Tracker rotators mentioned earlier are for light- to medium-duty use. You cannot use this system for 20-ft large-scale arrays; these motors are not up to the job. The Orb-Tracker is a relatively inexpensive means to set up a computer-controlled satellite tracking antenna system.

Figure 15-3 Orb-Tracker chassis panel.

Orb-Tracker Orbit 360 Parts List

U1	STAMP 2 microcontroller (original)
U2	LTC1298 A/D converter IC
U3	LM7805 5-V regulator IC
R1	1-kΩ $^1/_4$-W resistor
R2, R3	50-Ω 10-W resistor
C1, C2	0.1-μF 35-V disk capacitor
C3	10-μF 35-V electrolytic capacitor
C4, C5	1000-μF 50-V electrolytic capacitor
C6	470-μF 35-V electrolytic capacitor
D1, D2, D3, D4	1N1002 silicon diode
D5, D6	Full-wave diode bridge
Z1, Z2, Z3, Z4	Potter & Brumfield 5-V low-current relay (JWD-107-5)
T1	25-V center-tapped 2-A transformer
SW1, SW2	SPST switch (toggle)
F1	120-V 1.5-A fuse
J1	DB-9 RS-232 jack
Miscellaneous	Circuit board, header, wire, etc.

Orb-Tracker Yaesu 5500 Parts List

U1	STAMP 2 microcontroller (original)
U2	LTC1298 A/D converter IC
U3	LM7805 5-V regulator IC
R1	1-kΩ $^1/_4$-W resistor
C1, C2	0.1-μF 35-V disk capacitor
C3	10-μF 35-V electrolytic capacitor
C4	470-μF 35-V electrolytic capacitor
D1, D2, D3, D4	1N1002 silicon diode
Z1, Z2, Z3, Z4	Potter & Brumfield 5-V low-current relay (JWD-107-5)
T1	6-V 500-mA transformer
SW1, SW2	SPST switch (toggle)

F1	120-V 1.5-A fuse
J1	DB-9 RS-232 jack
Miscellaneous	Circuit board, header, wire, etc.

```
'  SST.BS2
'  G Brigman, KC4SA & M Hammond, N8MH/
'  Driver for Gemini/Orbit 360 antenna rotators.
'  Any +/- DC voltage driven rotator
'I/O Assignments
'       A/D Chip Select.................P0 - out
'       A/D Clock.......................P1 - out
'       A/D I/O pin.....................P2 - in
'       AZ CCW Relay....................P3 - out
'       AZ CW Relay.....................P4 - out
'       EL CCW Relay....................P5 - out
'       EL CW Relay.....................P6 - out
'       Serial Data Input...............P16 - out
'       Radio TTL output................P13 - out
'       LCD display output..............P12 - out
'    (255/360)*100=141 Convert 360 degs/rev to binary
' Data input expected to conform to WISP format:  AZxxx.x ELxx.x
'DECLARATIONS
StartPOS:
'Assign VARIABLES here.
    globvar     var   word    'Global variable
    azpos       var   word    'Actual azimuth position
    newazpos    var   word    'Requested new azimuth
    elpos       var   word    'Actual elevation position
    newelpos    var   word    'Requested new elevation
    direction   var   byte    'Rotor direction
    brake       var   byte    'Opposite Rotor Direction
    ad          var   word    '12 bit A/D value
    config      var   nib     'Configuration bits for A/D
    startb      var   config.bit0 'Startbit
    sgldif      var   config.bit1 'mode select
    oddsign     var   config.bit2 'channel select
    msbf        var   config.bit3 '0's after data xfer complete
'Assign CONSTANTS here.
    portout     con   15      'Define serial freq. output port
    portin      con   16      'Define serial input port
    lcd         con   12      'Define serial display line
    I           con   254     'Instruction prefix
    ClrLCD      con   1       'Clear LCD instruction
    N96N        con   $4054   '9600 baud, inverted, no parity
    degree      con   113     '40950/360=113 steps/degree
    minstep     con   2       'Minimum step in degrees
    cs          con   0       'A/D chip select. 0 = active
    clk         con   1       'A/D clock. Out=rising, In=falling
    dio         con   2       'A/D data I/O pin number
    azccw       con   3       'counter clockwise
    azcw        con   4       'clockwise
    elccw       con   5       'counter clockwise
    elcw        con   6       'clockwise
'  INITIALIZE I/O PINS
```

Listing 15-1 Orb-tracker program—Orbit 360 rotator.

```
InitPorts:
     high cs                    'Deselect the A/D
     high dio                   'Prepare A/D I/O line
     for globvar=3 to 6
          low globvar
          pause 100
     next
     low lcd                    'make pin 12 low
InitLCD:
     pause 1000
     serout 0,N96N,[I,ClrLCD]       'clear LCD display
     serout lcd,N96N,[I,128,"SAEBRTrack  N8MH"] 'send upper text line to LCD
'    serout lcd,N96N,[I,200,"/"] 'send lower text to LCD
     pause 5
     gosub DoAz                 'Get initial rotor position Az
     gosub DoEl                 'Get initial rotor position El
     gosub WriteAz              'Update the display
     gosub WriteEl              'Update the display
GetData:
'    gosub DoAz                 'Get initial rotor position Az
'    gosub DoEl                 'Get initial rotor position El
'Read the decimal azimuth data.
     serin portin,84,[wait("AZ"),dec newazpos,wait(".")]
'Read the decimal elevation data.
     serin portin,84,2000,GetData,[wait("EL"),sdec newelpos,wait(".")]
'convert the values to binary
'    newazpos = (newazpos*degree)/10
'    newelpos = (newelpos*degree)/10
     if newelpos < 0 then QGetData
'UNCOMMENT this line if you don't want 360 degree elevation control
'I found it useful when working on the array.
     if newelpos < 0 or newelpos > 90 then QGetData
AzWork:
     if abs(azpos-newazpos) <= minstep then ElWork
     direction=azccw
     brake=azcw
     if newazpos<azpos then AzGoCCW
     direction=azcw
     brake=azccw
AzGoCW:
     high direction     'turn on the appropriate line
TestAzCW:
     pause 100
     gosub DoAz
     gosub WriteAz
     if (azpos+minstep)<newazpos then TestAzCW
     low direction
     gosub SetBrake
     goto ElWork
AzGoCCW:
     high direction     'turn on the appropriate line
TestAzCCW:
     pause 100
     gosub DoAz
     gosub WriteAz
```

Listing 15-1 Orb-Tracker program—Orbit 360 rotator (*Continued*).

```
     if (azpos-minstep)>newazpos then TestAzCCW
     gosub SetBrake
     low direction
ElWork:
     if abs(elpos-newelpos) <= minstep then QGetData
     direction=elccw
     brake=elcw
     if newelpos<elpos then ElGoCCW
     direction=elcw
     brake=elccw
ElGoCW:
     high direction       'turn on the appropriate line
TestElCW:
     pause 100
     gosub DoEl
     gosub WriteEl
     if (elpos+minstep)<newelpos then TestElCW
     low direction
     gosub SetBrake
     goto QGetData
ElGoCCW:
     high direction         'turn on the appropriate line
TestElCCW:
     pause 100
     gosub DoEl
     gosub WriteEl
     if (elpos-minstep)>newelpos then TestElCCW
     low direction
     gosub SetBrake
QGetData:
     goto GetData           'Loop through the process
SetBrake:
' UNCOMMENT THE FOLLING LINE TO DISABLE THE BRAKE
'  goto QSetBrake
     pause 1
     high brake
     pause 100
     low brake
QSetBrake:
     return
DoAz:
     config=config "PIXymbolsPCx">| %1011   'Set config array
     oddsign=0
     gosub convert
     azpos=abs(((ad*10)/degree)-360) 'convert to decimal and invert
QDoAz:
     return

WriteAz:
     serout lcd,N96N,[I,193,"AZ     "] 'clear old characters on LCD
     pause 1
     serout lcd,N96N,[I,193,"AZ ",dec azpos,223] 'send azimuth to LCD
     pause 1
QWriteAz:
     return
DoEl:
```

Listing 15-1 Orb-Tracker program—Orbit 360 rotator (*Continued*).

```
     config=config "PIXymbolsPCx">| %1011     'Set config array
     oddsign=1
     gosub convert
     elpos=abs(((ad*10)/degree)-360) 'convert to decimal and invert
QDoEl:
     return
WriteEl:
     serout lcd,N96N,[I,202,"EL      "] 'clear old characters on LCD
     pause 1
     serout lcd,N96N,[I,202,"EL ",dec elpos,223] 'send azimuth to LCD
     pause 1
QWriteEl:
     return
Convert:
     low cs              'Activate the A/D.
     shiftout dio,clk,lsbfirst,[config<\\>4]  'Send config bits
     shiftin dio,clk,msbpost,[ad<\\>12]   'Get data bits
     high cs             'Deactivate the A/D.
QConvert:
     return
'   End of program
```

Listing 15-1 Orb-Tracker program—Orbit 360 rotator (*Continued*).

```
'YAESU.BS2
' G Brigman, KC4SA & M. Hammond N8MH/
' Driver for Yaesu G-5500 antenna rotators.
' Azimuth + Elevation
' 3.4.2001
'  I/O Assignments
'         A/D Chip Select.................. P0 - out
'         A/D Clock.........................P1 - out
'         A/D I/O pin.......................P2 - in
'         AZ CCW Relay..................... P3 - out
(         AZ CW Relay.......................P4 - out
'         EL CCW Relay......................P5 - out
'         EL CW Relay.......................P6 - out
'         Serial Data Input.................P16 - out
'         Radio TTL output................. P13 - out
'         LCD display output................P12 - out
' Data input expected to conform to WISP format:  AZxxx.x ELxx.x
'DECLARATIONS
StartPOS:
'Assign VARIABLES here.
     globvar      var    word   'Global variable
     azpos        var    word   'Actual azimuth position
     newazpos     var    word   'Requested new azimuth
     elpos        var    word   'Actual elevation position
     newelpos     var    word   'Requested new elevation
     direction    var    byte   'Rotor direction
     brake        var    byte   'Opposite Rotor Direction
     ad           var    word   '12 bit A/D value
     config       var    nib    'Configuration bits for A/D
     startb       var    config.bit0 'Startbit
     sgldif       var    config.bit1 'mode select
     oddsign      var    config.bit2 'channel select
```

Listing 15-2 Orb-Tracker program—Yaesu 5500 series rotator.

```
    msbf        var    config.bit3 '0's after data xfer complete
'Assign CONSTANTS here.
    portout     con    15   'Define serial freq. output port
    portin      con    16   'Define serial input port
    lcd         con    12   'Define serial display line
    I           con    254  'Instruction prefix
    ClrLCD      con    1    'Clear LCD instruction
    N96N        con    $4054  '9600 baud, inverted, no parity
    azdegree    con    96   '40950/450= 91 steps/degree; adjust for best correlation
    eldegree    con    229  '40950/180=228 steps/degree; adjust for best correlation
    minstep     con    2    'Minimum step in degrees
    cs          con    0    'A/D chip select. 0 = active
    clk         con    1    'A/D clock. Out=rising, In=falling
    dio         con    2    'A/D data I/O pin number
    azccw       con    3    'counter clockwise (AZ LEFT)
    azcw        con    4    'clockwise (AZ RIGHT)
    elccw       con    5    'counter clockwise (EL DOWN)
    elcw        con    6    'clockwise (EL UP)
'INITIALIZE I/O PINS
InitPorts:
    high cs          'Deselect the A/D
    high dio         'Prepare A/D I/O line
    for globvar=3 to 6
        low globvar
        pause 100
    next
    low lcd                                     'make pin 12 low
InitLCD:
    pause 1000
    serout 0,N96N,[I,ClrLCD]                     'clear LCD display
    serout lcd,N96N,[I,128,"SAEBRTrack  N8MH"]  'send text to LCD
    serout lcd,N96N,[I,200,"/"]                  'send 2nd line of text to LCD
    pause 5
    gosub DoAz                                   'Get initial rotor position Az
    gosub DoEl                                   'Get initial rotor position El
    pause 200
    gosub WriteAz                                'Update the display
    gosub WriteEl                                'Update the display
GetData:
'Read the decimal azimuth data.
    serin portin,84,[wait("AZ"),dec newazpos,wait(".")]
'Read the decimal elevation data.
    serin portin,84,2000,GetData,[wait("EL"),sdec newelpos,wait(".")]
 'UNCOMMENT this line if you don't want 360 degree elevation control
'I found it useful when working on the array.

'    if newelpos < 0 or newelpos > 180 then QGetData
'the following line keeps elevation between 0 and 90 degrees
    if newelpos < 0 or newelpos > 90 then QGetData
AzWork:
    if abs(azpos-newazpos) <= minstep then ElWork
    direction=azccw
    brake=azcw
```

Listing 15-2 Orb-Tracker program—Yaesu 5500 series rotator (*Continued*).

```
     if newazpos<azpos then AzGoCCW
     direction=azcw
     brake=azccw
AzGoCW:
     high direction      'turn on the appropriate line
TestAzCW:
pause 100
     gosub DoAz
     gosub WriteAz
     if (azpos+minstep)<newazpos then TestAzCW
     low direction
     gosub SetBrake
     goto ElWork
AzGoCCW:
     high direction     'turn on the appropriate line
TestAzCCW:
     pause 100
     gosub DoAz
     gosub WriteAz
     if (azpos-minstep)>newazpos then TestAzCCW
     gosub SetBrake
     low direction
ElWork:
     if abs(elpos-newelpos) <= minstep then QGetData
     direction=elccw
     brake=elcw
     if newelpos<elpos then ElGoCCW
     direction=elcw
     brake=elccw
ElGoCW:
     high direction     'turn on the appropriate line
TestElCW:
     pause 100
     gosub DoEl
     gosub WriteEl
     if (elpos+minstep)<newelpos then TestElCW
     low direction
     gosub SetBrake
     goto QGetData
ElGoCCW:
     high direction     'turn on the appropriate line
TestElCCW:
     pause 100
     gosub DoEl
     gosub WriteEl
     if (elpos-minstep)>newelpos then TestElCCW
     low direction
     gosub SetBrake
QGetData:

     goto GetData       'Loop through the process
SetBrake:
'   COMMENT THE FOLLING LINE TO ENABLE THE BRAKE
     goto QSetBrake  'don't use brake for Yaesu G-5500
     pause 1
     high brake
```

Listing 15-2 Orb-Tracker program—Yaesu 5500 series rotator (*Continued*).

```
     pause 100
     low brake
QSetBrake:
     return
DoAz:
     config=config "PIXymbolsPCx">| %1011   'Set config array
     oddsign=0                'select AZ line on A/D chip
     gosub convert
'    azpos=(ad*10)/azdegree  'convert to decimal
     azpos=((ad*10)-400)/azdegree  'convert/correct for non-Zero volts at 0deg
QDoAz:
     return
WriteAz:
     serout lcd,N96N,[I,193,"AZ  "] 'clear old characters on LCD
     pause 1
     serout lcd,N96N,[I,193,"AZ ",dec azpos,223] 'send azimuth to LCD
     pause 1
QWriteAz:
     return
DoEl:
     config=config "PIXymbolsPCx">| %1011      'Set config array
     oddsign=1                   'select EL line on A/D chip
     gosub convert
'    elpos=(ad*10)/eldegree     'convert to decimal
     elpos=((ad*10)-600)/eldegree  'convert/correct for non-Zero volts at 0deg
QDoEl:
     return
WriteEl:
     serout lcd,N96N,[I,202,"EL "] 'clear old characters on LCD
     pause 1
     serout lcd,N96N,[I,202,"EL ",dec elpos,223] 'send azimuth to LCD
     pause 1
QWriteEl:
     return
Convert:
     low cs            'Activate the A/D.
     shiftout dio,clk,lsbfirst,[config<\\>4]  'Send config bits
     shiftin dio,clk,msbpost,[ad<\\>12]   'Get data bits
     high cs           'Deactivate the A/D.
QConvert:
     return
'   End of program
```

Listing 15-2 Orb-Tracker program—Yaesu 5500 series rotator (*Continued*).

16

HAM RADIO REPEATER

CONTENTS AT A GLANCE

The amateur radio repeater is most often the center or stabilizing force of a local ham radio community. The now ubiquitous VHF/UHF repeater had its humble roots in the late 1950s and has been going strong ever since. A ham repeater, for those nonamateur folks out there, is a means of extending the range of a radio system. Ham radio repeaters are often set up on hill or mountain tops as shown in Fig. 16-1, and can significantly increase communication range between to VHF/UHF stations. A mobile radio operator in city A, for example, can now talk to a mobile operator or base station in city B. Repeaters are not limited to amateur radio operators but are also used by public service departments such as police and fire companies every day in almost every community. Repeaters are usually placed in high locations to support a wide area or range. Radio repeaters consist of a transmitter on one specific frequency, a separate receiver operating on another frequency, a repeater controller, a power supply, and an antenna, as shown in Fig. 16-2.

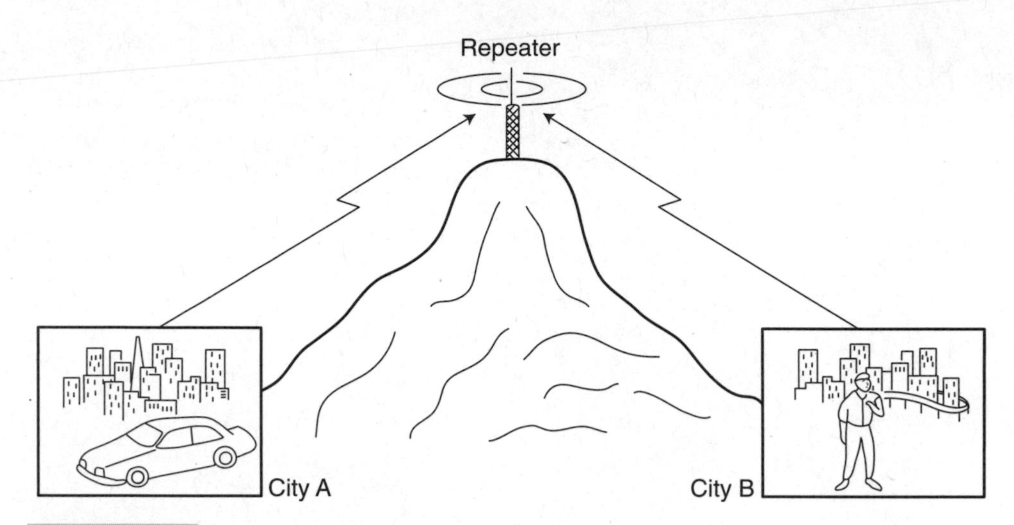

Figure 16-1 **Repeater coverage.**

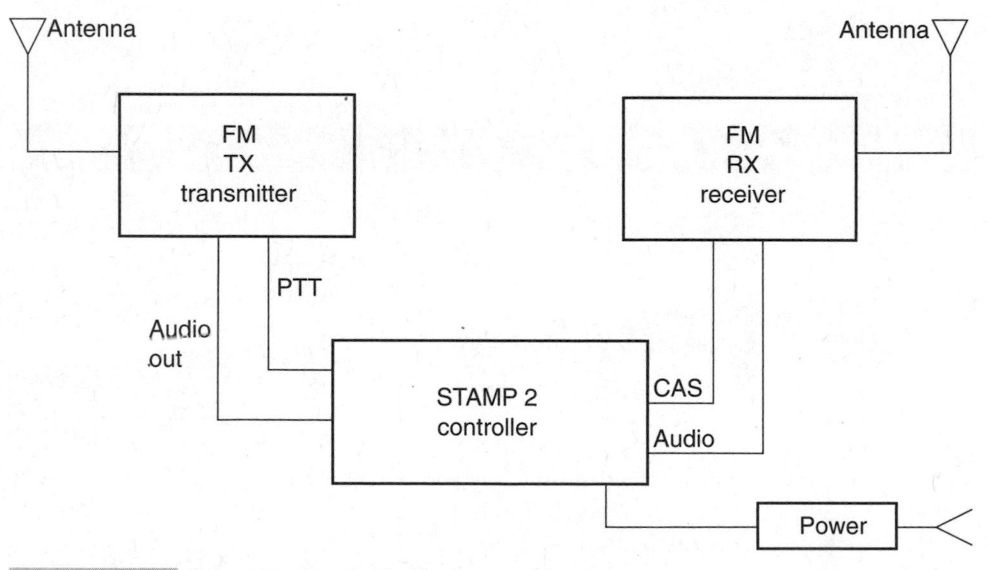

Figure 16-2 **Repeater block diagram.**

Repeater Function

The repeater controller is the glue that holds the repeater system together. It can consist of a simple logic circuit and a relay or it can be as complex as an entire computer system. The receiver audio and a carrier-operated relay (COR) are utilized to control the transmitter via the controller. When an FM receiver detects a signal via the COR detector, it tells the controller a signal is present and that the controller should turn on the FM transmitter and rebroadcast the receiver's audio, thus repeating the signal over a wide area. Usually the transmitter and receiver are on two different frequencies; i.e., on the 2-m amateur band, the receiver and transmitter are separated by 600 kHz, so as not to interfere with each other. Repeaters are usually coordinated with a governing body to assure that other repeaters do not interfere with each other and to control overlap between repeaters.

Once again, the BASIC STAMP 2 comes to our rescue in designing a ham radio repeater controller. The amateur radio repeater in this chapter can perform a number of repeater functions. The repeater controller contains a touch-tone decoder that can be used to remotely control the system. The repeater also can act as a beacon and send out an ID signal to identify itself every 5 minutes. The system also has a time-out timer, roger-beep, and burglar alarm feature, and an optional continuous tone coded squelch system (CTCSS) control feature with the optional TS-64 board from Communications Specialists.

Basic Components

The ham radio repeater controller is shown in Figs. 16-3 and 16-4. The diagram in Fig. 16-4 shows the brain of the repeater controller, i.e., the original BASIC STAMP 2 microprocessor. Fifteen data I/Os are used in the ham radio repeater project. The receiver audio output is connected to pin 4 of J3. The audio input from receiver section is fed to both the analog switch at U1 via R1, C1, and C4 and then to then touch-tone decoder chip input at U2. The audio to both U1 and U2 is controlled via two 20-kΩ potentiometers.

The Mitel 8870 touch-tone decoder chip is a complete DTMF receiver combining both a band-split filter and digital decoder, as shown in Fig. 16-5. The filter section uses switched capacitor techniques for the high and low filter groups. The 18-pin decoder chip uses a digital

Figure 16-3 Radio repeater circuit board.

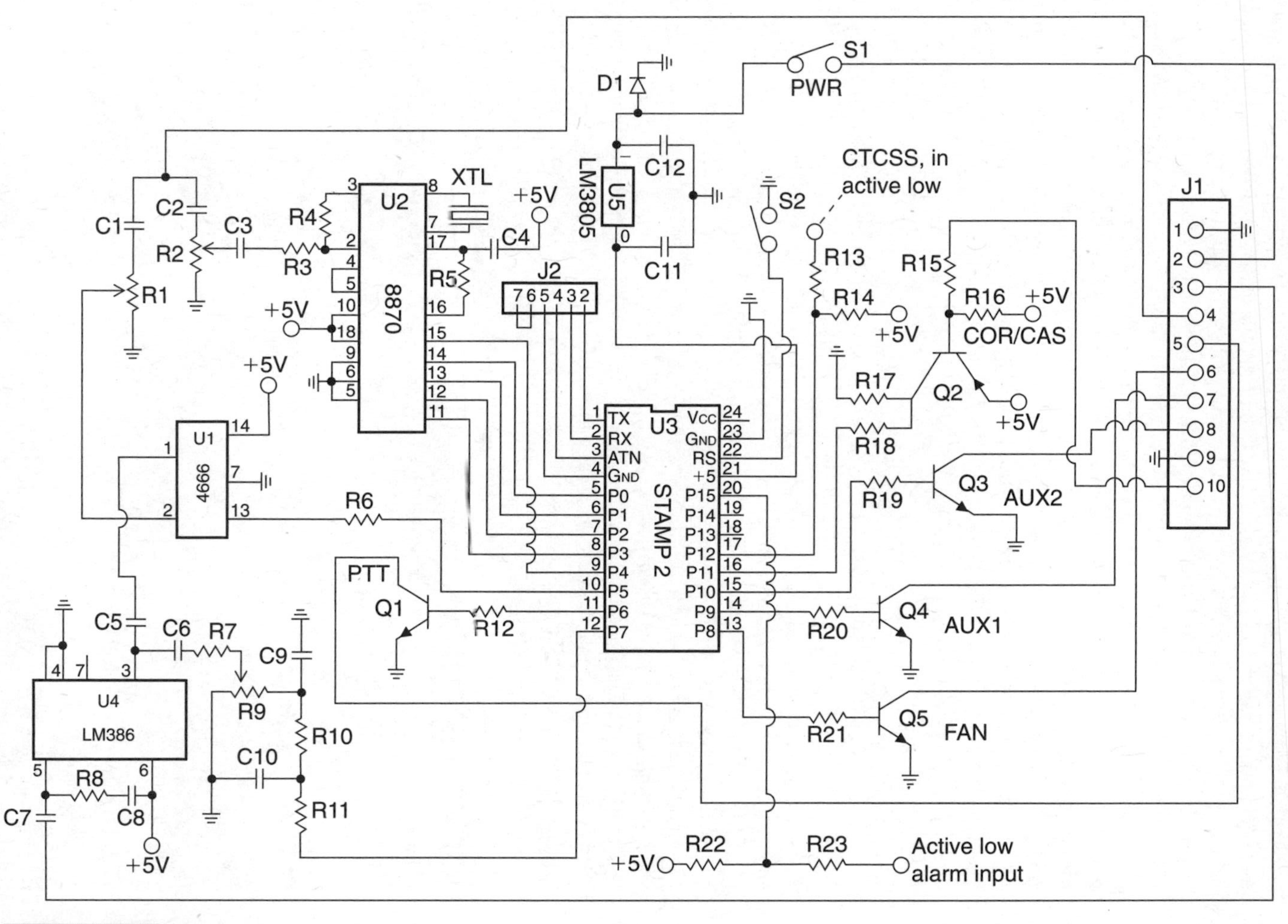

Figure 16-4 Ham repeater schematic.

code converter and latch, in concert with a digital detection algorithm to detect and decode all 16 DTMF tone pairs into a 4-bit code; see Table 16-1. The M8870 touch-tone decoder requires only a few components, and the chip pinouts are shown in Table 16-2. The touch-tone decoder receives audio control information from the receiver and processes it, through the STAMP 2. The binary digital outputs from the touch-tone decoder, i.e., Q1 through Q4, are fed directly to the STAMP 2 controller on pins 11, 12, 13, and 14. The 8870 touch-tone decoder also outputs a delayed steering output signal (StD) on pin 15. The delayed steering logic output presents a logic high signal when a received tone pair has been registered and the output latch updated. This output returns to logic low when the voltage on St/GT falls. Note the StGT is the steering input/guard time; see data sheets for the 8870 decoder in App. 1 on the CD-ROM.

The audio output to the repeater's transmitter is coupled to J3 on pin 3. The output from the STAMP 2 on pin P7 is fed to the network of R9, R10, R11, and C11, and the output from the analog switch on U1 is sent to the LM386 along with the attenuable signal from the STAMP 2. The output of the LM386 is then input to J3-3. The STAMP 2 controls the transmitter via the PTT control transistor at Q1. Table 16-3 lists input and output connections at the J3 pins.

The COR/CAS or carried operated relay output from the repeater receiver is connected to J3-10. The COR/CAS input is also tied to the STAMP 2 on P6. The PTT (push to talk) line to the repeater transmitter is controlled by the output of the STAMP 2 at P7, which controls the transistor at Q3. The repeater controller also has two auxiliary control functions handled by AUX1 and AUX2.

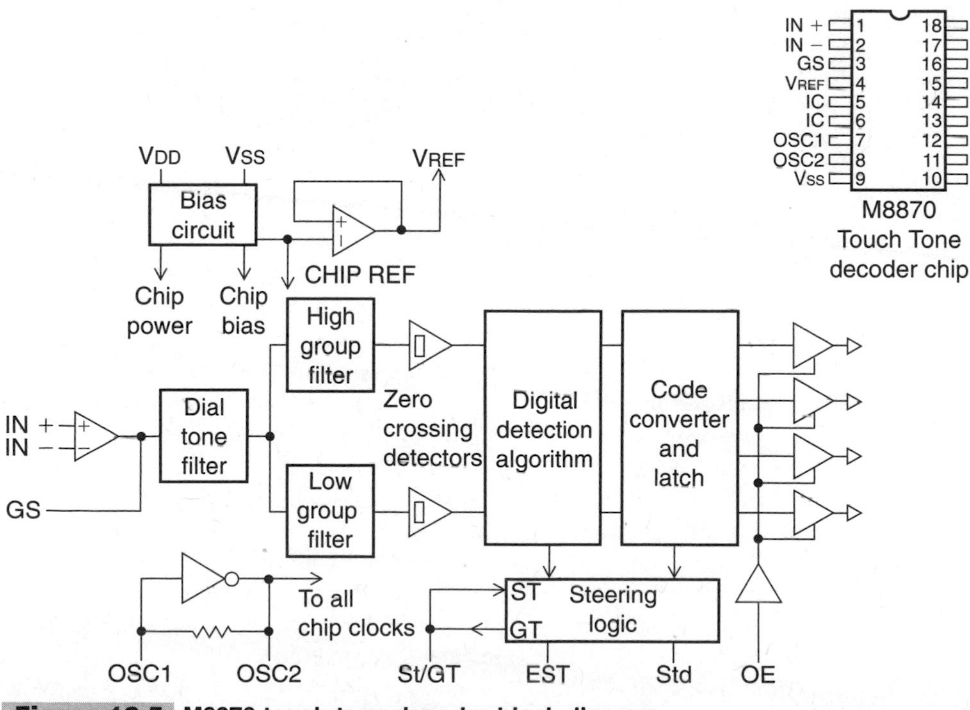

Figure 16-5 M8870 touch-tone decoder block diagram.

TABLE 16-1 TOUCH-TONE DECODING

FLOW	FHIGH	KEY	OE	Q4	Q3	Q2	Q1
697	1209	1	H	0	0	0	1
697	1336	2	H	0	0	1	0
697	1477	3	H	0	0	1	1
770	1209	4	H	0	1	0	0
770	1336	5	H	0	1	0	1
770	1477	6	H	0	1	1	0
852	1209	7	H	0	1	1	1
852	1336	8	H	1	0	0	0
852	1477	9	H	1	0	0	1
941	1336	0	H	1	0	1	0
941	1209	*	H	1	0	1	1
941	1477	#	H	1	1	0	0
697	1633	A	H	1	1	0	1
770	1633	B	H	1	1	1	0
852	1633	C	H	1	1	1	1
941	1633	D	H	0	0	0	0
Any	Any	Any	L	Z	Z	Z	Z

One possible use for AUX1 might be to utilize an ISD 2590 voice message to announce the repeater highlights or special notes about the repeater. The diagram in Fig. 16-6 illustrates a simple circuit for using the voice message chip. The audio output from the voice message chip would have to be interfaced to the transmit audio J3-3, so that the voice message would go out over the air. Voice messages are recorded in the voice chip by using the REC pushbutton switch and the microphone at pins 17 and 18. The play pushbutton is tied to transistor Q4, through a small relay.

The repeater controller will also accept a 5-V signal through an "alarm" switch contact to notify the control operator of someone possibly tampering with the repeater. Flood or water sensors could also be employed as an alarm input source.

An optional external CTCSS or subaudible decoding board can be integrated into the repeater. A Circuit Specialist's TS-64 board is shown in the schematic at Fig. 16-7. The output of the CTCSS board can be connected to the repeater controller via R14 on pin P12 of the STAMP 2 microprocessor. Additional connections on the subaudible board are the 12-V power source and ground. The yellow or the PL wire from the TS-64 board is connected according to the manufacturer of your transmitter. Note, JP7 must be bridged on the TS-64 board while JP1 through JP6 must be configured to the proper PL tone for your application.

A normally open pushbutton system reset switch is shown at S2. This switch is connected to the reset or RST pin of the STAMP 2 in the event of a system "lock-up problem," should it ever occur.

TABLE 16-2 M8870 TOUCH-TONE DECODER CHIP PINOUTS

NAME	FUNCTION	DESCRIPTION
IN+	Noninverting input	Connections to the front-end amplifier
IN-	Inverting input	Connections to the front-end amplifier
GS	Gain select	Access to output of front-end amp for feedback resistor
Vref	Reference volts output	Nominal $V_{dd}/2$, may be used to bias the input at midrail
IC	Internal connection	Must be tied to V_{ss}
IC	Internal connection	Must be tied to V_{ss}
OCS1	Clock input	3.579-MHz-crystal between these pins/internal oscillator
OSC2	Clock output	3.579 MHz crystal between these pins/internal oscillator
Vss	Negative power	Normally connected to 0 V
OE	3-state output enable	Logic high enables the outputs Q1–Q4/Internal pull-up
Q1, Q2	3-state outputs	When enabled by OE, provides code to last valid digit pair
Q3, Q4	3-state outputs	When enabled by OE, provides code to last valid digit pair
StD	Delayed steering logic	Logic high when received tone pair registered, output latch updated; returns to logic low when voltage on St/Gt falls below VTSt
Est	Early steering logic	Logic high when the digital algorithm detects a recognizable tone pair; momentary loss of signal will cause Est to go to logic low
St/GT	Steering/guard time	Steering input/guard time output; a voltage greater than VTSt detected at St causes the device to register the detected tone pair and update output latch. A voltage less than VTSt frees the device to accept a new tone pair. The GT output acts to reset the external steering time constant.

The STAMP 2 controller is powered via pin 21 and an LM7805 regulator IC was used to supply 5 volts to the repeater controller chips. A 300-Ω series resistor was placed in the input to the regulator. The input to the regulator chip can be a +12-V power supply or trickle-charged battery. A 12-V "wall wart" could alternatively be used to power the system.

TABLE 16-3 J3 INPUT/OUTPUT CONNECTION LEGEND

J3 PIN NUMBER	REPEATER FUNCTION
1	Ground connection
2	+ 12–13.6-V input voltage to repeater controller
3	TX output (transmitter)
4	RX input (receiver)
5	PTT (push to talk)
6	Cabinet fan
7	AUX1 (auxillary function 1)
8	AUX2 (auxillary function 2)
9	Ground connection
10	COR/CAS carrier-operated relay connection to receiver

The ham radio repeater was constructed on a 4- by 6-in single-sided circuit board and then placed in a metal enclosure. When constructing the repeater controller, be sure to use integrated circuit sockets for all the ICs in the event of a possible component failure at a later date. Be sure to carefully observe the proper polarity of the capacitors and diodes when installing them. When installing the integrated circuits, be sure to take care to properly orient them in the sockets, to avoid damaging the ICs on power-up. When you are finished building the ham radio repeater circuit board, you can then locate a suitable enclosure in which to mount the repeater controller. You can simply place the controller board in an aluminum enclosure box, or you can install the controller circuit board in a larger "rack" panel that can house the receiver, transmitter, and controller for later expansion.

In order to program the STAMP 2 with the repeater controller software, you will need to make up a programming cable. A 4-pin 0.010-in male header is mounted on the circuit board, to mate with the programming cable. Next you will need to construct a programming cable, using a 4-pin female header at one end and a DB-9 female RS-232 connector at the other end. Header P1 is then connected to a 9-pin EIA female, which is connected to your PC in order to download programs to the STAMP 2.

Programming the Repeater

Once the ham radio repeater circuit board is completed and mounted, you are ready to connect up your programming cable, apply power, and download the ham radio repeater software. First locate the STAMP 2 Windows editor program called STAMPW.EXE. Finally, locate and download the repeater controller program titled RC1.BS2 (Listing 16-1). Now remove power from the repeater controller and connect your transmitter and receiver, as well as the antennas.

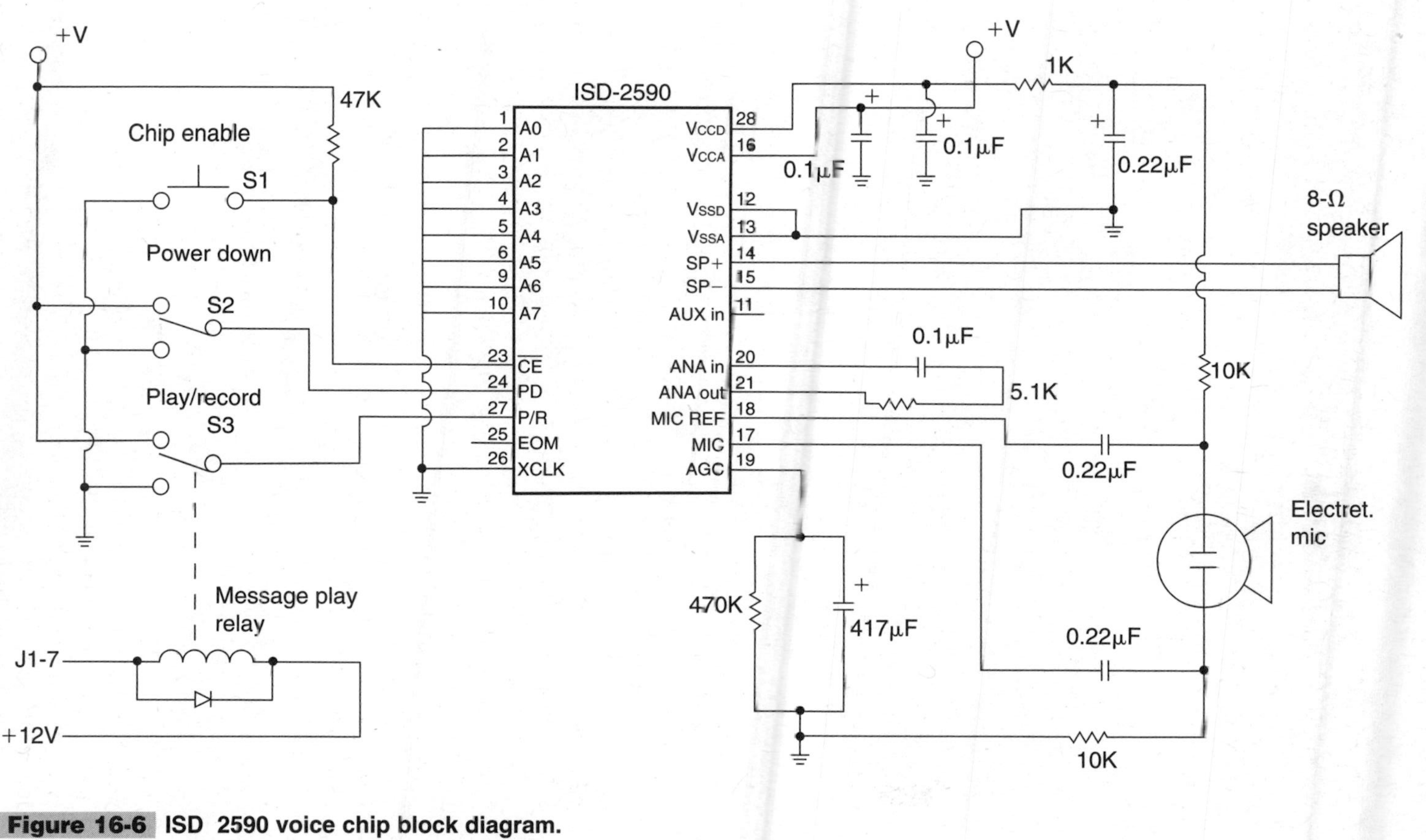

Figure 16-6 ISD 2590 voice chip block diagram.

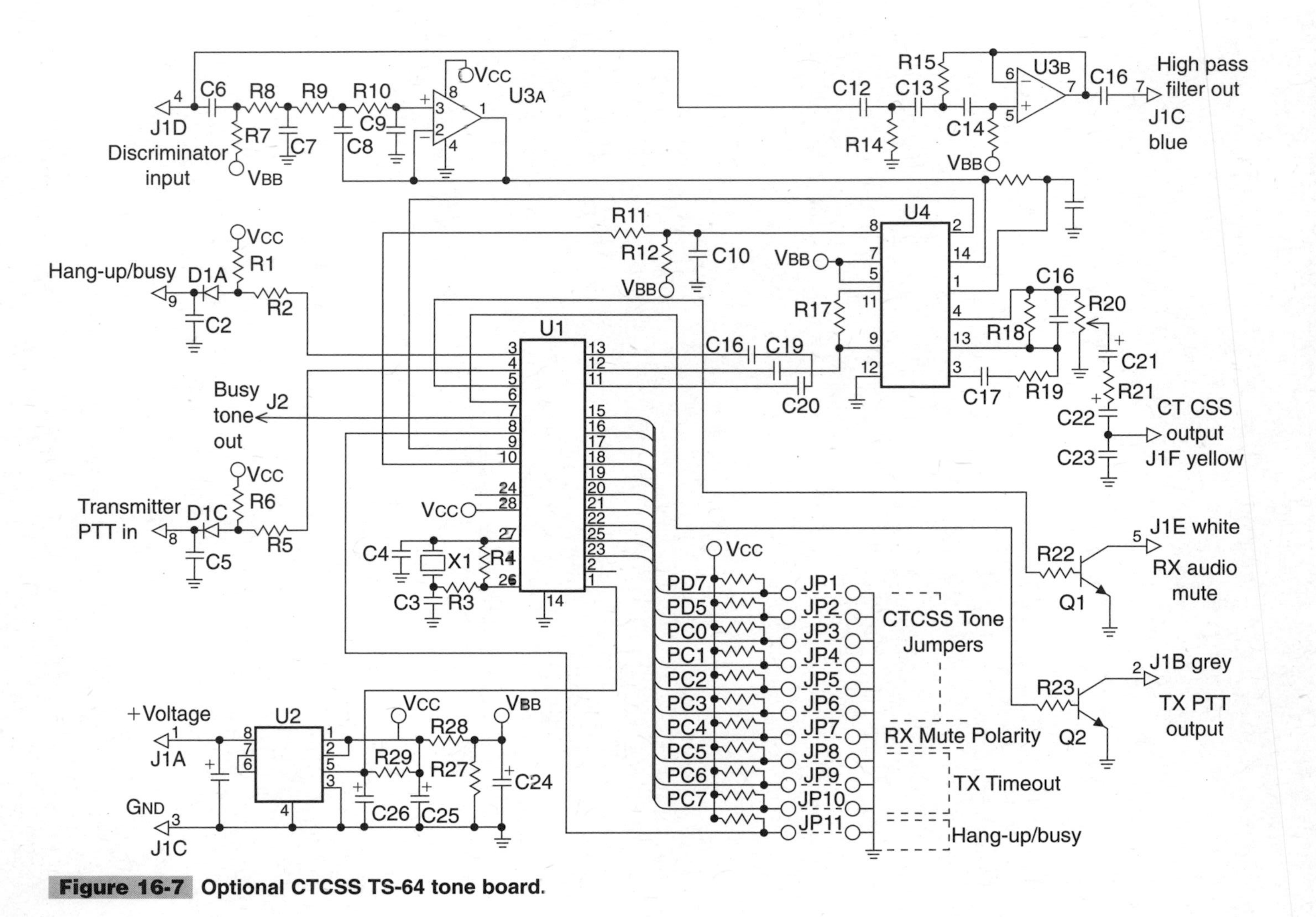

Figure 16-7 Optional CTCSS TS-64 tone board.

The ham radio repeater controller will respond to a number of preprogrammed touch-tone functions, illustrated in the beginning of the software listing.

Apply power once again when all the repeater components are in place and your ham radio repeater controller is ready to serve you and your radio friends!

```
RC2.BS2
''''''''''''''''''''''''''''''''
'USE AT YOUR OWN RISK          '
'WRITTEN BY ROGER CAMERON      '
''''''''''''''''''''''''''''''''
'400 PEYTON                    '
'CLINTON MO 64735              '
'www.iland.net/~noyox647       '
'                              '
'REPEATER CONTROLLER           '
'WITH 3 DIGIT DTMF             '
'ENABLE AND DISABLE,           '
'TWO AUX OUTPUT, FAN           '
'CONTROL,BURGLARALARM,         '
'BEACON MODE  AND MORE         '
'                              '
'IF YOU DECIDE YOU USE THIS    '
'CODE PLEASE SEND N0YOX        '
'$20.00 AT THAT TIME I WILL    '
'EMAIL YOU THE LATEST VERSION  '
'                              '
''''''''''''''''''''''''''''''''


'''''''''''''''''''''''''DTMF CODES'''''''''''''''

'210 CTCSS ON REQUIRES EXTERNAL CTCSS BOARD
'211 CTCSS OFF REQUIRES EXTERNAL CTCSS BOARD
'531 TOGGLES DTMF ENABLE/DISABLE
'222  REPEATER OFF
'123  REPEATER ON
'189 AUX 1 OFF
'188 AUX 1 ON
'199 AUX 2 OFF
'198 AUX 2 ON
'123 FORCE ID
'255 TEST ALARM MUST FIRST TURN ALARM ON
'788 TOGGLES ALARM
'888 BEACON ON
'999 BEACON OFF
''''''''''''''''''''''''''''''''''''''''''''''''''
'2=DAH 1=DIT 0=SPACE
DATA @0,2,1,0,2,2,2,2,2,0,2,1,2,2,0,2,2,2,0,2,1,1,2,0,2,1,1,2,1,0,1,2,1
DATA @60,0
TONE CON 700
''''''''''''''''''''''''VARIABLES'''''''''''''
IDTIME VAR WORD
FANTIME VAR WORD
TOTIMER VAR WORD
CODE VAR WORD
I VAR BYTE
DTMF VAR BYTE
BCOUNT VAR NIB
ALARMC VAR NIB
CHARACTER VAR NIB
```

Listing 16-1 Ham radio repeater program.

```
FANST VAR BIT
TAILRQ VAR BIT
REPEATER VAR BIT
BEACONMODE VAR BIT
ALARMST VAR BIT
CTCSS VAR BIT
CORSTATE VAR BIT
PROGMODE VAR BIT
PROGMODE = 0
CTCSS = 0
RESTART:
      READ 60,REPEATER
      IF REPEATER = 1 THEN REPOFF
      TOTIMER = 0
      PAUSE 2000
IDER:
      IF ALARMC > 7 THEN REALARMC 'AFTER  ALARMC > 7 T1S RESETS ALARMC TO 0
      HIGH 11 'TURNS TRANSMITTER ON
      IDTIME = 0
      FOR I = 0 TO 32 '32 IS THE NUMBER OF ELEMENTS - 1 IN THE DATA STATEMENT
      READ I, CHARACTER 'READS CW OUT OF DATA AREA
      GOSUB MORSE
      NEXT
      IF BEACONMODE = 1 THEN BEACON
      FOR I = 1 TO 100
      PAUSE 10
      GOSUB CHECKINS
      IF CORSTATE = 1 THEN CORE
      NEXT
WAAIT:
      IF REPEATER = 1 THEN REPOFF
      LOW 9
      LOW 11
      LOW 5 'turn off ac to power supply
      FANST = 0
WAAITT:
      IF REPEATER = 1 THEN REPOFF
      IF IN15 = 0 THEN ALARM
      FANTIME = FANTIME + 1
      IF FANTIME > 9000 THEN WFAN
      IDTIME = IDTIME + 1
      IDTIME = IDTIME MAX 5450
      GOSUB CHECKINS
      IF CORSTATE = 1 THEN CORE
      PAUSE 100
      GOTO WAAITT
CORE:
      IF REPEATER = 1 THEN REPOFF
      GOSUB CHECKAUDIO
      IF IDTIME > 5500 THEN IDER  'ID S IF NEEDED
      PAUSE 100
      IDTIME = IDTIME + 1 'ID COUNTER
      IF IN4 = 1 THEN DIG1 'GOES TO DTMF SUB IF TONE IS RECONIZED
      GOSUB CHECKINS
      IF CORSTATE = 1 THEN PTT
      IF TAILRQ = 1 THEN TAIL
      LOW 9 'TURNS OFF AUDIO
      LOW 11 'TURNS OFF  PTT
      GOTO CORE
PTT: 'THIS SUB KEYS TRANSMITTER
      TAILRQ = 1 'SETS TAIL REQUIRED BIT TO NEEDED
      HIGH 11 'KEYS TRANSMITTER
      IF IN15 = 0 THEN ALARM
```

Listing 16-1 Ham radio repeater program (*Continued*).

```
        GOSUB CHECKAUDIO
        GOSUB FAN 'TURNS FAN ON
        IF TOTIMER > 1800 THEN  TIMOUT
        TOTIMER = TOTIMER + 1 'TIMEOUT COUNTER
        IF IN4 = 1 THEN DIG1 'GOES TO DTMF SUB IF TONE IS RECONIZED'
        GOTO CORE
TAIL:
        GOSUB CHECKAUDIO
        FOR I = 1 TO 35
        PAUSE 20
        GOSUB CHECKINS
        IF CORSTATE = 1 THEN CORE
        NEXT
        TOTIMER = 0
        TAILRQ = 0
        FREQOUT 7,200,900
        FREQOUT 7,100,400
        FOR I = 1 TO 100
        PAUSE 20
        GOSUB CHECKINS
        IF CORSTATE = 1 THEN CORE
        NEXT
        LOW 11
        GOTO CORE
WFAN:
        IF FANST = 1 THEN WFAN2
        HIGH 5 'turn on ac to power supply
        FANST = 1
        FANTIME = 8000
        GOTO WAAITT
WFAN2:
        LOW 5 'turn off ac to power supply
        FANST = 0
        FANTIME = 0
        GOTO WAAITT
TIMOUT:
        HIGH 11
        LOW 9 'AUDIO OFF
'       1 = DAH /// 2 = DIT ////3 = CHARACTER SPACE
        FOR I = 0 to 4
        LOOKUP I, [2,0,2,2,2],CHARACTER
        GOSUB MORSE
        NEXT
        PAUSE 200
        LOW 11
TMOUT:
        GOSUB CHECKINS
        IF CORSTATE = 1 THEN TMOUT
        PAUSE 3500
        TOTIMER = 0
        TAILRQ = 0
        HIGH 11
        PAUSE 300
'       1 = DAH /// 2 = DIT ////3 = CHARACTER SPACE
        FOR I = 0 to 4
        LOOKUP I, [2,0,2,2,2],CHARACTER
        GOSUB MORSE
        NEXT
```

Listing 16-1 Ham radio repeater program (*Continued*).

```
PAUSE 100
      GOSUB CHECKINS
      IF CORSTATE = 1 THEN CORE
      LOW 9
      LOW 11
      GOTO WAAIT
FAN:
      HIGH 5 'turn on ac to power supply
      FANST = 1
      RETURN
REPOFF:
      LOW 5
      LOW 9
      LOW 11
      REPEATER = 1
      WRITE 60,1
      REPW:
      PAUSE 30
      IF IN15 = 0 THEN ALARM
      IF IN4 = 1 THEN DIG1 'GOES TO DTMF SUB IF TONE IS RECONIZED
      PAUSE 30
      GOTO REPW
REPON:
      WRITE 60,0
      PAUSE 1000
      GOTO RESTART
AUX1:
      PAUSE 1900
      GOSUB CONFIRMON
      HIGH 6
      GOTO CORE
AUX1OFF:
      PAUSE 1900
      GOSUB CONFIRMOFF
      LOW 6
      GOTO CORE
AUX2:
      PAUSE 1900
      GOSUB CONFIRMON
      HIGH 8
      GOTO CORE
AUX2OFF:
      PAUSE 1900
      GOSUB CONFIRMOFF
      LOW 8
      GOTO CORE
CHECKAUDIO:
      GOSUB CHECKINS
      IF CORSTATE = 1 THEN AUDION
      IF CORSTATE = 0 THEN AUDIOFF
AUDION:
      HIGH 9
      RETURN
AUDIOFF:
      LOW 9
      RETURN
CONFIRMON:
      HIGH 11
      PAUSE 600
```

Listing 16-1 Ham radio repeater program (*Continued*).

```
'      2 = DAH /// 1 = DIT ////3 = CHARACTER SPACE
       FOR I = 0 to 5
       LOOKUP I, [2,2,2,0,2,1],CHARACTER
       GOSUB MORSE
       NEXT
       PAUSE 100
       RETURN
CONFIRMOFF:
       HIGH 11
       PAUSE 600
'      2 = DAH /// 1 = DIT ////3 = CHARACTER SPACE
       FOR I = 0 to 12
       LOOKUP I, [2,2,2,0,1,1,2,1,0,1,1,2,1],CHARACTER
       GOSUB MORSE
       NEXT
       PAUSE 100
       RETURN


DTMFER:
       DTMF = INL & %00001111
       LOOKUP DTMF,["D84#206B195A3*7C-"],DTMF  'STRAIGHT PINS
       'LOOKUP DTMF,["D1234567890*#ABC-"],DTMF  'CROSSED PINS
       IF DTMF = "1" THEN DTMF1
       IF DTMF = "2" THEN DTMF2
       IF DTMF = "3" THEN DTMF3
       IF DTMF = "4" THEN DTMF4
       IF DTMF = "5" THEN DTMF5
       IF DTMF = "6" THEN DTMF6
       IF DTMF = "7" THEN DTMF7
       IF DTMF = "8" THEN DTMF8
       IF DTMF = "9" THEN DTMF9
       IF DTMF = "0" THEN DTMF0
       IF DTMF = "*" THEN OTHER
       IF DTMF = "#" THEN OTHER
       IF DTMF = "A" THEN OTHER
       IF DTMF = "B" THEN OTHER
       IF DTMF = "C" THEN OTHER
       IF DTMF = "D" THEN OTHER
       RETURN
DTMF1:
       DTMF = 1
       RETURN
       DTMF2:
       DTMF = 2
       RETURN
       DTMF3:
       DTMF = 3
       RETURN
       DTMF4:
       DTMF = 4
       RETURN
       DTMF5:
       DTMF = 5
       RETURN
       DTMF6:
       DTMF = 6
       RETURN
       DTMF7:
```

Listing 16-1 Ham radio repeater program (*Continued*).

```
      DTMF = 7
      RETURN
      DTMF8:
      DTMF = 8
      RETURN
      DTMF9:
      DTMF = 9
      RETURN
      DTMF0:
      DTMF = 0
      RETURN
REALARMC:' THIS RESETS ALARM COUNTER FOR NEXT ALARM
      ALARMC = 0
      GOTO IDER
OTHER:
      IF IN4 = 1 THEN OTHER
      IF REPEATER = 1 THEN REPOFF
      GOTO CORE
      RETURN
MORSE:
      BRANCH CHARACTER,[SPACE,DIT,DAH]
      RETURN
DAH:
      GOSUB CHECKAUDIO
      FREQOUT 7,216,TONE
      GOSUB CHECKAUDIO
      FREQOUT 7,72,0
      RETURN
DIT:
      GOSUB CHECKAUDIO
      FREQOUT 7,72,TONE
      GOSUB CHECKAUDIO
      FREQOUT 7,72,0 'SMALL SPACE
      RETURN
SPACE: 'SPACE BETWEEN CHARACTERS
      FREQOUT 7,144,0
      RETURN
'___________________________


      ''''''''''''''''''''DTMF SUB'''''''''''''''''''''''''''''

DIG1:
      PAUSE 100
      IF IN4 = 0 THEN CORE
      I = 0
      CODE = 0
      LOW 9
      GOSUB DTMFER 'GETS DIGIT AND CONVERTS IT TO A DIGIT
      CODE = DTMF * 100
      GOSUB CHECKDTMF ' WAITS FOR DTMF TO DISAPPEAR AND REAPPEAR AND GETS NEXT
DIGIT
DIG2:
      CODE = CODE + (DTMF*10)
      GOSUB CHECKDTMF ' WAITS FOR DTMF TO DISAPPEAR AND REAPPEAR AND GETS NEXT
DIGIT
DIG3:
      CODE = CODE + DTMF
      PAUSE 750 'GIVES YOU TIME TO UNKEY AND HEAR RESPONSE
```

Listing 16-1 Ham radio repeater program (*Continued*).

```
      GOTO CHECKCODE 'CHECKS FOR VALID CODE
CHECKDTMF:'THIS SUB WAITS UNTIL DTMF DISAPEARS
            FOR I = 1 TO 100
            PAUSE 10
            IF IN4 = 0 THEN ENDFOR '0 TO LOW
            NEXT
            IF REPEATER = 1 THEN REPOFF
            GOTO CORE
ENDFOR:' THIS SUB WAITS FOR NEXT DTMF TO APPEAR
            FOR I = 1 TO 100
            PAUSE 10
            IF IN4 = 1 THEN ENDFOR2
            NEXT
            IF REPEATER = 1 THEN REPOFF
            GOTO CORE
ENDFOR2:
            GOSUB DTMFER
            RETURN
CHECKCODE:
      IF CODE = 123 THEN REPON
      IF REPEATER = 1 THEN REPOFF
      IF CODE = 531 THEN PROGMTOG
      IF PROGMODE = 0 THEN CORE
      IF CODE = 222 THEN REPOFF
      IF CODE = 210 THEN CTCSSON
      IF CODE = 211 THEN CTCSSOFF
      IF CODE = 188 THEN AUX1
      IF CODE = 189 THEN AUX1OFF
      IF CODE = 198 THEN AUX2
      IF CODE = 199 THEN AUX2OFF
      IF CODE = 255 THEN ALARM
      IF CODE = 788 THEN ALARMTOG
      IF CODE = 888 THEN BEACONON
      IF CODE = 999 THEN BEACONOFF
      GOTO CORE
CTCSSON:
      CTCSS = 1
      GOSUB CONFIRMON
      GOTO CORE
CTCSSOFF:
      CTCSS = 0
      GOSUB CONFIRMOFF
      GOTO CORE
PROGMTOG: 'TURNS DTMF OFF/ON
      IF PROGMODE = 0 THEN PROGON
      PROGMODE = 0
      GOSUB CONFIRMOFF
      GOTO CORE
PROGON:
      PROGMODE = 1
      GOSUB CONFIRMON
      GOTO CORE
ALARMTOG:
      IF ALARMST = 0 THEN ALARMON
      ALARMST = 0
      GOSUB CONFIRMOFF
      GOTO CORE
ALARMON:
      ALARMST = 1
```

Listing 16-1 **Ham radio repeater program (*Continued*).**

```
       GOSUB CONFIRMON
       GOTO CORE
ALARM:
       IF ALARMST = 0 AND CODE = 255 THEN TESTALARM
       IF ALARMST = 0 THEN CORE
TESTALARM:
       ALARMC = ALARMC + 1
       IF ALARMC > 10 THEN ALARMTO
       LOW 9 'MUTES AUDIO
       GOSUB FAN
       FOR I = 0 TO 100
       FREQOUT 7,(100),TONE
       FREQOUT 7,(100),TONE+200
       NEXT
       LOW 11
       IF CODE = 255 THEN CORE
       PAUSE 10000
       GOTO CORE
BEACONON:
       GOSUB CONFIRMON
BEACON:
       BEACONMODE = 1
       BCOUNT = BCOUNT + 1
       IF BCOUNT > 7 THEN BEACONOFF
       GOTO CORE
BEACONOFF:
       GOSUB CONFIRMOFF
       BEACONMODE = 0
       BCOUNT = 0
       GOTO WAAIT
ALARMTO:
       IF IN15 = 0 THEN ALARMTO
       IDTIME = 0 ' T1S CAUSES IT TO TAKE 10 MINUTES UNTIL THE NEXT POSSIBLE ALARM
       GOTO CORE
CHECKINS:
       CORSTATE = 0
       IF CTCSS = 1 AND IN10 = 0 AND IN12 = 0 THEN CORSET
       IF CTCSS = 0 AND IN10 = 0 THEN CORSET
CKIN:
       RETURN
CORSET:
       CORSTATE = 1
       GOTO CKIN
```

Listing 16-1 **Ham radio repeater program (*Continued*).**

Ham Radio Repeater Parts List

U1	CD4066 analog switch IC
U2	M8870 touch-tone decoder chip
U3	BASIC STAMP 2 (original)
U4	LM358 linear op-amp

U5	LM7805 5-V regulator
Q1, Q3, Q4, Q5	2N2222 npn transistor
Q2	2N3906 pnp transistor
R1, R2, R9	20-kΩ potentiomer, 10 turn
R3, R4, R14, R16, R17, R22	100-kΩ 1/4-W resistor
R5	470-kΩ 1/4-W resistor
R6, R13, R15, R18, R23	10-kΩ 1/4-W resistor
R12, R19, R20, R21	10-kΩ 1/4-W resistor
R7, R10, R11	4.7-kΩ 1/4-W resistor
R8	10-Ω 1/4-W resistor
R13	300-Ω 1/4-W resistor
C1, C2, C3, C4, C5	0.1-μF 35-V disk capacitor
C6, C7, C10, C11	0.1-μF 35-V disk capacitor
C8	2.2-μF 35-V electrolytic capacitor
C9	100-pF 35-V Mylar capacitor
C12	220-μF 35-V electrolytic capacitor
C13	4.7-μF 35-V electrolytic capacitor
D1	1N4002 silicon diode
XTL	3.579-MHz color burst crystal
S1	SPST power toggle switch
S2	Pushbutton switch SPST (normally open)
J1	4-pin male header (plug)
J2	DB-9 female RS-232 connector
J3	10-position dual screw terminal
P1	4-pin female header (jack)
Miscellaneous	PC circuit board, IC sockets, wire, enclosure, etc.

Optional TS-64 CTCSS Tone Board Parts List

U1	TS-64 microprocessor
U2	LP2951CM low-dropout regulator

U3	LM385D dual op-amp
U4	MF6CWM-50 low-pass filter
D1	IMN10 triple diode array
Q1	MMBTA42LT npn silicon transistor
Q2	BCX56 npn SOT-89 transistor
X1	AT38 3.58-MHz crystal
C24	6.8-μF 4-V 20% tantalum capacitor
C21, C22	2.2-μF 16-V 20% tantalum capacitor
C25, C26	1-μF 20-V 20% tantalum capacitor
C1, C6	0.1-μF 20-V 20% tantalum capacitor
C8, C15, C17	0.01-μF 50-V capacitor
C7, C10	0.01-μF 50-V capacitor
C11, C20	6800-pF 50-V 10% monochip capacitor
C12, C13	4700-pF 50-V 10% monochip capacitor
C14, C19, C16, C18	3300-pF 50-V 10% monochip capacitor
C2, C5, C9, C23	180-pF NPO 50-V monochip capacitor
C3, C4	22-pF NPO 50-V monochip capacitor
R21, R23	2.2-kΩ $^1/_8$-W 5% chip resistor
R15	3.6-kΩ $^1/_8$-W 5% chip resistor
R22, R27	4.7-kΩ $^1/_8$-W 5% chip resistor
R28	7.5-kΩ $^1/_8$-W 5% chip resistor
R3, R17	11-kΩ $^1/_8$-W 5% chip resistor
R11, R12, R14	33-kΩ $^1/_8$-W 5% chip resistor
R1, R2, R5, R6	47-kΩ $^1/_8$-W 5% chip resistor
R6, R24, R25, R26	47-kΩ $^1/_8$-W 5% chip resistor
R19	82-kΩ $^1/_8$-W 5% chip resistor
R7, R18	91-kΩ $^1/_8$-W 5% chip resistor
R9, R10, R29	160-kΩ $^1/_8$-W 5% chip resistor
R8	200-kΩ $^1/_8$-W 5% chip resistor
R13	680-kΩ $^1/_9$-W 5% chip resistor
R4, R16	2.7-MΩ $^1/_8$-W 5% chip resistor
R20	5-kΩ 3MM chip trimmer pot
RN1	4.7-kΩ × 8 chip resistor
J1	9-pin subminiature header
Miscellaneous	9-pin plug-in cable assembly, TS-64 PC board

17

BALLOON TELEMETRY CONTROLLER

CONTENTS AT A GLANCE

The balloon telemetry project is a project that is both unusual and fun that you can build using a BASIC STAMP 2 processor. Many ham radio enthusiasts combine their love for radio with their interest in ballooning. This project combines radio, ballooning, and telemetry all in one encompassing project. This project might also appeal to hobbyists interested in rocketry, who may wish to measure temperature and pressure at high altitudes for altitude versus temperature or temperature versus pressure studies. While this project is intended for hams, nonhams can also build the project by using a GMRS or FRS radio instead of VHF/UHF amateur radio equipment. This project is designed to be interfaced with a radio transmitter capable of voice transmission. The balloon telemetry project monitors both pressure and temperature and sends the data over radio in Morse code via the microphone input of the radio used.

Basic Components

This project centers around the use of a BASIC STAMP 2 microprocessor as the main component in this project, as illustrated in Fig. 17-1. The original BASIC STAMP 2 is used for this project, but the alternative STAMP 2 could easily be substituted if you wish to save some money. This project begins with a Motorola MPX5100A pressure sensor. The pressure sensor is a 6-lead, 5-V device, with a 0- to 14.5-psi range. The MPX5100A is a single-element thin-film piezoresistive transducer designed for microcontroller operation. The pressure transducer is connected to a calibration resistor/capacitor network shown at C1/R1. The output from the sensor is next coupled to a voltage divider network at Ra/Rb. The divider is a resistor-ratio network formed by a 100-kΩ and a 10-kΩ resistor. The output from the divider is then fed to the input of a Maxim MAX187 analog-to-digital converter. The MAX187 is an 8-pin single-channel converter ideally suited to microprocessor control.

The STAMP 2 controls pins 7 and 8, respectively, the chip select (CS) and SCLK pins. The digital output of the A/D converter is shown at pin 6, and is fed to P1 of the BASIC STAMP 2. Two thermistor temperature sensors are connected to the STAMP 2 via pins P6 and P7, through capacitors C2 and C3. Three output ports are used in this project. Pin 3 on the STAMP 2 provides an audio output to the radio transmitter, through a coupling capacitor. The STAMP 2 pin P4 is used to control a relay, which provides contact closure for turning a transmitter or oscillator off and on. Pin P5 of the STAMP 2 is used to control an optional servomotor, which controls a mirror. The enhanced camera feature controls a mirror that is used to change camera angles, in order to allow the on-board camera to see different balloon or rocket aerial views. This alternative aspect of the project could utilize a small live mini TV camera, with a mini ATV transmitter to broadcast balloon pictures back to earth.

Construction of the Controller

The balloon telemetry controller project was constructed on a small printed circuit board with the STAMP 2, the A/D converter, the pressure sensor, and the 5-V regulator. Integrated circuit sockets for the STAMP 2 and A/D converter are highly recommended for this project. A socket helps to protect the STAMP 2 from excessive heat when soldering and also facilitates easy removal in the event of a system failure down the road. When building the balloon telemetry controller, be sure to observe the correct polarity while installing capacitors, diodes, and transistors. When installing the ICs, pay particular attention to the orientation of the pressure transducer and analog-to-digital converter chip. Taking time for these details will prevent unpleasant surprises on power-up. Note that power switch S1 does double duty both to apply power and to shut down the A/D converter chip. A momentary pushbutton is connected to ground at pin 22, which provides a system reset button in the event of the processor lockup. The STAMP 2 can be powered by a 9- to 12-Vdc power source. A rechargeable battery pack is recommended for this project. The 9- to 12-V supply is also fed to the regulator at U3, which provides 5 V to the pressure sensor and A/D converter. A system ground connection is applied to pin 23 of the

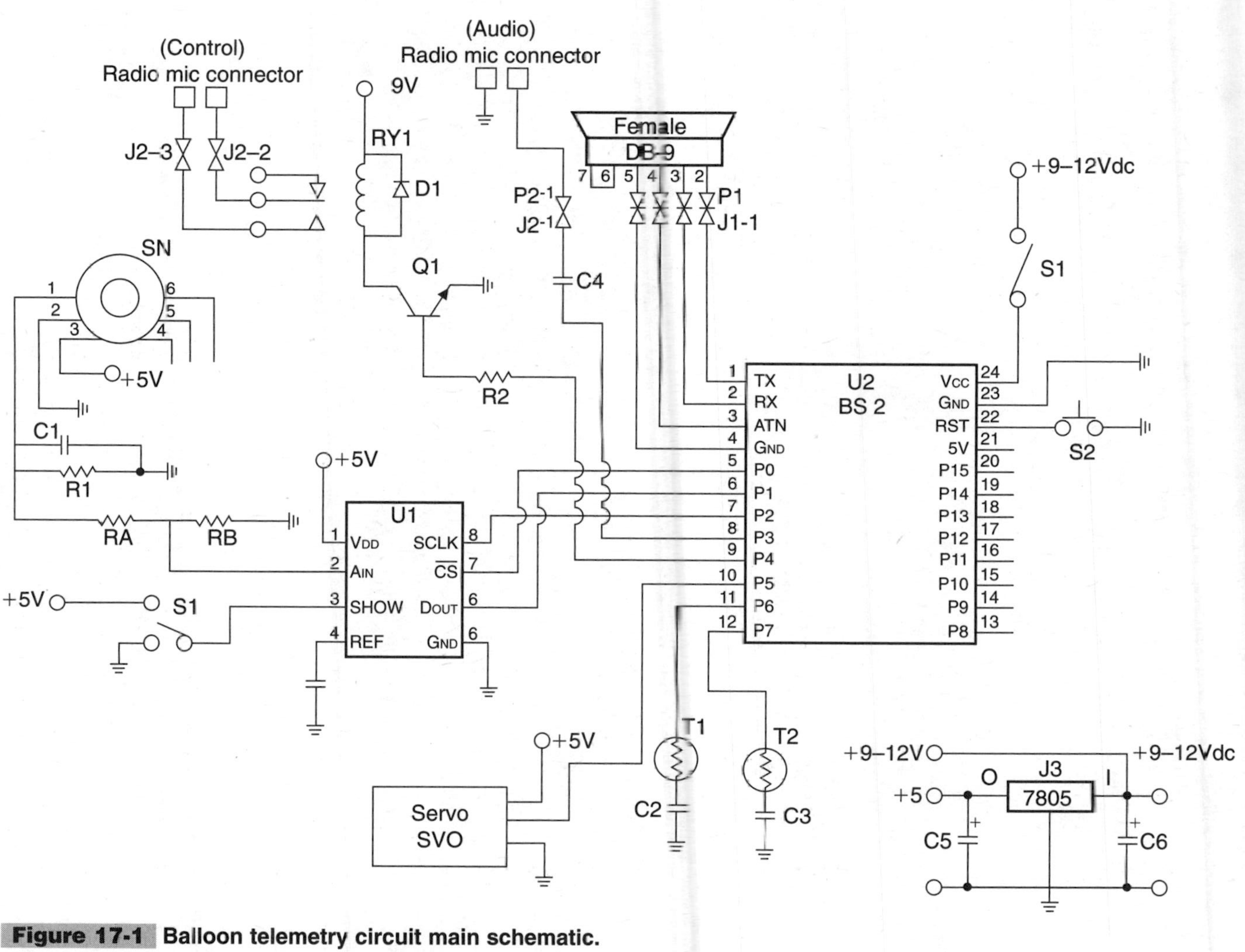

Figure 17-1 Balloon telemetry circuit main schematic.

original STAMP 2. Install two male circuit board headers, one for the programming cable J1 and another for the audio/control cable J2. A cable can now be constructed between P1 and a 9 pin RS-232 female connector to allow you to download the balloon telemetry program. Next, make up a cable from P2 to a connector leading to your transmitter unit, which will provide control and audio connections from your controller to your radio.

Once the circuit has been constructed and mounted in a small plastic box, Styrofoam can be placed around the plastic box to provide a safety enclosure for the circuit board. You will be ready to download the balloon telemetry program into the STAMP 2. Apply power to the circuit and download the program. Once the program is installed and running you can test out the system initially. You may wish to program your own call letters into the system. Take a look at the program and scroll down to the line: lookup index1, (130,13,180,99,163),Character 'N4YWK. You will have to change the numerical data to reflect your own call sign.

Two Configurations

The balloon telemetry system is shown here in two major configurations, but the system is flexible and will allow many different setups. If you choose to use the balloon telemetry project to simply send temperature and pressure readings, you can use the relay RY1 to key a 146- or 440-MHz transmitter. The line from P3 can be used to feed the audio output from the STAMP 2 to the microphone input of the telemetry transmitter used; see Fig. 17-2.

If you elect to use the camera option with a mini TV camera and an ATV transmitter, you can feed the audio output of the STAMP 2 at pin P3 to the audio input to the ATV transmitter. Using this method, you can transmit both camera pictures and telemetry audio on the same transmitter. The relay control at P4 can then be used to turn the ATV transmitter on and off. If you are going to use a still-picture film camera with a time lapse feature and a small FRS or 2-m amateur radio transmitter, then you will have to send the audio from P4 to the microphone input of the transmitter you wish to use; see Fig. 17-3.

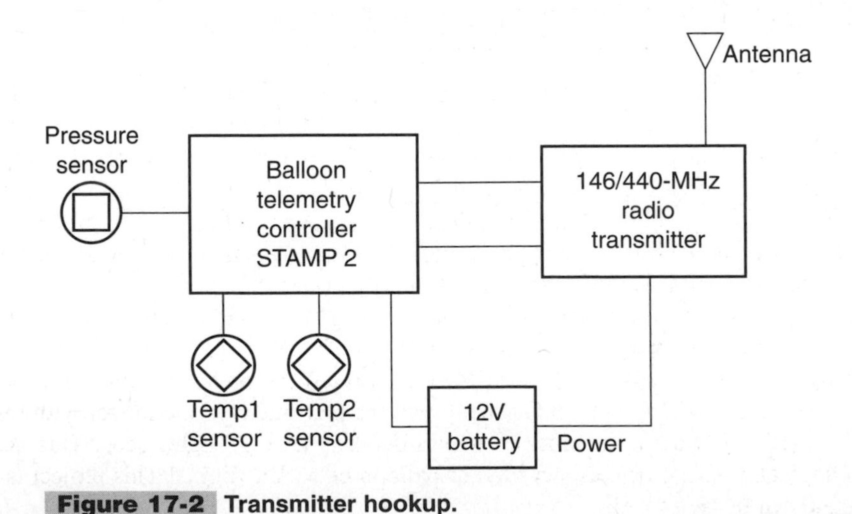

Figure 17-2 Transmitter hookup.

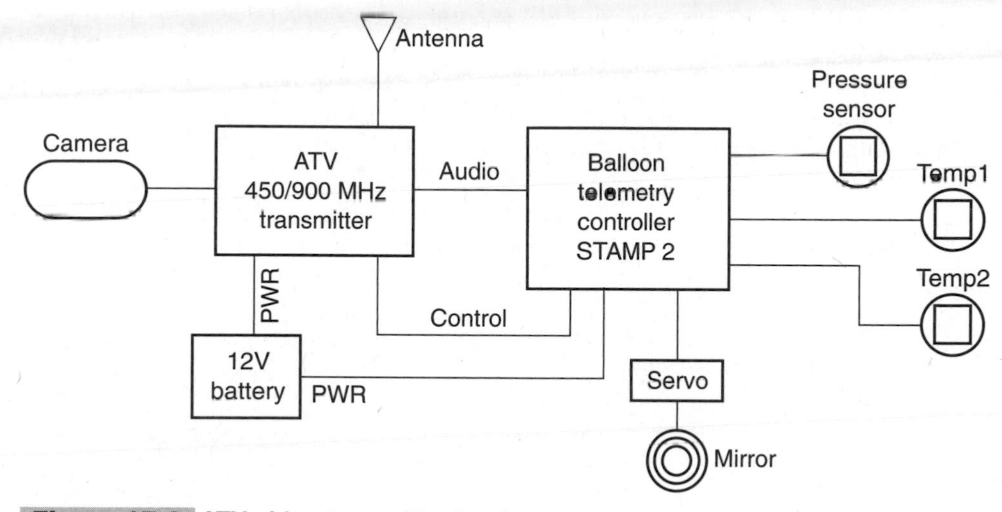

Figure 17-3 **ATV video transmitter hookup.**

Now that your balloon telemetry circuit is complete, you will need to locate the STAMP 2 Windows editor program titled STAMPW.EXE. Next, connect up your programing cable and apply power to the circuit. Finally, you will need to download the balloon telemetry program named BTELEMETRY.BS2 (Listing 17-1).

With power applied to the balloon telemetry circuit and a transmitter connected to the system, you can make a few tests. The system will begin transmitting an ID tone and then will begin sending data in the form of CW or Morse code. Next, apply heat/cold to one of the thermistors and when the transmitter begins sending CW again on its next cycle, the CW message will reflect a different value for the changed temperature reading from the heated or cooled thermistor.

Controller Operation

In operation the system will first send the call sign of the radio operator. Next, it will send raw data in decimal values of the current pressure reading. The system will next send the two thermistor readings. The telemetry unit will send data by first sending the thousands value, then the hundreds, next the tens, and finally the ones. The system will cycle through the pressure and temperature values and then move the mirror, so that the camera will get a particular view. Once the new cycle begins anew, the new pressure and temperature readings will be followed by a mirror movement.

If you elect to use the camera option, you will need to obtain a model airplane hobby servo. The servo is mounted to a small mirror. The STAMP 2 controls the servo, allowing two mirror positions, so the system will cycle between the two positions, allowing two camera views. You will need to experiment with the placement of the mirror with respect to your small TV camera, in order to maximize the viewing angles seen. This camera option adds an exciting new aspect to your balloon or rocket flights! This project is a lot of fun and can be addictive!

```
BTELEMETRY.BS2
(Will Payne N4YWK
Symbol CS     = 0         'pin 0, to A/D convertor chip select
Symbol AD     = pin1      'pin 1, to A/D convertor data
Symbol CLK    = 2         'pin 2, to A/D convertor clock
Symbol CW_aud    = 3      'pin 3, CW audio tone output
Symbol CW_key    = 4      'pin 4, CW key output
Symbol Servo   = 5        'pin 5, servo motor PWM control
Symbol Therm1    = 6      'pin 6, thermistor RC
Symbol Therm2    = 7      'pin 7, thermistor RC
Symbol Character = b0     'variable address, CW character
Symbol Elements = b1      'variable address, CW element count
Symbol Index1  = b2       'variable address, loop index
Symbol Index2  = b3       'variable address, loop index
Symbol SrvPos = b10       'servo position register
Symbol ADresult = w2      'variable address, A/D convertor data
Symbol ADnormal = w3      'variable address, altitude and temp normal
Symbol DFTone  = 120      'DF tone pitch
Symbol Tone  = 112        'CW tone pitch
Symbol Quiet = 0          'CW silence
Symbol Dit_length = 6     'CW rate constant, dit time
Symbol Dah_length = 18    'CW rate constant, dah time
Symbol Wrd_length = 36    'CW rate constant, word space time
Symbol Chr_length = 24    'CW rate constant, char space time
Symbol PScale  = 46       'scale constant for thermistor RC network
Symbol PosA    = 100      'servo position A, alternates with below
Symbol PosB    = 200      'servo position B, alternates with above
   let pins = 007         'set first three pins high
   let dirs = %11111101 'set all pins to output except pin 1
Top:
   for index1 = 0 to 4  'count thru CW characters to send
        lookup index1,(130,13,180,99,163),Character      'N4YWK
        gosub Morse      'send it
        next
   gosub Word_sp             'send space before continue
Telemetry:                   'send telemetry in decimal CW
gosub ADconv                 'acquire 12 bit A/D static pressure
gosub Send_4_Digits          'send raw data in decimal CW
pot Therm1,PScale,ADresult 'pot read thermistor 1
gosub Send_3_Digits          'send raw data in decimal CW


pot Therm2,PScale,ADresult 'pot read thermistor 2
gosub Send_3_Digits          'send raw data in decimal CW
sound CW_aud,(DFtone,255)  'sound a long audio tone
gosub Move_It                'move something
high CW_key                  'key down
for Index1 = 1 to 100        'wait 100 counts, without sleeping.
     pause 100               'each count is 100 mS.
     next                    'stay awake so servo will not sleepwalk.
low CW_key                   'key up
    'sleep 10                     '10 sec battery saver sleep
'gosub Srvo_pulse            'reposition servo if it sleepwalks.
goto Top                     'loop forever
Move_It:             'move a servo motor using PWM control, 20 mS cycle 10-90%
                     'cycles the servo between two positions, PosA and PosB,
```

Listing 17-1 Balloon telemetry program.

```
                'alternating every execution cycle.
 if SrvPos = PosA then Srvo_skip        'if servo is at PosA skip down
let SrvPos = PosA                       'else next servo command to PosA
 goto Srvo_pulse                        'go command it
     Srvo_skip:                         'if it was already at PosA
         let SrvPos = PosB              'servo will command to PosB


     Srvo_pulse:
     for Index1 = 1 to 100              'Send 100 pulses, about 2 sec.
           pulsout Servo,SrvPos         'Form the pulse, i * 10 uS wide.
           pause 18                     'Wait 20 mS.
           next                         'Keep cycling.
        return
Send_4_Digits:
    Character=ADresult/1000//10         'get thousands digit
    gosub CWval                         'look up and send CW code
Send_3_Digits:
    Character=ADresult/100//10          'get hundreds digit
    gosub CWval                         'look up and send CW code
    Character=ADresult/10//10           'get tens digit
    gosub CWval                         'look up and send CW code
    Character=ADresult//10              'get ones digit
    gosub CWval                         'look up and send CW code
    gosub Word_sp                       'send space before return
    debug ADresult                      'test port
    return
Morse:                                  'send a CW character, one byte of
packed bits
                                        '5 MSBs are 0=dit and 1=dah
                                        '3 LSBs are number of elements in
character
 let Elements = Character & %00000111     'mask off element count
     Bang_Key:
     for index2 = 1 to Elements              'count thru elements
          if Character >= 128 then Dah       '1 = dah
          goto Dit                           '0 = dit
     Reenter:
         let Character = Character * 2       'shift to next element
         next
     sound CW_aud,(Quiet,Chr_length)         'space after character
     return                                  'one character was sent
       Dit:                                  'send a Dit
           high CW_key                       'key down
           sound CW_aud,(Tone,Dit_length)    'sound an audio tone
           low CW_key                        'key up
           sound CW_aud,(Quiet,Dit_length)   'the sound of silence
           goto Reenter                      'one Dit was sent
       Dah:                                  'send a Dah
           high CW_key                       'key down
           sound CW_aud,(Tone,Dah_length)    'sound an audio tone
           low CW_key                        'key up
           sound CW_aud,(Quiet,Dit_length)   'the sound of silence
           goto Reenter                      'one Dah was sent
Char_sp:                                     'send space between
characters
    sound CW_aud,(Quiet,Dah_length)
    return

Word_sp:                                     'send space between words
```

Listing 17-1 Balloon telemetry program (*Continued*).

```
    sound CW_aud,(Quiet,Wrd_length)
    return
CWval:                                              'look up CW code for
decimal digits 0-9
   lookup Character,(253,125,61,29,13,5,133,197,229,245),Character
   gosub Morse
   return
ADconv:                                     'acquire result from MAX187 12
bit serial A/D convertor
   low CLK                                  'CLK low before start
   low CS                                   'select the convertor
   pulsout CLK,1                            'pulse CLK to start
   let ADresult = 0                         'clear the result register
   for Index2 = 1 to 12                     'count the bits in
        let ADresult = ADresult * 2         'shift the result
        pulsout CLK,1                           'clock the convertor
        let ADresult = ADresult + AD            'add the data bit
        next                                    'until complete
   high CS                                      'deselect the convertor
   return                                       'go home with ADresult
end
```

Listing 17-1 Balloon telemetry program (*Continued*).

Balloon Telemetry Project Parts List

SN	Motorola MPX5100A pressure sensor
T1, T2	Thermistors, Keystone NTC (Digikey KC002G)
U1	Maxim MAX 187 A/D converter
U2	BASIC STAMP 2 (original)
U3	LM7805 5-V regulator
R1	51-kΩ $^1/_4$-W resistor
R2	1-kΩ $^1/_4$-W resistor
Ra	100-kΩ $^1/_4$-W resistor
Rb	10-kΩ $^1/_4$-W resistor
C1	50-pF 35-V capacitor
C2, C3	0.1-μF 35-V capacitor
C4	0.47-μF 35-V capacitor
C5	1-μF 50-V electrolytic capacitor
C6	0.2-μF 50-V capacitor
D1	1N4002 silicon diode
Q1	2N2222 npn transistor
RY1	6–9-V minirelay (RadioShack)

S1	Jumper or slide switch, SPDT
S2	Normally open pushbutton switch
SVO	Hobby servomotor, 3-wire type
BATT	9- to 12-V rechargeable battery pack
Miscellaneous	Connectors, wire, jack, plugs, DB-9, etc.

18

R/C KITE/GLIDER DIGITAL CAMERA SYSTEM

CONTENTS AT A GLANCE

Photography is a powerful medium in capturing our lives, hobbies, and world around us. Aerial photography adds a whole dimension to seeing the world from a different perspective. Combining aerial photography and model building and remote control, the R/C digital camera control system combines the best of three hobbies in one project. The pictures you can capture with the R/C digital camera system are simply fascinating; see Fig. 18-1. In the past, film cameras produced only about 5 percent useful results because of motion blur, but with the digital camera system almost every picture is simply beautiful. The remote control digital camera system will work with a model R/C aeroglider or a nylon parafoil (lifting kite). The remote control digital camera system is built around the BASIC STAMP 2 microcomputer as the interface/controller between a model aircraft remote control system receiver and a digital camera. Once the controller is constructed, it's simply a matter of connecting cables between the receiver and the controller and the controller and a digital camera.

Figure 18-1 Kite camera view.

Many new digital cameras support a serial input/output jack that will easily interface with the STAMP 2 for serial control. The STAMP 2 accepts a single-channel output from a multichannel remote control receiver and uses the receiver pulses for sensing. The STAMP 2 in turn acts as an interface to the digital camera, snapping pictures as determined by you remotely. The R/C camera system can also be utilized to snap pictures on a timed interval if desired.

Basic Components

The R/C digital camera control system, as mentioned, revolves around the BASIC STAMP 2 microprocessor. It is shown in Fig. 18-2. The first eight I/O pins P0 through P7 are used for mode programing functions, via a bank of in-line DIP switches. The opposite side of the switch bank is tied to the +5-V supply. Each I/O input pin is also coupled to ground via 10-kΩ pull-down resistors. The three program functions are shown in Table 18-1. The first mode is the single-shot R/C mode: toggling the transmitter control from off to on. The second mode is the time lapse R/C mode: the TX control is left on. Note that trigger intervals are made in variable steps between 10 seconds to an hour. The third mode is the stand-alone mode: the camera performs in the local time lapse mode, without intervention.

The two LEDs on the controller circuit board are status indicators. The red error lamp is flashed if an acknowledgment (Ack) is not received from the camera following any command in the shutter-firing sequence, while the green lamp is to confirm that a shutter command has been successfully received by the camera. It is flashed after an (Ack)

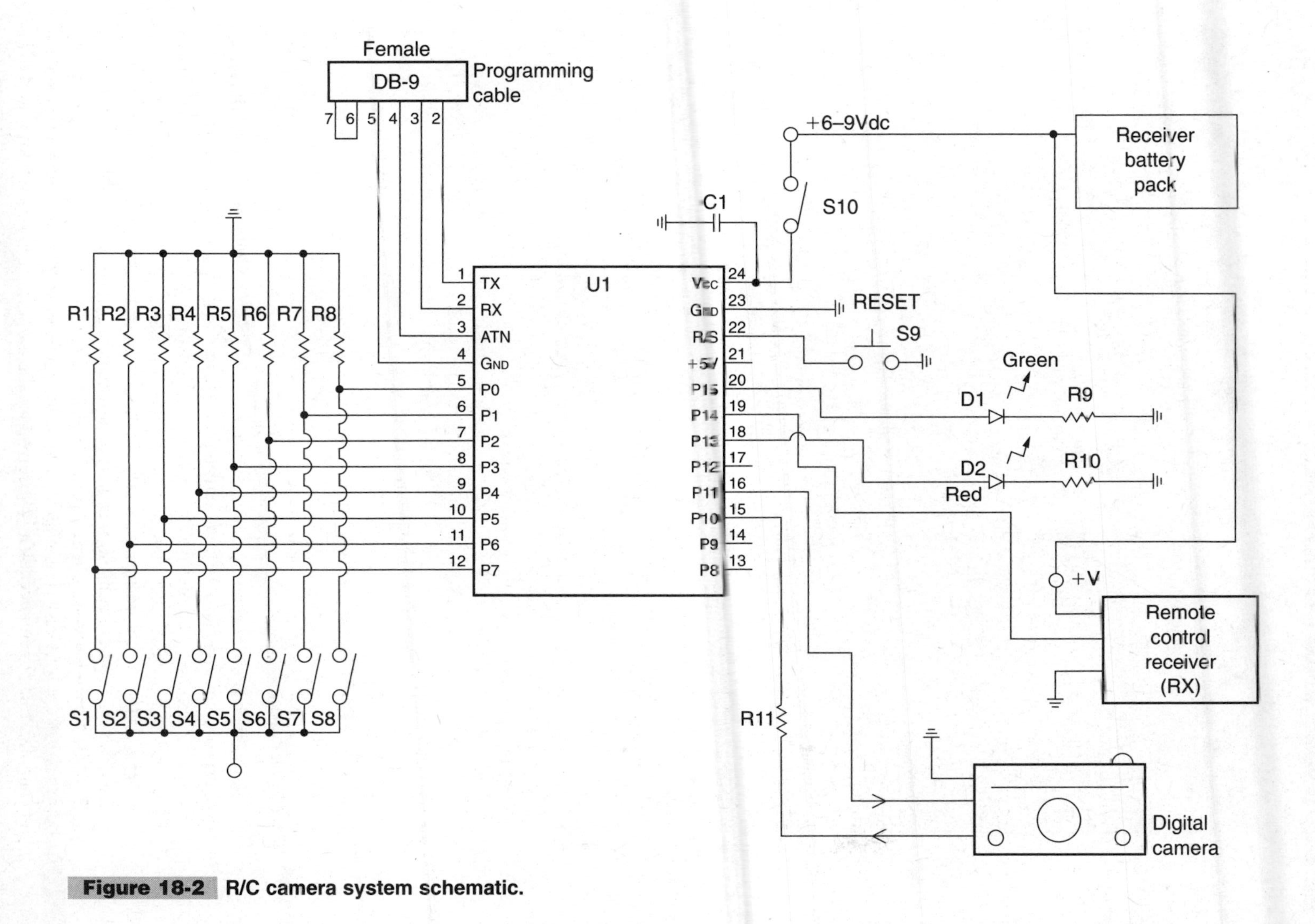

Figure 18-2 R/C camera system schematic.

TABLE 18-1

I/O PIN	FUNCTION
0 through 5	TimeoutVal 0–127; pin 0 is LS bit; all switches OFF = 0; all switches ON = 63
6	Multiplier for TimeoutVal (pins 0–5); on: timeout = TimeoutVal minutes; off: timeout = TimeoutVal × 10 seconds
7	Mode; off = R/C; on = stand-alone

is received after the last command in the shutter-firing sequence (except in continuous mode).

The STAMP 2 controller is shown powered by the remote control receiver, using a 6-V battery pack. The battery system also powers the interface and other servos used for aileron and elevator functions. Programming the STAMP 2 controller is facilitated by using pins 1 through 4. These four pins are brought out to a 4-pin single-row female header, mounted on the circuit or perfboard. You will need to make a programming cable, with a male header at one end and a DB-9 female connector at the other end for initial programming.

Construction of the Interface

Construction of the R/C digital photography interface is quite simple, and it can be built on a small perfboard or circuit board. RadioShack has a number of suitable inexpensive perfboards that could be used for the controller/interface circuit, or you could elect to construct your own circuit board if desired; see Fig. 18-3. The layout of the circuit is quite simple. The prototype was constructed using an integrated circuit socket for the STAMP 2. When installing components on the circuit board, be sure to observe the correct polarity for the capacitors and LEDs. When installing the BASIC STAMP 2, be careful to orient the controller correctly when installing it to the IC socket. On the low pin number side of the STAMP 2, two 4-position DIP switch assemblies are mounted close to the STAMP 2. Each I/O input, P0 through P7, also requires a resistor on each input pin. You can choose either single resistors or an SIP multiresistor package to do the job. The multiresistor SIP packs are easy to work with but a bit more difficult to locate than conventional resistors. On the high I/O pin number side of the STAMP 2, i.e., pins 8 through 15, you will need to install the status indicator LEDs and the RS-232 input/outputs, as well as the reset pushbutton. The reset switch, a normally open pushbutton switch, is used in the event of a system lockup. A 4-pin single-row female header makes an ideal serial interface connector for the RS-232 input/output to the digital camera on pins P10 and P11. Remember to also include a ground connection for the serial I/O connection that is available from STAMP 2 pin 23. A second single-row female header should also be installed for the receiver input on pin P14. Remember to connect a ground connection to the header. Next you will have to make up a

Figure 18-3 **R/C camera system PC board.**

camera cable between the camera and the RS-232 header and an R/C receiver-to-header cable between the receiver and the RX input header. The smaller and lighter the circuit board the better, to lessen the total system weight.

Installation

The prototype of the R/C digital camera project was installed underneath the sailplane or glider. Initial flight tests used a cardboard box instead of a real camera. The test box should be weighted with the approximate weight of the camera/interface assembly to establish a center of gravity and to see the effect on the aerodynamics or handling characteristics. The box was taped under the sailplane, and was tested with different box positions. Two configurations seemed to work best. First, the test box was used with the simulated camera pointing downward. In this test, it was determined that there was a reduction in glide but handling was safe; see Fig. 18-4. In the next best configuration, the simulated camera was pointed sideways. This produced good handling and good glide performance; see Fig. 18-5.

A Futaba five-channel R/C control system is used for this project; other systems such as Hitec or Multiplex could be used. The prototype R/C system is used to control ailerons, elevator, spoiler, rudder, and the camera interface. You could also get by with a four-channel system, with a bit less control. Simply refer to the receiver documentation to determine the pinouts of the receiver. Once a free servo port channel is determined, you will need to

Figure 18-4 Glider with camera pointing downward.

Figure 18-5 Glider with camera pointing sideways.

obtain a connector to mate the free servo port channel output in order to make a cable from the receiver to the STAMP 2 interface. This camera control channel is then fed to P14, so the STAMP 2 can interpret the incoming receiver pulses. The other receiver channels are coupled directly to the respective servos for airframe control.

The R/C digital camera project was tested with three different digital cameras, but other cameras will work. The prototype was tested with the Olympus models C400L, D340, and C840L cameras, all of which are based on the Fujitsu chip set. When wiring the cameras, be sure to take time to carefully determine the correct wiring to avoid damage to the camera. Most digital cameras use the 3-wire serial system for remote control, i.e., serial TX, serial RX, and ground connections. Other digital cameras based on the Fujitsu chip set can be seen at http://photopc.sourceforge.net/.

Connecting the Systems

After constructing your camera controller interface, you are now ready to wire up the rest of the system. The diagram in Fig. 18-6 illustrates the interconnection between the controller and the receiver and controller and the digital camera. The receiver is connected to the BASIC STAMP 2 controller via header/plug J1/P1 on the controller board. The interface controller is wired to the digital camera via header/plug J2/P2. The diagrams in Figs. 18-7 and 18-8 depict two popular serial camera connections. Note that the cameras are powered from their own internal batteries and only serial signal leads connect the camera and controller via three wires, i.e., TX, RX, and ground. The R/C digital remote control camera system can be powered in two ways. The receiver battery pack is connected to the receiver module and a jumper set consisting of the plus lead and ground is wired from the receiver to the controller at pin 24, the V_{cc} lead. Powering the controller in this fashion utilizes the STAMP 2's internal regulator to power just the controller. Alternatively, the STAMP 2 could be powered by applying regulated +5 V directly to pin 21.

Once the digital R/C camera controller system is wired up and power is applied, it is now time to test out the system. First, you will need to locate the STAMP 2 editor program titled STAMPW.EXE; next, you will need to download the RCAMERA.BS2 (Listing 18-1) program into the STAMP 2 controller. Finally, you will need to make some final tests to make sure your system will be airworthy, before flying your camera glide/kite. Now is a good time to also test the mode DIP switches on the system controller for R/C control. It is recommended that the

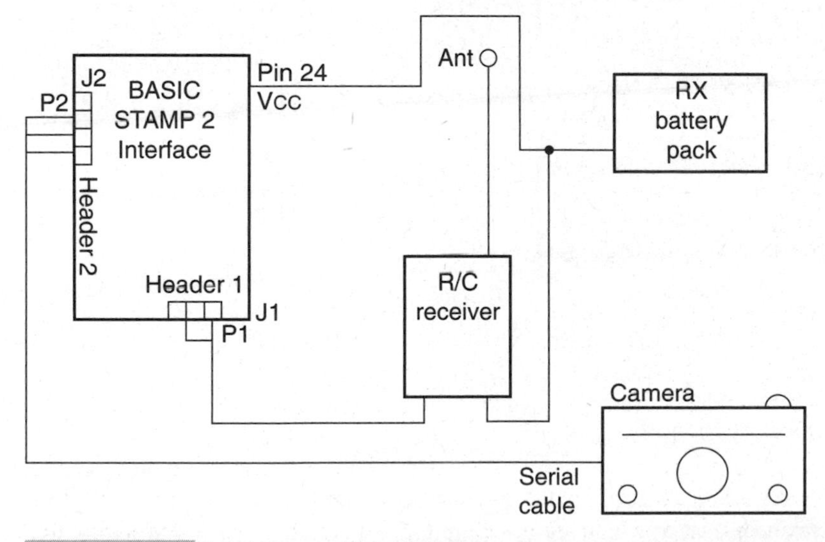

Figure 18-6 Interconnection diagram.

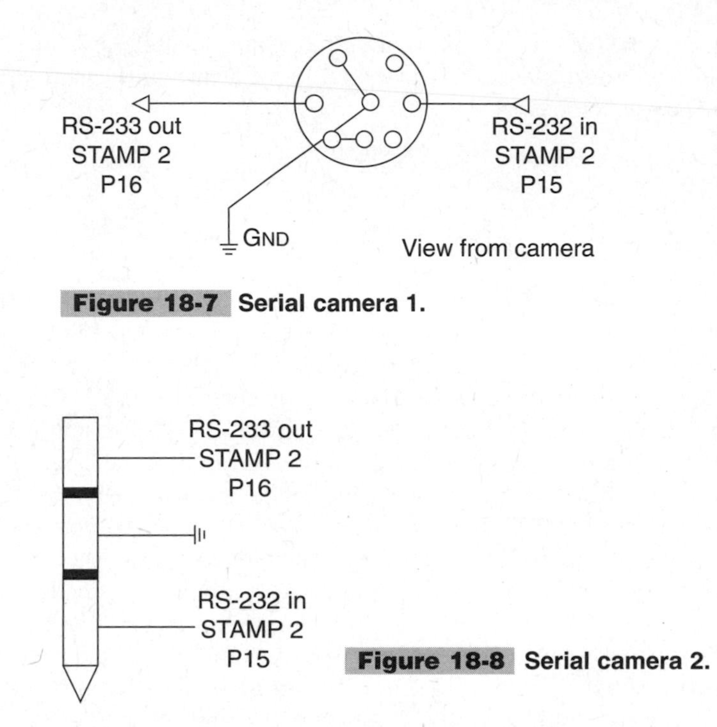

Figure 18-7 Serial camera 1.

Figure 18-8 Serial camera 2.

receiver/controller/camera be wired up completely before applying power. Next, turn on the R/C transmitter section and test to make sure all systems are working properly. Test the remote servos to ensure they are all functioning. Lastly, test the camera controller functions to ensure the camera shutter is working. Now that the whole system is working, you are ready to go fly a glider or kite!

R/C Digital Camera Control System Parts List

U1	BASIC STAMP 2 (original)
U2	LM7805 5-V regulator
R1, R2, R3, R4	10-kΩ $^1/_4$-W resistor
R5, R6, R7, R8	10-kΩ $^1/_4$-W resistor
R9, R10	330-Ω $^1/_4$-W resistor
R11	22-kΩ $^1/_4$-W resistor
C1	4.7-μF 50-V electrolytic capacitor
D1	LED (green)

D2	LED (red)
PL	DB-9/female RS-232 connector
S1, S2, S3, S4	Mini DIP switches
S5, S6, S7, S8	Mini DIP switches
S9	Momentary pushbutton SPST (normally open)
S10	SPST toggle switch (on-off)
Miscellaneous	Headers, sockets, PC board, wire, jumpers, camera cable
Battery	6–9-V battery pack
R/C receiver servos	Futaba, JR, Hitec, Multiplex, etc.—any receiver using regular R/C
R/C transmitter	Futaba, JR, Hitec, Multiplex, etc.
Camera	Olympus C400, D340, C840L cameras work great (*Note:* other cameras based on the Fujitsu chip set will work)

```
'RCAMERA.BS2
' RADIO CONTROLLED DIGITAL CAMERA INTERFACE
'
'   Copyright M. Shellim 1999-2002
'    Author M. Shellim
'    Mods and fixes: Gary Thurmond

'    Monitors one channel of a Futaba receiver
'    and controls Olympus C400L digital
'    camera via serial i/o port configured for
'    RS232 send/receive

'    RC mode dip switch sets repeat in 10 sec interval.
'    RC mode dip switch set to 00000(65535).
'    Repeat interval will be as short as the camera cycle time.
'    D-340R, std. = 3.0 sec, HQ = 5.1 sec.
'    Changes not yet subjected to rigorous testing.
'    Seems to work but haven't tried hard to break it (Ditto - MS).
'
'    Note that there are debug statements and maybe
'    some rem's. Hope I cleaned out other edits I made
'    that do not address the RC interval subject.
'    This is not a final version, just an example
'    of speeding up the interval in the RC mode.
'    The subroutines that I have hacked at are -
'    PUSH_SHUTTER, see comments there.
'    AWAIT_WITH_TIMEOUT, see comments there.

' Flashing LED parameters

flash_len VAR WORD
flash_num VAR BYTE
flash_pin VAR BYTE
```

Listing 18-1 Radio-controlled digital camera interface program.

```
BAUD_MODE CON 16416 ' 19200 inverted

' I/O pin configuration

' Masks          111111
'               5432109876543210
'               ________
TIMEOUT_MASK   CON %0000000000111111 ' Pins 0-5
SEC10_OR_MIN   CON %0000000001000000  ' Pin 6
MODE_MASK      CON %0000000010000000  ' Pin 7
SERIN_PIN        CON 10
SEROUT_PIN       CON 11
RED_LED_PIN      CON 13
RX_PIN           CON 14
GREEN_LED_PIN    CON 15
' Pins - dev board
'RX_PIN          CON  8
'SEROUT_PIN      CON 11
'GREEN_LED_P     CON 12
'RED_LED_PIN     CON 15
'SERIN_PIN       CON 14

' Misc Constants
pictime CON 43  ' D-340R in std Q mode (640x480)
'pictime CON 63 ' D-340R in HQ mode (1280x960)
'TIMEOUT_MULT CON 10
'SHUTTER_DURATION CON 1

' Misc vars

serData    VAR BYTE
prevPulse VAR BIT
currPulse VAR BIT
iloop     VAR BYTE
musecs     VAR WORD
timeout    VAR WORD
wMask      VAR WORD

' Program Begins
gosub flash_roger
if INS & MODE_MASK = 0  then main_rc
if INS & MODE_MASK <> 0 then main_self_time

' Subroutine main_self_time

main_self_time:

    ' Press shutter

    gosub push_shutter

    ' Wait for timeout.
    ' time-out period (secs)  = 9.6 * timeout - 1.77

    gosub read_timeout
    sleep timeout * 10 -2
```

Listing 18-1 **Radio-controlled digital camera interface programs (*Continued*).**

```
    ' Repeat
    goto main_self_time:

' Subroutine main_rc
main_rc:
    currPulse = 1 ' To prevent initial triggering

wait_for_toggle_on:

    ' Wait for the user to toggle shutter control off to on
    gosub rc_await_0_to_1

debug "0 to 1",cr

press_shutter_and_wait:

     ' Press shutter, and wait till user resets Tx shutter button
     ' or timeout occurs

     gosub push_shutter
     gosub await_0_with_timeout

     ' If the shutter control was left ON (i.e. timed out), press
     ' shutter again and wait

     if CurrPulse = 1 then press_shutter_and_wait

     ' Otherwise, the shutter control was switched to off, so
     ' wait for it to be on.

debug " ",cr
debug "W T on",cr

    goto wait_for_toggle_on

' Subroutine rc_await_0_to_1
rc_await_0_to_1:
    PrevPulse = CurrPulse
    gosub get_pulse
    if PrevPulse = 1 or CurrPulse = 0 then rc_await_0_to_1
return

' Subroutine await_0_with_timeout
await_0_with_timeout:
     intLoop var WORD

     ' Read timeout value from pins into "timeout"

     gosub read_timeout

     ' 100 Loops = ten seconds
     ' pictime is time from wakeup to pic finish

     for intLoop = 1 to (timeout * 100)- pictime
           gosub get_pulse
           if CurrPulse = 0 then await_0_end
           pause 85
     next
```

Listing 18-1 Radio-controlled digital camera interface program (*Continued*).

```
     ' Timeout !

await_0_end
     return

' Subroutine PUSH_SHUTTER

PUSH_SHUTTER:
debug " ",cr
debug "attn",cr
    ' ATTN
    ' Camera attn. takes >,0.82 sec, >,1.1 sec if it went to sleep,
    ' the Oly 180 sec. internal timeout.
    Serout SEROUT_PIN, BAUD_MODE, [$00]
    Serin  SERIN_PIN,  BAUD_MODE, 1300,serin_timeout,[WAIT ($15)]

debug "baud 19200",cr
    ' Set baud = 19200
    Serout SEROUT_PIN, BAUD_MODE,
[$1B,$53,$06,$00,$00,$11,$02,$00,$00,$00,$13,$00]
    Serin  SERIN_PIN,  BAUD_MODE, 500,serin_timeout,[WAIT ($06)]

debug "flash off",cr
    ' Force flash OFF
    Serout SEROUT_PIN, BAUD_MODE,
[$1B,$43,$06,$00,$00,$07,$02,$00,$00,$00,$09,$00]
    Serin  SERIN_PIN,  BAUD_MODE, 500, serin_timeout,[WAIT ($06)]

fire_shutter:
debug "fire",cr
    ' Fire shutter looks at dip switch and get_pulse after fist pic.
    ' If dip switch is zero (65535) and get_pulse is one then take
    ' another pic with all delays bypassed.

    Serout SEROUT_PIN, BAUD_MODE, [$1B,$43,$03,$00,$02,$02,$00,$04,$00]
    Serin  SERIN_PIN,  BAUD_MODE, 500,serin_timeout,[WAIT ($06)]
debug "wait 06",cr

    Serin  SERIN_PIN,  BAUD_MODE, [WAIT ($05)]
debug "wait 05",cr

    gosub read_timeout
    if timeout <> 65535 then baud_sleep
    gosub get_pulse
    if CurrPulse = 1 then fire_shutter

baud_sleep:
debug "baud 0",cr
    ' Set baud = 0
    Serout SEROUT_PIN, BAUD_MODE,
[$1B,$53,$06,$00,$00,$11,$00,$00,$00,$00,$11,$00]

    gosub flash_roger

done:   return
serin_timeout:

debug "timeout",cr
```

Listing 18-1 Radio-controlled digital camera interface program (*Continued*).

```
    gosub flash_error
    goto done

' Subroutine FLASH
' flashes a LED for a caller
' specified count and length of time.

FLASH:
   ' make flash_pin an output pin

    wMask = DCD flash_pin
    DIRS = DIRS | wMask

    ' Loop to output a 1 then 0.
    for iloop = 1 to flash_num

          ' Output a 1

          OUTS = OUTS | wMask
          pause flash_len

          ' Output a 0

          OUTS = OUTS & (~wMask)
          if iLoop = flash_num then flash_end
          pause flash_len
     next
flash_end:
     return

' subroutine GET_PULSE
' Sets currPulse to 0 or 1 depending
' on the length of the receiver pulse
' being monitored on i/o pin 8.

get_pulse:
     pulsin RX_PIN,1,musecs
     if musecs < 575 then pulse_off  '2x575=1150 mS
     if musecs > 925 then pulse_on   '2x925=1850 mS
     return
pulse_on
     currPulse = 1
     goto pulse_end
pulse_off:
     currPulse = 0
pulse_end:
     return

' subroutine read_timeout
'  variable "timeout" is assigned the
'  number of 10 second periods to wait.
'  (10 seconds is the smallest practical
'  duration between shutter presses.)

read_timeout:
    flMinutes VAR WORD

    ' Read i/o pins specifying timeout and units
```

Listing 18-1 Radio-controlled digital camera interface program (*Continued*).

```
    timeout = INS & TIMEOUT_MASK
    flMinutes = INS & SEC10_OR_MIN

    ' Treat zero timeout as max

    if timeout <> 0 then timeout_apply_mult
    timeout = 65535
    goto timeout_end

timeout_apply_mult:

    ' If minute units specified apply  multiplier
    ' to convert to 10sec units.

    if flMinutes = 0 then timeout_end
    timeout = timeout * 6

timeout_end:
    return
' Subroutine FLASH_ERROR
flash_error:
     flash_pin = RED_LED_PIN
     flash_num = 1
     flash_len = 500
     gosub flash
     return

' Subroutine FLASH_ROGER
flash_roger:
     flash_pin = GREEN_LED_PIN
     flash_num = 3
     flash_len = 100
     gosub flash
     return
```

Listing 18-1 **Radio-controlled digital camera interface program (*Continued*).**

19

TELE-ALERT

CONTENTS AT A GLANCE

The Tele-Alert is a unique new low-cost, multichannel, microprocessor-controlled remote event/alarm reporting system that will immediately notify you via your cellular phone (Fig. 19-1).The Tele-Alert can monitor up to four different alarm or event conditions from a host of different types of sensors, as shown in Fig. 19-2, and report the particular channel that was activated. The Tele-Alert can be configured to monitor voltage levels, temperature changes, movement, contact closures, windows, doors, safes, and perimeters as well as computer equipment. The Tele-Alert can be utilized to protect your home, office, shop, or vacation home.

The Tele-Alert is used with your existing telephone line; therefore, no charges are added to your monthly phone bill. The Tele-Alert can be programmed to call you on your cellular phone, or to call a friend, neighbor, co-worker, or a relative if desired. The Tele-Alert is expandable for future applications.

The Tele-Alert is compact and easy to use as well as expandable. Simply plug in a 9-V power supply, connect up your telephone line and the input connections, and you are ready

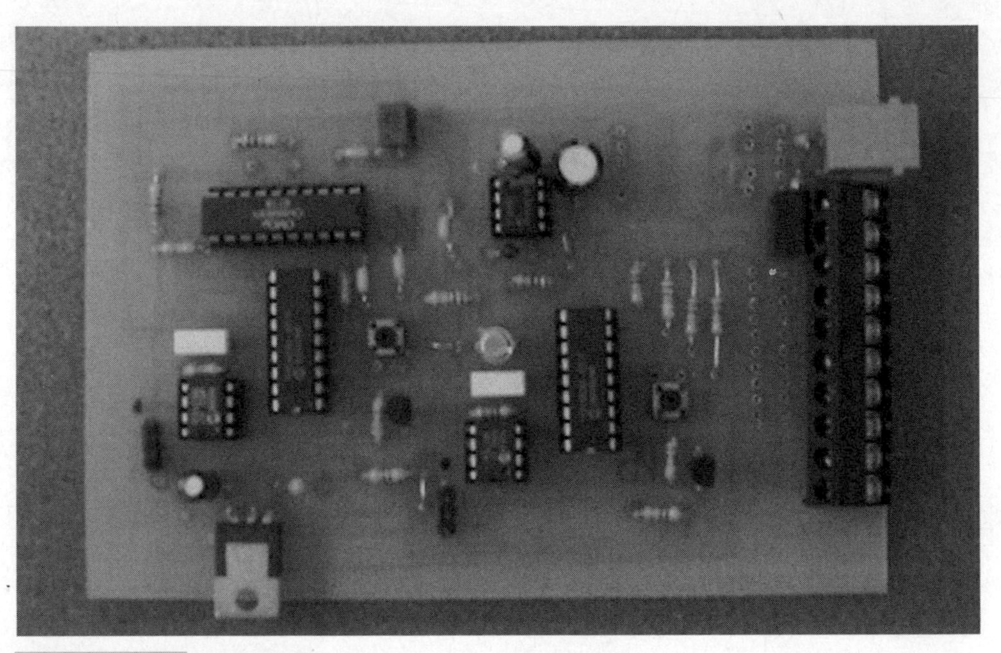

Figure 19-1 Tele-Alert.

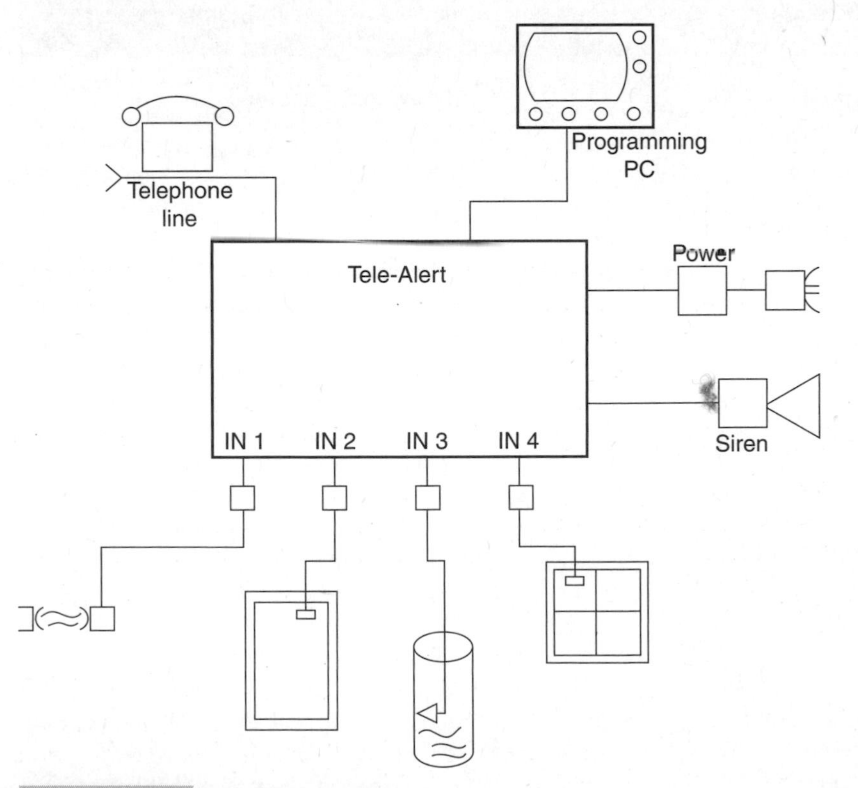

Figure 19-2 Tele-Alert system.

to go. The optional motion module allows the Tele-Alert to detect moving persons and will trigger the Tele-Alert; in addition the motion module contains two normally open and two normally closed alarm loop channels, so a number of alarm sensors/switches can also be used. The optional temperature/voltage level module allows your Tele-Alert to monitor up to four temperature or voltage level presets, and report the changes via your cellular phone or pager. The optional listen-in module will allow you to listen in to the area being protected for up to 2 minutes, if desired. The optional enhancement modules can turn your Tele-Alert into a complete alarm system to protect your home, cottage, or office (see Chap. 22, "Input Sensor Modules").

Basic Components

The Tele-Alert multichannel alarm reporting system begins with the circuit diagram shown in Fig. 19-3. The heart of the Tele-Alert is the PIC 16C57 chip, which emulates the BASIC STAMP 2 computer. The PIC 16C57 is preloaded with a BASIC interpreter much like the Parallax BASIC STAMP 2 (BS2). The PIC 16C57, however, is a much less expensive approach to solving a problem, but requires a few more external parts than the BASIC STAMP 2. The PIC 16C57 requires very little in the way of support chips to make it function—only two, a 24LC16B EEPROM for storage and a MAX232 for communication. Only a few extra components are needed to form a functional microprocessor with serial input/output. A 20-MHz ceramic resonator, a few resistors, and a diode are all that are needed to use the processor in its most simple form. The 28-pin PIC 16C57 is a very capable and versatile little microprocessor that can perform many tasks.

In order to turn the BS2 alternative into a Tele-Alert project, a few more parts must be added to the basic configuration. First, you need to configure pins 10 through 17 as inputs, for a total of 8 inputs that can be used to sense events or alarm conditions. The first four input pins are brought out to the terminal block for easy access. The second four input pins are shown at IN5 through IN8. Note the resistors placed across each of the input pins; these inputs are reserved for future use. Pins 18 through 25 are configured as output pins. Pin 18 is used to activate or enable the microphone in the listen-in module, if utilized. Pin 19 is used to drive the solid-state relay at U4. The microprocessor activates the LED in U4, which in turn closes the relay contacts at pins 4 and 6 of the relay; this essentially shorts out blocking capacitor C11 and allows the phone to go on hook and begin dialing your cell phone. Pins 20 and 21 of U1 are audio output pins. Pin 20 of U1 outputs the touch-tone signals needed to dial the phone when an alarm condition is sensed, while Pin 21 is used to output the tone sequences that indicate via your cell phone which event/alarm channel has been activated. The capacitors at C9 and C10 are used to couple the touch-tones and the alarm sequence tones to the coupling transformer network formed by T1 and the associated components. Capacitor C10 is placed across T1 while zener diodes D2 and D3 are used as voltage clamps at the input of the transformer.

Transformer T1 is used to couple the audio from the Tele-Alert to the phone line, once the unit has been triggered. Resistor R9 is used to couple one side of the transformer to the phone line at J6, and also acts to hold the phone line. The other secondary transformer lead is coupled to the phone line via C11. C11 is a blocking capacitor that keeps the phone off

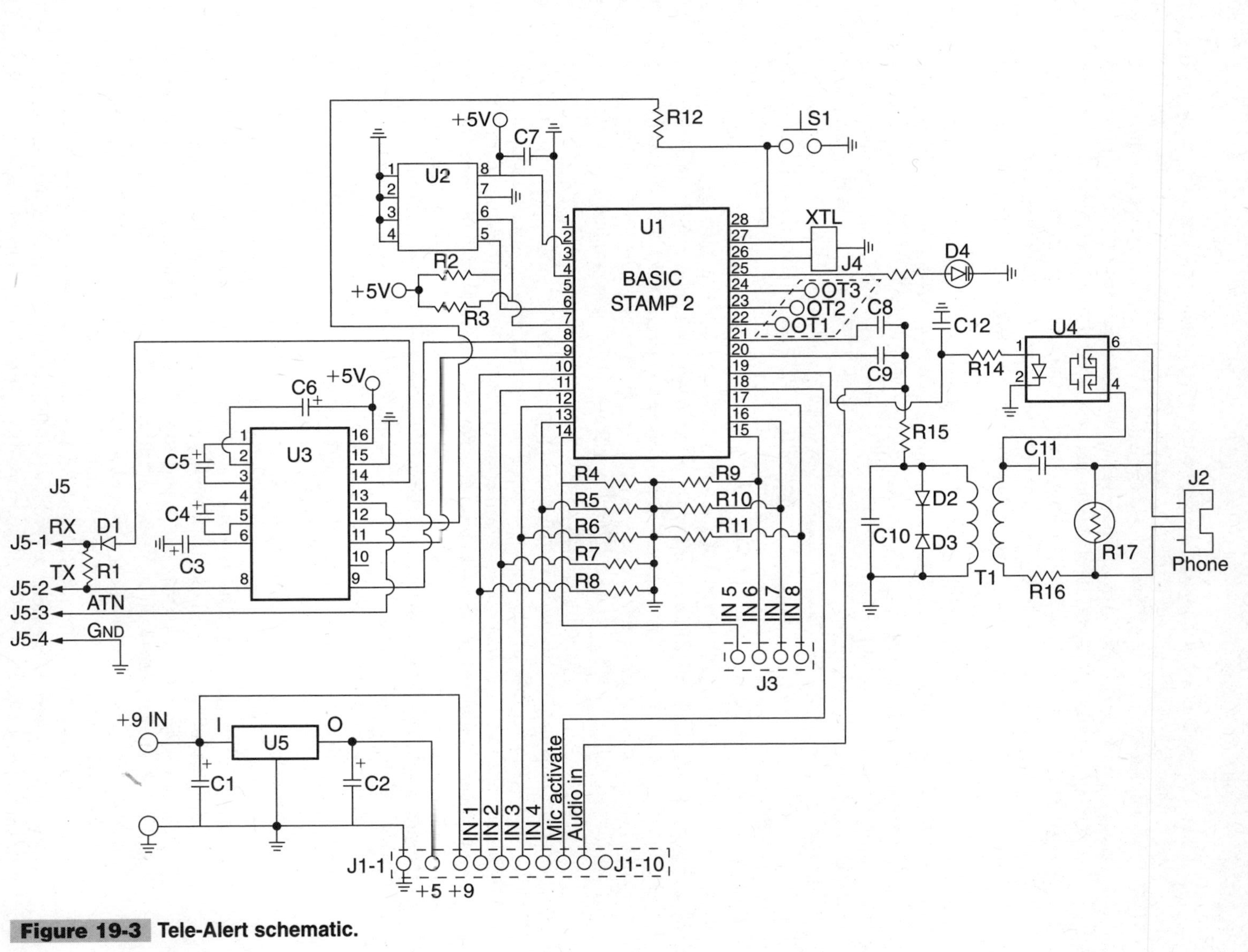

Figure 19-3 Tele-Alert schematic.

hook until an alarm condition occurs. The solid-state relay at U4 acts to short out C11 when an event or alarm condition occurs, thus coupling T1 to the phone line. Resistor R16 is a metal oxide varistor (MOV) and is used to protect the circuit from high voltage spikes. Pins 22 and 23 may be utilized as auxiliary outputs to drive local alarm sirens or outdoor lamps if desired, while pin 24 is left for further expansion. Pin 25 of U1 is used as a status indicator. The 20-MHz ceramic resonator is connected between pins 26 and 27 to establish the clock reference for the BASIC interpreter chip at U1. Pin 28 is utilized to reset the microprocessor, via S1, if the system locks up. The serial 16-kbyte EEPROM memory is coupled to the microprocessor via pins 6 and 7.

The microprocessor communicates via U3, a MAX232 serial communication chip. The MAX232 is coupled to the microprocessor through pins 8 and 9, very simply an input and an output pin. Four capacitors are all that are required to animate the MAX232 serial communication chip. These capacitors are required as a charge pump to create a minus voltage for the serial chip. The MAX232 is coupled to 9-pin serial connector for serial communication with a laptop or personal computer for programming purposes.

The Tele-Alert circuit is powered via the regulator at U5, which provides 5 V to U1, U2, and U3. A 9-Vdc "wall wart," or wall cube, power supply is used to provide power to the Tele-Alert circuit. The 9-V source is also used to provide power to optional enhancement modules.

Constructing Tele-Alert

Construction of the Tele-Alert is quite straightforward. The Tele-Alert prototype is constructed on a dual-sided glass-epoxy circuit board. Be sure the circuit board has the component side facing up toward you, as you begin placing parts into the circuit board. This is a *very* important step, so take your time and make sure you are placing the components on the correct side of the circuit board, before you begin soldering parts to the double-sided circuit board.

You can begin by placing the resistors and capacitors on the board. Be sure to observe the correct polarity when installing the capacitors or the circuit will not work correctly. Integrated circuit sockets were installed for the ICs in the event the circuit needs to be serviced at some point in time. These sockets can now be installed. Next, you can install the diodes and the LED. Once again, be sure to observe the correct polarity of the diodes when installing. Now install the ceramic resonator and the MOV, followed by the transformer and regulator U5. Please note the orientation of the semiconductors before installing them. The integrated circuits usually have either a small circular cutout or a notch at the top of the chip. Generally, pin 1 is just to the left of the circle or cutout. Next, you can install the male communication header J5, and the two optional enhancement female headers, J1 and J2. Last, install the reset switch. The Tele-Alert circuit board measures 4 by $2^1/_2$ in and can be housed in a suitable plastic enclosure. Note, the basic Tele-Alert is only about $^1/_2$ in high, but if you intend to add optional modules at a later time, you should consider an enclosure that has more height to accommodate the optional circuit boards.

Operating Tele-Alert

Operation of the Tele-Alert is simple, First you will need to connect a 9-V wall wart power supply to the circuit via the two power input pins at the top of the board. Next you will need to make up a serial communication or programming cable that connects the Tele-Alert header J5 to the serial port of your programming computer; see Fig. 19-4. Now you will need to fire up you PC and load the supplied disk into your computer. Make a directory called TELE and dump the contents of the disk into that directory. Once the communication cable has been attached, you will need to load the STAMPW.EXE program. This is the STAMP 2 editor program you will use to load the Tele-Alert program into the microprocessor. You should also see the CELL.BS2 (Listing 19-1) program, which is used to operate the Tele-Alert system. Start the STAMPW.EXE program, and then scroll through the list and highlight the Tele-Alert program, titled CELL.BS2. Press "enter" and the program should be now displayed on the screen. Scroll down the displayed program, and look for the simulated phone number and replace it with *your own phone number.* Remember, a digit 10 is programmed by a zero. If you want to dial a long-distance number, just add the full sequence of numbers. If you wish to dial out from a PBX telephone system, you will need to first put in the access number such as an 8 or 9 followed by a comma and the actual phone number. Next, press ALT-R to load the program into the Tele-Alert circuit. The program produces different sequences of tone, which are transmitted to your remote cell phone. The tone sequences can be changed for your particular needs. You can change these sequences to songs if you choose.

Your Tele-Alert is now programmed and ready to operate. At last, you can connect your event or alarm inputs to the set of solder pads at the far edge of the circuit board, just below the phone line connection (see Table 19-1). Remember, you must use normally open circuit switches for the alarm inputs unless you are using the optional motion module, which provides both normally open and normally closed inputs. Connect your phone line via the phone jack and your Tele-Alert is now ready to serve you.

Now you are ready to simulate an alarm condition, to see if your Tele-Alert functions. You can test the circuit in one of two ways. A simulated approach is to connect a crystal headphone across the secondary of T1 at J6, with the Tele-Alert disconnected from the

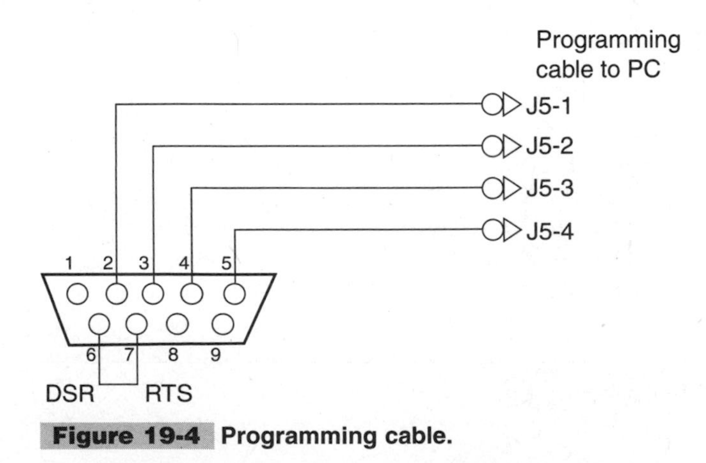

Figure 19-4 Programming cable.

```
'CELL2.BS2
'inputs go to +5 volts for activation
flash           Var   byte
new_io_state    Var   byte
old_io_state    Var   byte
call_state      Var   byte            'Set up conditions
io              Var   byte
timer           Var   byte
i               Var   byte
j               Var   word
TXpin           Con   10
LedPin          Con   15
Lamp            Con   13
Siren           Con   12
Tele            Con   9
MIC             Con   8
recall_delay    Con   120
C               Con   3000   'C - 523    'Tones used for annunciation
D               Con   2000   'D - 587
E               Con   1000   'E - 659
F               Con   500    'F - 698
G               Con   250    'G - 783
A               Con   150    'A - 880
B               Con   100    'B - 987
R               Con   0

init:                                            'Init conditions
DIRL = %00000000
DIRH = %11111111
new_io_state = %00000000
old_io_state = %00000000

for flash = 1 to 3                           'indicate status conditions
 high LedPin
 pause 1500
 low LedPin
 pause 1500
next

main:                                        'main start routine
high LedPin
new_io_state=INL
pause 100
low LedPin
pause 100
if new_io_state <> %00000000 then dial_cell
after_io1:
old_io_state = new_io_state
goto main

delay_and_scan                               'check old/new compare input
old_io_state = new_io_state
 for timer = 1 to recall_delay
 high LEDPin
 new_io_state=INL
 pause 500
```

Listing 19-1 Tele-Alert program.

```
 low LedPin
 pause 500
 if new_io_state <> old_io_state then dial_cell
 old_io_state = new_io_state
 next
goto after_io1
dial_cell:                              'diaing routine
if new_io_state = %00000000 then main
high LedPin
high Tele
high Lamp
high Siren
pause 500
dtmfout TXPin, 600, 600, [8]
dtmfout TXPin, 500, 100, [7,2,9,4,8,9,0]
sleep 10
gosub send_msg

goto delay_and_scan

send_msg:
io = new_io_state

if io.bit0=1 then gosub_chan_1          'alarm channel select
if io.bit1=1 then gosub_chan_2
if io.bit2=1 then gosub_chan_3
if io.bit3=1 then gosub_chan_4
pause 100

gosub_chan_1:
for i = 1 to 14
lookup i,[E,D,E,D,E,D,E,D,E,D,E,D,E,D],j
freqout 11,750,j,(j-1) max 32768
next
pause 200
goto fin_1

gosub_chan_2:
for i = 1 to 18
lookup i,[E,D,C,E,D,C,E,D,C,E,D,C,E,D,C,E,D,C],j
freqout 11,500,j,(j-1) max 32768
next
goto fin_2

gosub_chan_3:
for i= 0 to 24
lookup i,[E,F,G,C,E,F,G,C,E,F,G,C,E,F,G,C,E,F,G,C,E,F,G,C],j
freqout 11,250,j,(j-1) max 32768
next
goto fin_3

gosub_chan_4:
for i= 0 to 28
lookup i, [C,D,E,G,C,D,E,G,C,D,E,G,C,D,E,G,C,D,E,G,C,D,E,G,C,D,E,G],j
freqout 11,150,j,(j-1) max 32768
next
goto fin_4
```

Listing 19-1 Tele-Alert program (*Continued*).

```
fin_1:fin_2:fin_3:fin_4:
pause 200
high MIC
pause 30000
low TELE
low LEDPin
low Siren
low LAMP
low MIC
pause 100
return
end
```

Listing 19-1 **Tele-Alert program (*Continued*).**

phone line. Apply power to the circuit, and connect a normally open switch across input IN1. Activate the switch at IN1 and you should begin hearing activity at the crystal headphone. First you should hear the touch-tone sequences followed by the alarm tone sequence. If this checks out, then you can move on to a real test by connecting the Tele-Alert to the phone line and repeating the same test.

The main Tele-Alert circuit board has two 10-position female header jacks, i.e., J1 and J2, near the input terminal solder pads. These two 10-pin headers allow you to add the optional enhancement modules. The listen-in module plugs in to the first header socket near the input terminals at J1, while the temperature/voltage module (TVL) or the motion module plugs into the second header row pins J2. These optional modules are described in Chap. 22.

Alarm Configurations

The Tele-Alert system is very flexible and can be used in a number of different alarm configurations. The Tele-Alert could be used to monitor safes, doors, windows, floor mats, computers, movement, temperature and voltage changes, and smoke and fire sensors. The Tele-Alert can even monitor existing local alarms by using the Tele-Alert as a multichannel dialer. With the Tele-Alert and the motion module you can easily protect an entire vacation house or cabin, using the existing phone line. You can even monitor an overheating computer or universal power supply (UPS) failure. The Tele-Alert can be utilized in almost any alarm configuration, from multizoned alarm systems to simple multievent annunciators, with a little imagination. The Tele-Alert can also drive a siren or flashing outdoor lights to create a local "noisy" alarm if desired. The diagram in Fig. 19-5 illustrates optional local alarm connections. Relay 1 can be used to drive small loads such as an electronic Sonalert, or, with the addition of relay 2, you can be drive larger loads, such as motor sirens or flashing outdoor lamps. You can drive two external loads from OT-1 and OT-2. Output OT-3 is left for future expansion. The Tele-Alert can be used day or night with your existing phone line; no additional phone lines or phone bills are incurred.

Now, with the Tele-Alert, you can be notified of an alarm condition when you are shopping, boating, driving, golfing, traveling, or working outdoors. The Tele-Alert can dial multiple phone numbers or repeatedly dial the same number, as desired. If a second party

TABLE 19-1 INPUT/OUTPUT TERMINAL CONNECTIONS

		ALARM INPUT TERMINALS AND HEADER BLOCK—J1/J2	
J1-1	J2-1	IT-1	+9 Vdc input
J1-2	J2-2	IT-2	Ground
J1-3	J2-3	IT-3	IN1 input channel 1
J1-4	J2-4	IT-4	IN2 input channel 2
J1-5	J2-5	IT-5	IN3 input channel 3
J1-6	J2-6	IT-6	IN4 input channel 4
J1-7	J2-7	IT-7	Microphone enable
J1-8	J2-8	IT-8	Audio input mic
J1-9	J2-9	IT-9	+5 Vdc
J1-10	J2-10	IT-10	N/C
		AUXILIARY INPUTS	
J3-1	IN5	Aux. input channels 5—optional Page-Alert use	
J3-2	IN6	Aux. input channel 6—optional Page-Alert use	
J3-3	IN7	Aux. input channel 7—optional Page-Alert use	
J3-4	IN8	Aux. input channel 8—optional Page-Alert use	
		AUXILIARY OUTPUTS	
J4-1	OT-1	Aux. output channel 1—for external siren	
J4-2	OT-2	Aux. output channel 2—for external lamp	
J4-3	OT-3	Aux. output channel 3—future use	
		SERIAL COMMUNICATION—PROGRAMMING CABLE	
J5-1	RX	Pin 2 DB-9 RS-232	
J5-2	TX	Pin 3 DB-9 RS-232	
J5-3	ATN	Pin 4 DB-9 RS-232	
J5-4	GND	Pin 5 DB-9 RS-232	
		PHONE LINE CONNECTIONS	
J6-1	L1	Ring—phone line connection	
J6-2	L2	Tip—phone line connection	

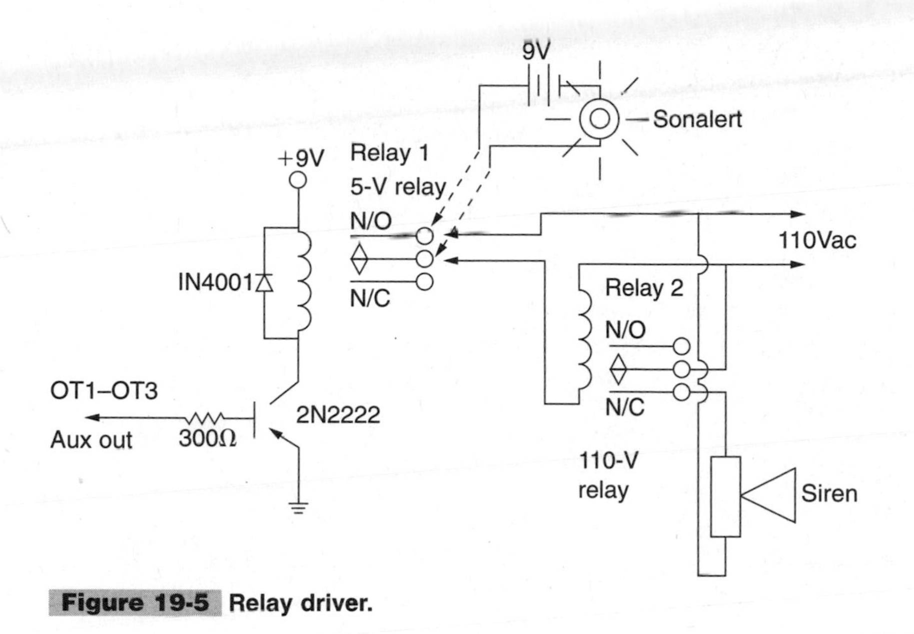

Figure 19-5 Relay driver.

is notified, the condition can be verified and then the police can be called, depending upon the severity of the alarm condition.

The Tele-Alert is always there to serve you, to free you, and to give you peace of mind! Why not build one for yourself?

Tele-Alert Parts List

R1, R2, R3, R12	4.7-kΩ 1/4-W resistor
R4, R5, R6, R7	10-kΩ 1/4-W resistor
R8, R9, R10, R11	10-kΩ 1/4-W resistor
R13	330-Ω 1/4-W resistor
R14	620-Ω 1/4-W resistor
R15	1-kΩ 1/4-W resistor
R16	150-Ω 1/4-W resistor
R17	130-V rms MOV
C1	10-μF 35-V electrolytic capacitor
C2, C7, C8, C9	0.1-μF 35-V capacitor
C3, C4, C5, C6	1-μF 35-V electrolytic capacitor
C10	0.001-μF 35-V capacitor

C11	0.1-μF 250-V Mylar capacitor
C12	47-μF 35-V electrolytic capacitor
D1	1N914 silicon diode
D2, D3	3.9-V zener diode
D4	LED
XTL	20-MHz ceramic resonator
T1	600-Ω audio transformer
J1, J2	10-position male header jacks
J3	3-position male header jack
J4, J5	4-position male header jack
U1	PIC 16C57 microprocessor
U2	24LC16B EEPROM memory
U3	MAX232 serial communication chip
U4	PVT412L MOS relay
U5	LM7805 5-Vdc regulator
Miscellaneous	PC board, wire, IC sockets

20 PAGE-ALERT

CONTENTS AT A GLANCE

The Page-Alert is a low-cost microprocessor-controlled alarm reporting system that can monitor up to four different alarm conditions and immediately report them to your numerical pager; see Fig. 20-1. The Page-Alert can be used to protect your home, office, shop, or vacation home while you are away. The Page-Alert can be configured to monitor voltage levels, temperature, movement, doors, windows, and electronic equipment problems such as computer failures; see Fig. 20-2. The Page-Alert can free you to be two places at once!

The optional enhancement modules, such as the pyroelectric motion module, can be used to sense body heat and provide an output that will activate the Page-Alert. The temperature/voltage level module can be utilized to monitor temperature fluctuations or voltage level changes, depending on the configuration, and report the problem to your pager.

The Page-Alert utilizes your existing telephone line, so it incurs no additional monthly phone bills, which are generally required by alarm companies. Simply plug in the power supply, connect it to a regular phone line, connect at least one sensor, and you ready to remotely monitor just about any alarm condition. On receiving a call from your Page-Alert, you can elect to respond yourself or call a neighbor, friend, or coworker to solve the problem. You could also elect to notify the police.

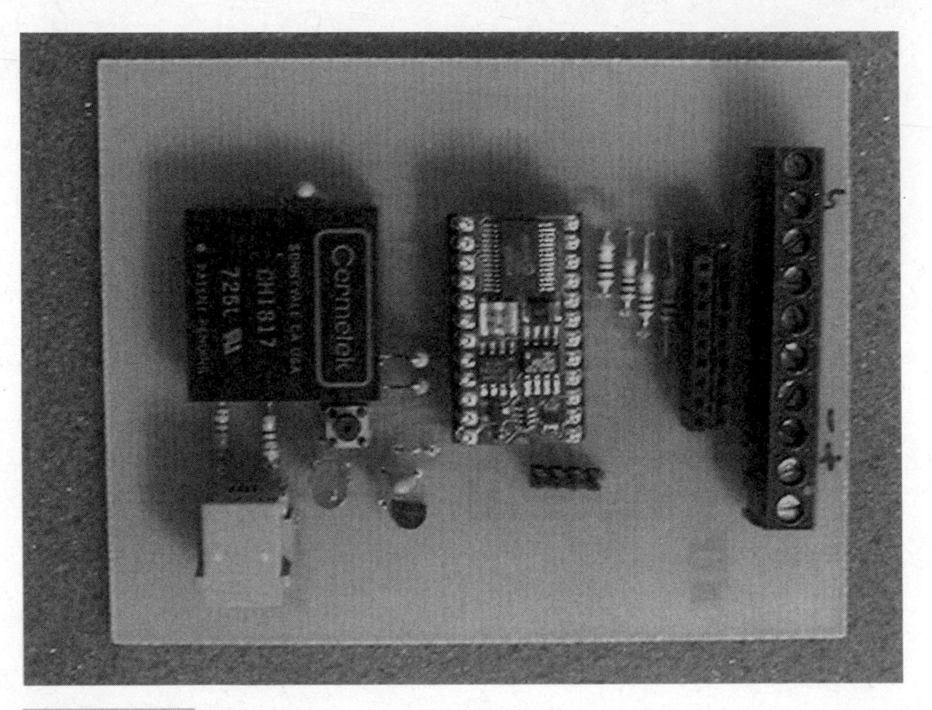

Figure 20-1 Page-Alert.

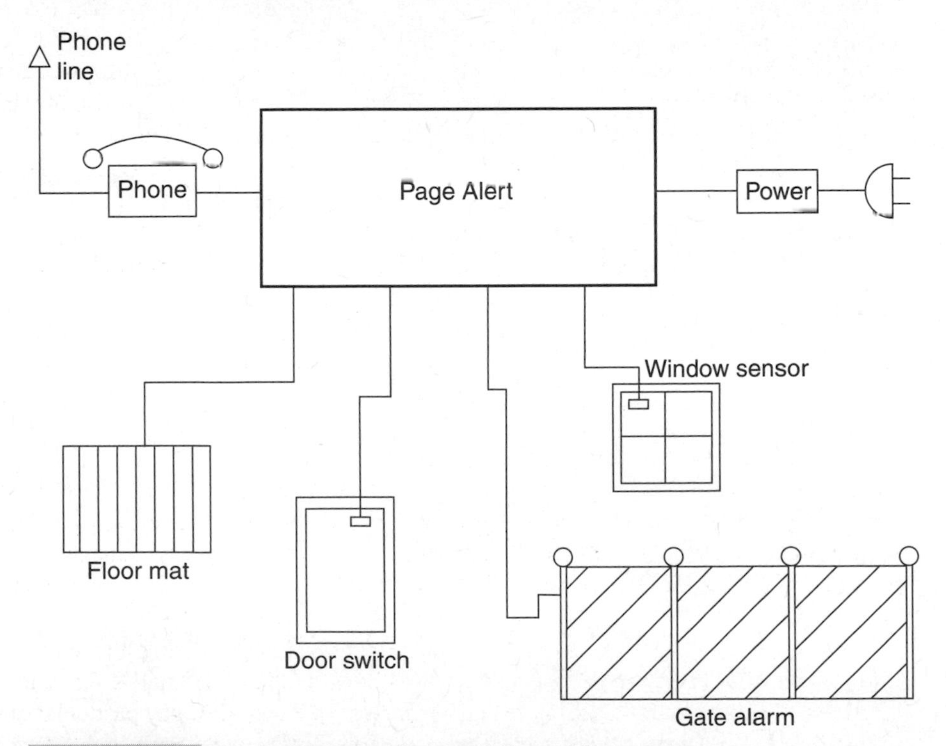

Figure 20-2 Block diagram of overall Page-Alert system.

Basic Components

The Page-Alert multichannel alarm reporting system centers around the original BASIC STAMP 2 (BS2) microcomputer at U2, as shown in Fig. 20-3. The BS2 microcomputer is a small but powerful computer capable of up to 4000 instructions per second. The BS2 microcomputer consists of main processor, memory, reset, regulators, and touch-tone/X-10 generator, all combined in a 24-pin chip carrier. The BS2 computer runs on an interpreted BASIC language. The BS2 has 16 input/output lines. Table 20-1 illustrates the pinout of the BS2 microcontroller, including the serial port connections to your programming computer. The BS2 microcomputer is generally used for specific or dedicated control application.

The Page-Alert scans up to four alarm input channels simultaneously. All four channels are configured as normally open inputs, with 10-Ω resistors across the inputs at pins P0 through P3. Table 20-2 illustrates the pin connections used on the BS2 for the Page-Alert project. The inputs to the Page-Alert are brought out to the screw terminals at J4 for easy connections to the outside world. Inputs P0 through P3 are used as input channels; P4 through P7 are not used in this project. The output of the microcontroller at P9 is utilized to drive the data access arrangement module (DAA) at M1. The DAA is the telephone-company-approved telephone interface that should be used to couple and isolate electronic circuits to the phone line; it provides the correct input/output level interfacing, relay, and protection circuits needed for interfacing (see Fig. 20-4). Once an alarm input is triggered, the Page-Alert activates pin 14, which in turn drives the OH pin in the DAA module; this allows the phone to go off hook. The microprocessor now begins the dialing sequence to call your pager.

The BS2 microprocessor contains a touch-tone generator, which is utilized to dial your pager and also generate the identifier codes. Pin 15 or P10 on the BS2 is used to drive the audio signal from the touch-tone output to the TX(+) pin on the DAA, which dials the phone. The TX(−) pin on the DAA is connected to ground via a 0.1-μF disk capacitor. Pin 7 of the DAA is the 5-V power connection, while pin 9 is the ground connection. The RCV and RI pins on the DAA are not used here. Pins 1 and 2 of the DAA are connected to the phone line's ring and tip lines via the Sidactor protection device, followed by two 100-Ω resistors, which are used to couple the DAA to the RJ11 phone plug.

A reset function is provided at pin 22 of U2 and is connected via R5 to pin 21 (V_{dd}). This pin is a brownout detector and reset device. A reset pushbutton is connected between pin 22 and ground. A bypass capacitor is coupled across the power leads at pins 23 and 24. The power input on pin 24 can accept 5 to 15 Vdc, which powers the internal regulators in the BS2. Pin 21 is the 5-V system power pin from the regulator.

The regulator at U3 is used to power the Page-Alert system. A 12- to 15-Vdc "wall wart" power cube can be used to provide input power to the Page-Alert board at J4-1. The regulator provides 5 V to pin 21 of the BS2 controller at U2. This 5-V source is also utilized to provide power to the sensor daughter boards, which can be plugged into the female header at J3.

Construction of the Page-Alert is quite straightforward, utilizing the circuit layout design provided. The Page-Alert prototype was constructed on a small 2- by 4-in single-sided glass-epoxy circuit board. When constructing the Page-Alert pay particular attention to the polarity of the capacitors, and semiconductors. Integrated circuit sockets are highly

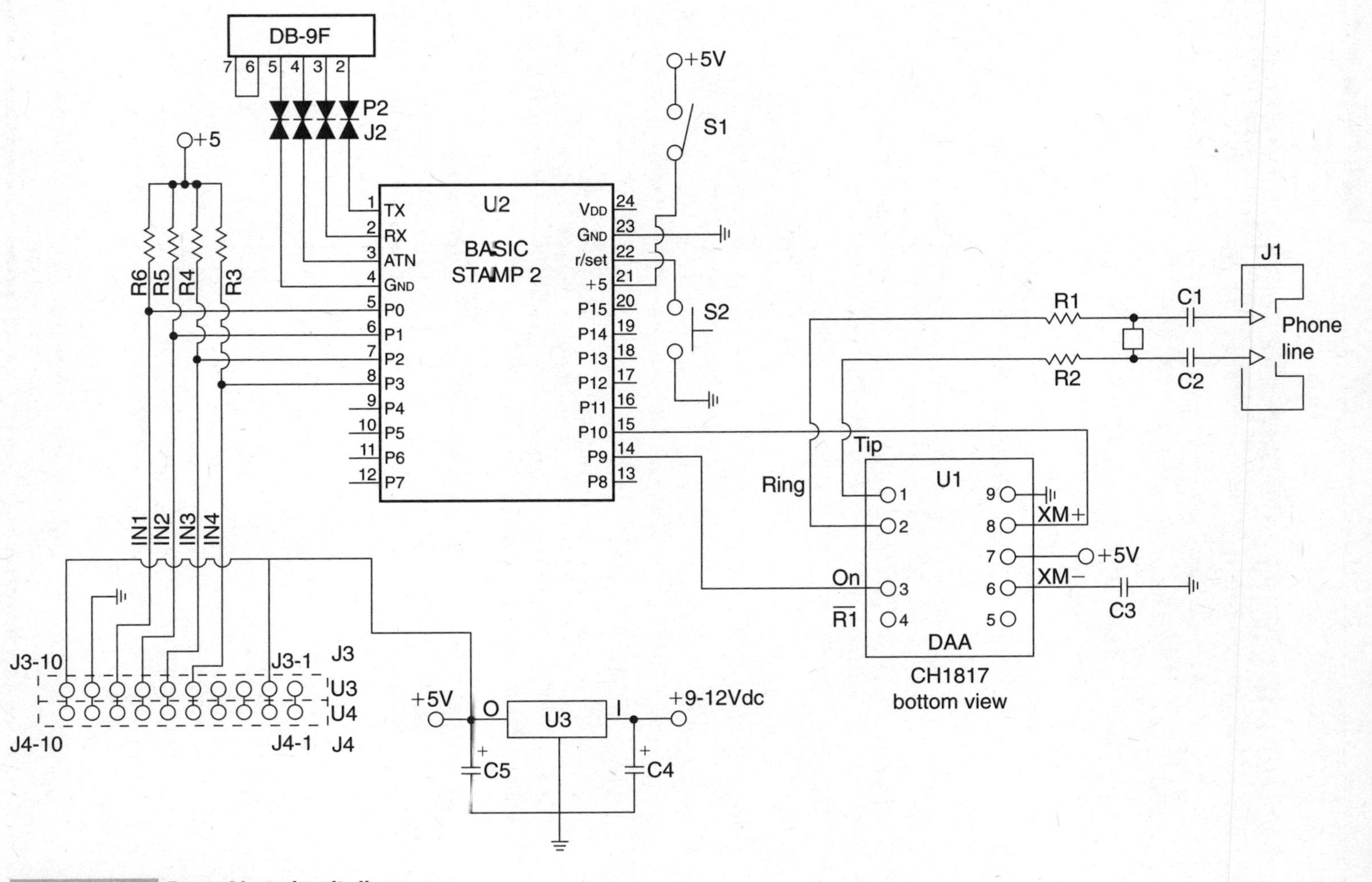

Figure 20-3 Page-Alert circuit diagram.

TABLE 20-1 STAMP 2 MICROPROCESSOR PINOUTS

PIN	NAME	FUNCTION	DESCRIPTION
1	SOU	Serial out	Temporarily connects to PC's Rx
2	SIN	Serial out	Temporarily connects to PC's Tx
3	ATN	Attention	Temporarily connects to PC's DTR
4	VSS	Ground	Temporarily connects to PC's ground
5	P0	User I/O 0	User ports that can be used for inputs or outputs
6	P1	User I/O 1	
7	P2	User I/O 2	
8	P3	User I/O 3	*Output mode:* Pins will source from VDD. Pins should not be allowed to source more than 20 mA or sink more than 25 mA. As groups, P0–P7 and P8–P15 should not be allowed to source more than 40 mA, or sink 50 mA.
9	P4	User I/O 4	
10	P5	User I/O 5	
11	P6	User I/O 6	
12	P7	User I/O 7	
13	P8	User I/O 8	*Input mode:* Pins are floating (less than 1-μA leakage).
14	P9	User I/O 9	The 0/logic threshold is approximately 1.4 V.
15	P10	User I/O 10	
16	P11	User I/O 11	*Note:* To realize low power consumption during sleep, make sure that no pins are floating, causing erratic power drain. Either drive them to VSS or VDD, or program them as outputs that don't have to source current.
17	P12	User I/O 12	
18	P13	User I/O 13	
19	P14	User I/O 14	
20	P15	User I/O 15	
21	VDD	Regulator out	Output from 5-V regulator (V_{in} powered). Should not be allowed to source more than 50 mA, including P0–P15 loads.
		Power in	Power input (V_{in} not powered). Accepts 4.5–5.5 V. Current consumption is dependent on run/sleep mode and I/O.
22	RES	Reset I/O	When low, all I/Os are inputs and program execution is suspended. When high, program executes from start. Goes low when VDD is less than 4 V or ATN is greater than 1.4 V. Pulled to VDD by a 4.7K resistor. May be monitored as a brownout/reset indicator. Can be pulled low externally (i.e., button to VSS) to force a reset. Do not drive high.
23	VSS	Ground	Ground. Located next to V_{in} for easy battery backup.
24	VIN	Regulator in	Input to 5-V regulator. Accepts 5.5 to 15 V. If power is applied directly to VDD, pin may be left unconnected.

TABLE 20-2 PAGE-ALERT PINOUTS USED

PIN	FUNCTION	DESCRIPTION
1	SOUT	Serial output programming pin
2	SIN	Serial input programming pin
3	ATN	Serial DTR line programming pin
4	VSS	Ground programming pin
		INPUTS
5	P0	IN1 alarm input 1
6	P1	IN2 alarm input 2
7	P2	IN3 alarm input 3
8	P3	IN4 alarm input 4
		OUTPUTS
14	P9	OH DAA relay drive
15	P10	TX DAA audio output pin
20	P15	LED status
21	VDD	+5 V from regulator
23	VSS	System ground
24	Vin	5–15-V input power (not used in project)

recommended for this project in the event of a component failure at a later date. Observe the phone line connector and the screw terminal strip at one edge of the circuit board. Note the 10-position screw terminal strip at J4, which is connected to a 10-pin female header at J3. The 10-pin header J3 is used to accept the motion module or the temperature voltage level (TVL) add-on modules, or other expansion devices. Table 20-3 illustrates the pinouts for the screw terminal strip and the 10-pin header connections. Once the Page-Alert has been completed, recheck the component placement and your solder connections. Now you are now ready to power your Page-Alert unit. Locate a 12–15-Vdc power supply to power your Page-Alert. You can readily elect to utilize the ubiquitous 12-Vdc wall wart to power your Page-Alert; usually these power supplies are quite economical. The Page-Alert prototype is housed in an economical 4- by 5-in Pactec enclosure.

Programming the Page-Alert

The Page-Alert is initially programmed via pins 1 through 4. These pins are the serial input/output connections used to program the BS2. The RX line is shown at pin 1, while the TX line is at pin 2. A DTR line is provided at pin 3, and the ground is at pin 4.

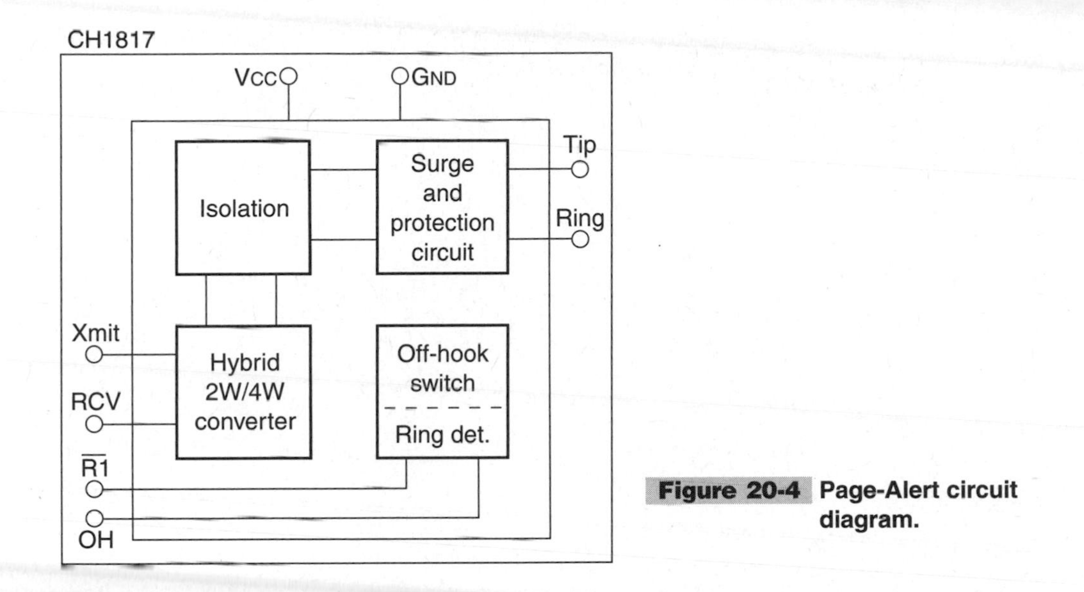

Figure 20-4 Page-Alert circuit diagram.

TABLE 20-3 ACCESSORY DAUGHTER BOARD I/O CONNECTIONS

J3—10-PIN HEADER—POWER AND I/O J4—10-PIN HEADER—SENSOR DAUGHTER BOARDS	
PIN	**PINOUT DESCRIPTION**
1	+5-V power pin
2	System ground
3	IN1 input pin
4	IN2 input pin
5	IN3 input pin
6	IN4 input pin
7	No connection
8	No connection
9	+5-V power
10	No connection

To program your Page-Alert, you simply apply power to the screw terminals and connect the Page-Alert via the serial cable to your personal computer and run the STAMP 2 editor program called STAMPW.EXE, provided on the CD-ROM supplied. The CD-ROM contains the editor program and the Page-Alert program. Once the STAMPW.EXE program is running, you simply load the PALERT.BS2 program (Listing 20-1) by pressing Alt-L. Once the PALERT.BS2 program appears on your computer screen, you will have to

```
'PALERT.BS2
' PAGE-ALERT
'identify variables
flash           VAR   byte
new_io_state    VAR   byte
old_io_state    VAR   byte
call_state      VAR   byte
io              VAR   byte
timer           VAR   byte

'identify constants
OHPin          CON   9
TxPin          CON   10
Siren          CON   12
Lamp           CON   13
LEDPin         CON   15
recall_delay   CON   120

'initialize variables and program
init:
DIRL = %00000000
DIRH = %11111111
new_io_state = %00000000
old_io_state = %00000000

for flash = 1 to 3               'status indicator
 high LEDPin
 pause 1500
 low LEDpin
 pause 1500
next

main:                            'main Page-alert routine
high LEDPin
new_io_state = INL
pause 100
low LEDPin
pause 100
if new_io_state <> %00000000 then dial_pager
after_io1:
old_io_state = new_io_state
goto main

delay_and_scan:                           'check old/new input status -compare
old_io_state = new_io_state
for timer = 1 to recall_delay
 high LEDPin
 new_io_state = INL
 pause 500
 low LEDPin
 pause 500
 if new_io_state <> old_io_state then dial_pager
 old_io_state = new_io_state
next
goto after_io1
```

Listing 20-1 Page-Alert program.

```
dial_pager:                                         'dial pager routine
if new_io_state = %00000000 then main
high LEDPin
high OHPin
high Siren
high Lamp
pause 500

dtmfout TxPin,500, 500, [8]
pause 200
dtmfout TxPin,400, 200, [7,9,9,6,6,5,8]
sleep 6
gosub send_msg

goto delay_and_scan                                 'turn alarm off and recycle
send_msg:
io = new_io_state
'debug ibin8 io, cr
dtmfout TxPin, 300, 200, [5,5,io.bit0,io.bit1,io.bit2,io.bit3,11]
pause 1000
low OHPin
low LEDPin
pause 30000
low Siren
low Lamp
pause 100
return
end
```

Listing 20-1 **Page-Alert program (*Continued*).**

enter your pager's number in the first phone number position and then you must enter the second number, or identifier, into the program. Next, press Alt-R to load/run the program in the BS2 microcomputer, and the Page-Alert is now ready to serve you. Once the program is loaded into the BS2 it remains there even if power is removed.

Using the Page-Alert

Now you are ready to utilize the Page-Alert. Connect a normally open alarm sensor to any one of the four input terminals and the 5-V terminal on the screw terminal. You can use any type of normally open type of alarm switch or sensor. Door or window switches could be used as well as any other sensor with a normally open set of contacts. Photoelectric, pyroelectric, or pressure sensors could also be utilized to trigger the Page-Alert, as well as the optional TVL and motion module daughter boards; see Chap. 22, "Input Sensor Modules."

In order to test-activate the Page-Alert, you could substitute a normally open pushbutton to start the Page-Alert unit. Once activated, the Page-Alert should come to life! First the status lamp will begin to flicker to indicate the program has started. Next, the OH line on the DAA is activated and the phone line goes off hook. Next the microprocessor begins touch-tone dialing your pager's phone number. The microprocessor then waits for a short interval, and the triggered alarm channel's ID, or identifier, is sent to your numerical pager.

Once alerted via your pager, you can respond yourself or you can elect to call a friend, neighbor, coworker, or even the police, depending on the severity of the problem. The Page-Alert uses your existing phone line, so there are no additional phone cost to use it. The Page-Alert can be used as a self-contained silent alarm using the motion module or TVL board or up to four alarm loops, consisting of normally open switch sensors. The Page-Alert also be used with an alarm controller box; in this way the Page-Alert becomes the phone dialer for up to four channels. A multizoned alarm system can be used to identify a particular channel or you can use the Page-Alert for both fire and burglar alarm applications at once.

The Page-Alert is a low-cost means to free you and alert you to intruders, equipment failures, or impending doom. Build one and have some fun!

Page-Alert Parts List

R1	240-Ω 1/4-W resistor
R2	5-kΩ trim pot
R3, R4, R5, R6	10k-Ω 1/4-W resistor
R7	330-Ω 1/4-W resistor
R8	4.7k-Ω 1/4-W resistor
R9, R10	100-Ω 1/4-W resistor
C1	1-μF 35-V electrolytic capacitor
C2	10-μF 35-V electrolytic capacitor
C3, C4	0.1-μF 35-V disk capacitor
D1	LED
D2	Sidactor P3002AB-ND
U1	LM7805 regulator (5 V)
U2	BS2 microcontroller (original)
M1	CH1817-D DAA module
S1	Momentary pushbutton switch
J1	RJ11 PC-mount telephone jack
J2	4-pin programming header—male
P2	4-pin programming header—female
J3	10-pin female header
J4	10-position screw terminal

21

DATA-ALERT

CONTENTS AT A GLANCE

The Data-Alert is a unique, low-cost, multichannel microprocessor-controlled remote event/alarm reporting system. It will immediately send an alarm notification or X10 commands via telephone to its Data-Term alarm display unit or a computer with a standard dial-up modem. The main Data-Alert board is shown in Fig. 21-1, and the modem board is depicted in Fig. 21-2. The Data-Alert can monitor up to four different alarm or event conditions from a host of different types of sensors and report the particular channel that was activated. The Data-Alert can be configured to monitor voltage levels, temperature changes, movement, contact closures, windows, doors, safes, and perimeters as well as computer equipment. The Data-Alert can be utilized to protect your home, office, shop, or vacation home.

The Data-Alert is used with your existing telephone line; therefore, no additional charges are added to your monthly phone bill. The Data-Alert is programmed to call you on your Data-Term display unit and display and/or perform commands. The Data-Alert can be expanded to monitor up to eight different input event or alarm conditions.

The optional motion module will allow the Data-Alert to detect moving persons in a protected space, while the optional temperature/voltage level module allows your Data-Alert to monitor up to four temperature or voltage level presets, and report the changes to your Data-Term.

Basic Components

The Data-Alert multichannel alarm reporting system begins with the main circuit diagram shown in Fig. 21-3. The heart of the Data-Alert is the PIC 16C57 chip, which emulates the BASIC STAMP 2 computer. The PIC 16C57 is preloaded with a BASIC interpreter much like the Parallax BASIC STAMP 2 (BS2). The PIC 16C57, however, is a much less expensive approach to solving a problem, but requires a few more external parts than the BASIC STAMP 2. The PIC 16C57 requires very little in the way of support chips to make it function. The PIC 16C57 basically requires only two support chips, a 24LC16B EEPROM for storage and a MAX232 for communication. Only a few extra components are needed to form a functional microprocessor with serial input/output. A 20-MHz ceramic resonator, a

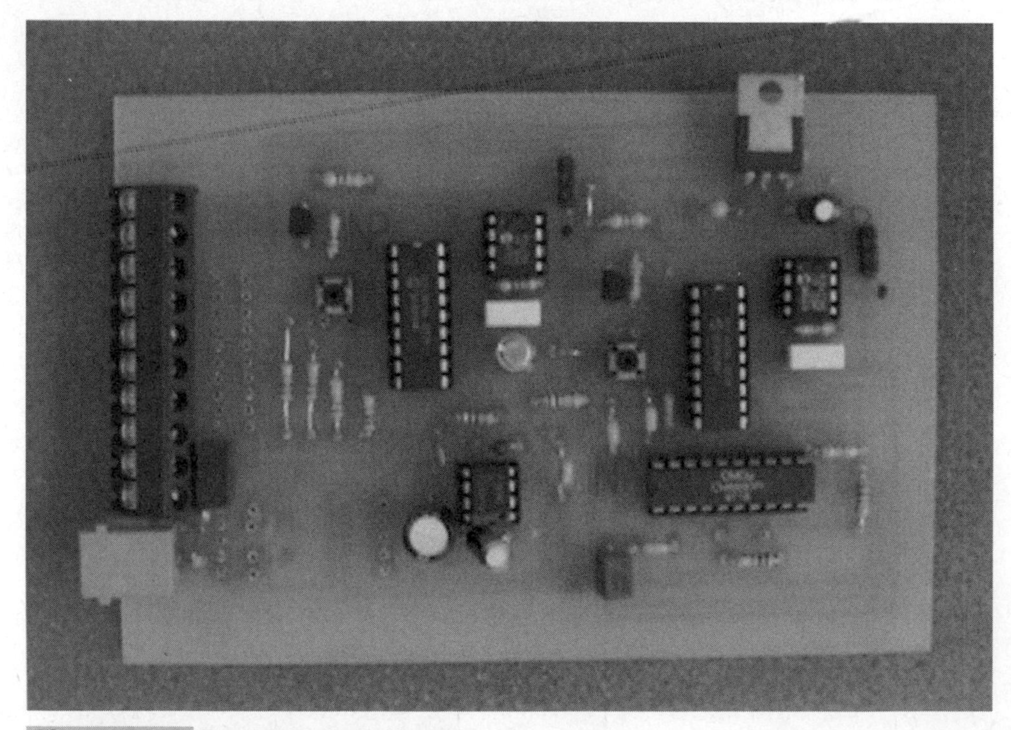

Figure 21-1 Data-Alert main board.

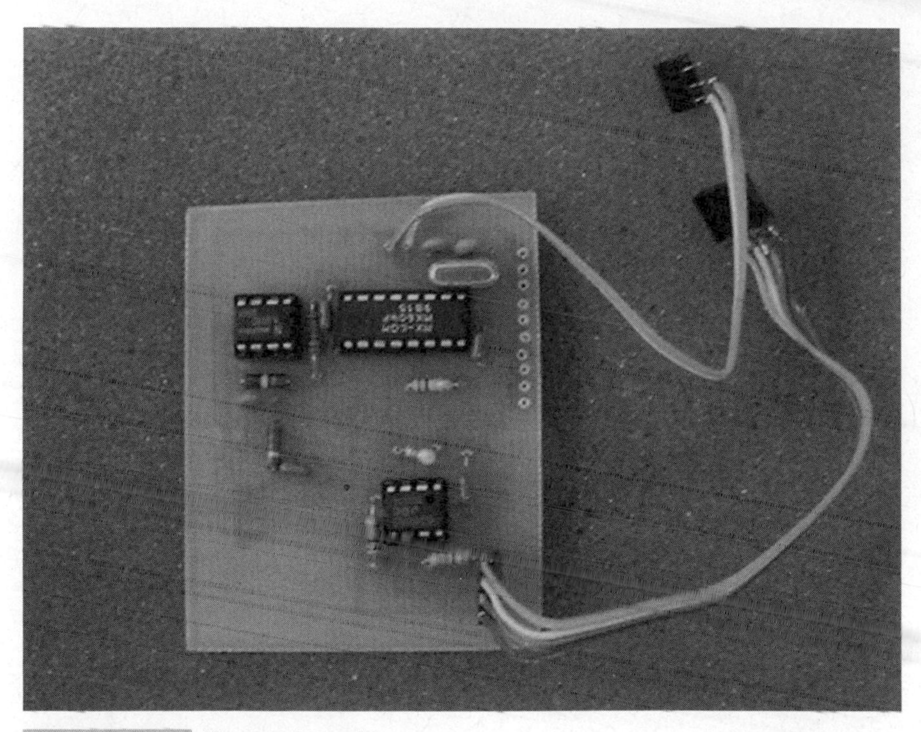

Figure 21-2 Modem board.

few resistors, and a diode are all that are needed to use the processor in its most simple form. The 28-pin PIC 16C57 is a very capable and versatile little microprocessor that can perform many tasks.

In order to turn the BS2 look-alike into a Data-Alert project, a few more parts must be added to the basic configuration. First, you need to configure pins 10 through 17 as inputs, for a total of eight inputs, that can be used to sense events or alarm conditions. The first four input pins are brought out to the terminal block for easy access. These first four inputs, IN1 through IN4, are used as inputs to the microprocessor. The second four input pins are IN5 through IN8; note the resistors placed across each of the input pins. Pins 18 through 25 are configured as output pins. Pin 19 is used to drive the solid-state relay at U4. The microprocessor activates the LED in U4, which in turn closes the relay contacts at pins 4 and 6 of the relay, essentially shorting out blocking capacitor C11 and allowing the phone to go on hook and begin dialing your cell phone or pager. Pins 20 and 21 of U1 are audio output pins. Pin 20 of U1 outputs the touch-tone signals needed to dial the phone when an alarm condition is sensed, while pin 21 is used to output the tone sequences that are used to indicate via your cell phone which event/alarm channel has been activated. The capacitors at C9 and C10 are used to couple the touch-tones and the alarm sequence tones to the coupling transformer network formed by T1 and the associated components. Capacitor C10 is placed across T1, while zener diodes D2 and D3 are used as voltage clamps diodes at the input of the transformer.

Transformer T1 is used to couple the audio from the Data-Alert to the phone line, once the unit has been triggered. Resistor R9 is used to couple one side of the transformer to the

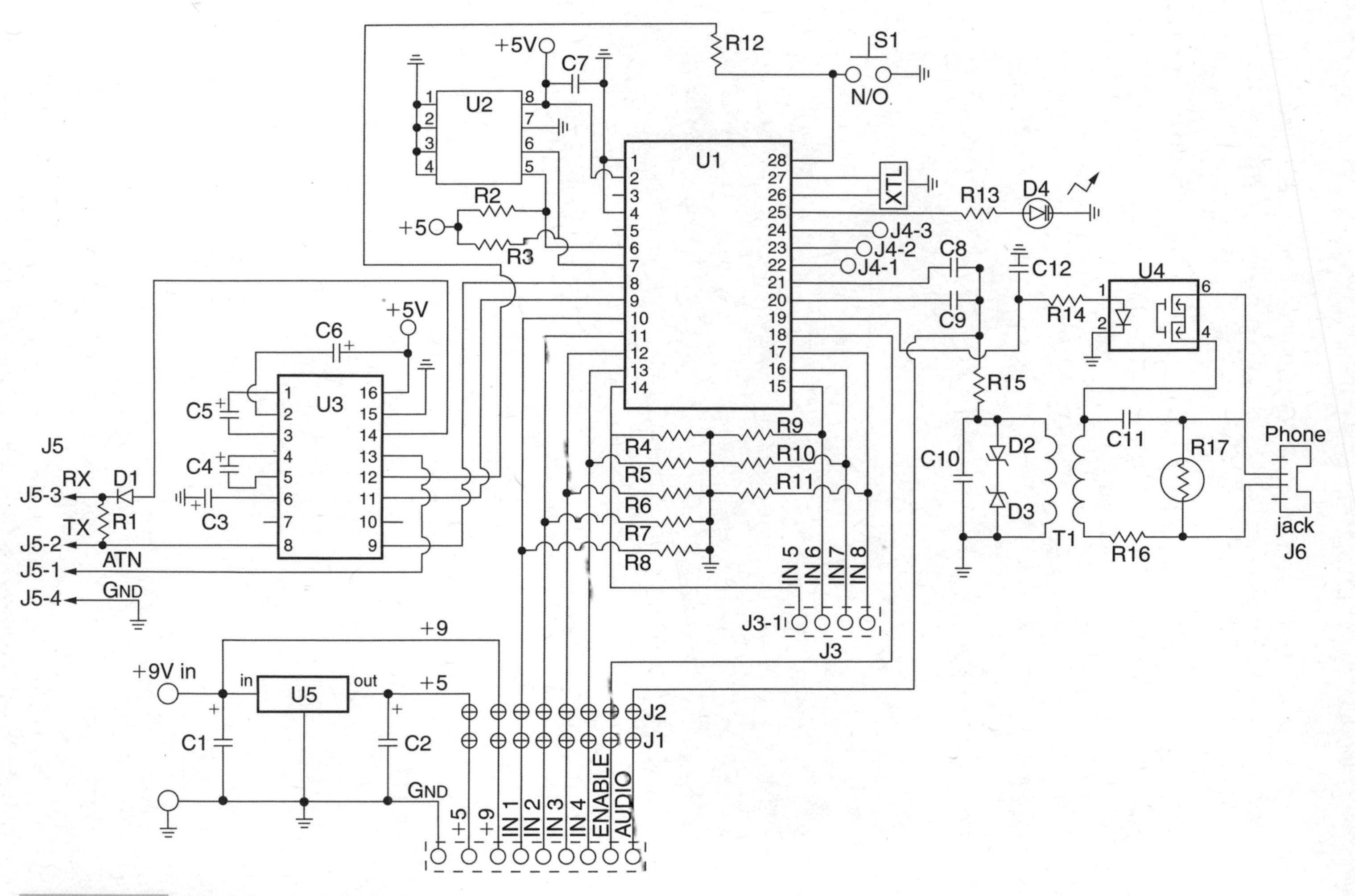

Figure 21-3 Data-Alert main board.

phone line at J6, and also acts to hold the phone line. The other secondary transformer lead is coupled to the phone line via C11. C11 is blocking capacitor that keeps the phone off hook until an alarm condition occurs. The solid-state relay at U4 acts to short out C11 when an event or alarm condition occurs, thus coupling the T1 to the phone line. Resistor R16 is a metal oxide varistor (MOV) and is used to protect the circuit from high-voltage spikes. Pins 22 and 23 may be utilized as auxiliary outputs to drive local alarm sirens or outdoor lamps if desired, while pin 24 is left for further expansion. Pin 25 of U1 is used as a status indicator. The 20-MHz ceramic resonator is connected between pins 26 and 27 to establish the clock reference for the BASIC interpreter chip at U1. Pin 28 is utilized to reset the microprocessor via S1 if the system locks up. The serial 16-kbyte EEPROM memory is coupled to the microprocessor via pins 6 and 7.

The microprocessor communicates via U3, a MAX232 serial communication chip. The MAX232 is coupled to the microprocessor through pins 8 and 9, very simply an input and output pin. Four capacitors are all that are required to animate the MAX232 serial communication chip. These capacitors are required as a charge pump to create a minus voltage for the serial chip. The MAX232 is coupled to 9-pin serial connector for serial communication with a laptop or personal computer for programming purposes.

The Data-Alert circuit is powered via the regulator at U5, which provides 5 V to U1, U2, and U3. A 9-Vdc "wall wart" or wall cube power supply is used to provide power to the Data-Alert circuit. The 9-V source is also used to provide power to optional enhancement modules.

The Data-Alert can also drive a siren or flash outdoor lights to create a local "noise" alarm if desired. The diagram in Fig. 21-4 illustrates optional local alarm connections. Relay 1 can be used to drive small loads such as an electronic Sonalert, or with the addition of Relay 2 you can be drive larger loads, such as motor sirens or flashing outdoor lamps. You can drive two external loads from OT-1 and OT-2. Output OT-3 is left for future expansion.

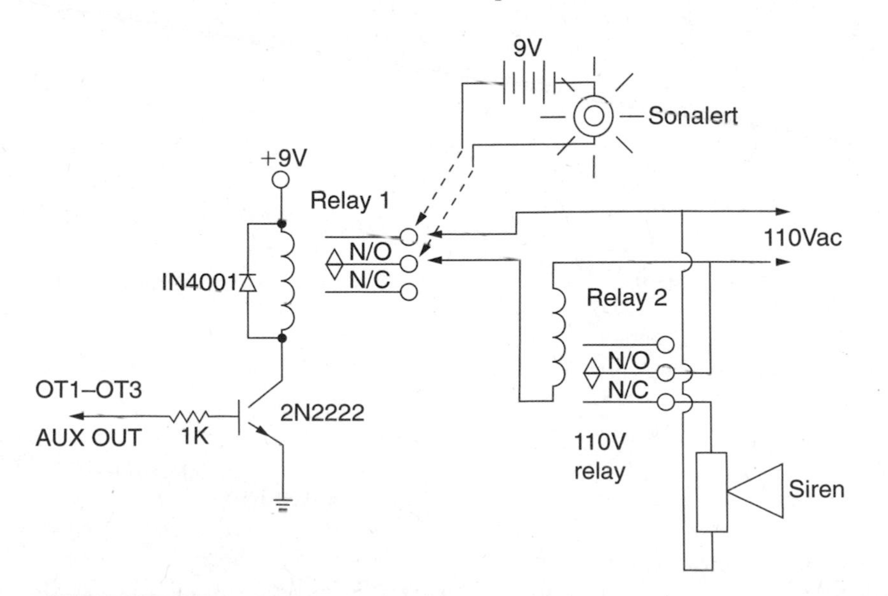

Figure 21-4 **High-power relay driver.**

Construction of the Data-Alert Board

Construction of the main Data-Alert board is quite straightforward. The Data-Alert prototype is built on a dual-sided glass-epoxy circuit board. Be sure the circuit board is facing you with the component side toward you as you begin placing parts into the circuit board. This is a *very* important step, so take your time and make sure you are placing the components on the correct side of the circuit board, before you begin soldering parts to the double-sided circuit board. You can begin by placing the resistors and capacitors on the board. Be sure to observe the correct polarity when installing the capacitors or the circuit will not work correctly. Integrated circuit sockets are installed for the ICs in the event the circuit needs to be serviced at some point in time. These sockets can now be installed. Next, you can install the diodes and the LED. Once again, be sure to observe the correct polarity of the diodes when installing. Now install the ceramic resonator and the MOV, followed by the transformer and regulator U5. Please note the orientation of the semiconductors before installing them. The integrated circuits usually have either a cutout or notch taken at the top of the chip or a small circle cutout at the top of the chip. Generally, pin 1 is just to the left of the circle or cutout. Next, you can install the male communication header J5 and the two optional enhancement female headers J1 and J2. Last, install the reset switch. The Data-Alert circuit board measures 4 by $2^1/_2$ in and can be housed in a suitable plastic enclosure. Note that the basic Data-Alert is only about $^1/_2$-in high, but if you intend to add optional modules at a later time, you should consider an enclosure that has more height to accommodate the optional circuit boards. Table 21-1 illustrates the Data-Alert main board pinout diagram.

Modem Module Board

The modem module allows your Data-Alert to report alarm conditions to the Data-Term display terminal. Each modem module uses a unique code that distinctly identifies itself to the your Data-Term unit. The modem module can also be used to report the exact time of the triggered event. Your Data-Term display unit can then log the various events from different Data-Alert systems located in different locations. The Data-Term display terminal also permits immediate alarm notification via X-10 ac power modules. Various sirens, lamps, etc., can be activated with the X-10 system. The Data-Alert is designed to monitor four different alarm channels or conditions such as motion monitors, temperature sensors, and door switches. Both the Data-Alert and Data-Term display terminal can be expanded to monitor more channels by modifying the software.

The modem module is a small add-on daughter board that plugs into the main Data-Alert system board. The modem module consists of three integrated circuits. The heart of the modem module is the MX604 1200-baud modem chip, which is animated by the BS2 controller in the Data-Alert. The second major component is the NJ6355 real-time clock chip, which is used to send the exact time of the event to the remote Data-Term display terminal, and an op-amp chip to amplify the outgoing signal from the modem chip. The MX604 modem chip requires only a few discrete components for operation. The modem chip is a full frequency-shift keying (FSK) modem with transmit and

TABLE 21-1 DATA-ALERT INPUT/OUTPUT CONNECTIONS

ALARM INPUT TERMINALS AND HEADER BLOCK—J1				
IT-1	J1-1	J2-1	+9-Vdc input	
IT-2	J1-2	J2-2	Ground	
IT-3	J1-3	J2-3	IN1 input channel 1	
IT-4	J1-4	J2-4	IN2 input channel 2	
IT-5	J1-5	J2-5	IN3 input channel 3	
IT-6	J1-6	J2-6	IN4 input channel 4	
IT-7	J1-7	J2-7	Microphone enable	
IT-8	J1-8	J2-8	Audio input mic	
IT-9	J1-9	J2-9	+5-Vdc output	
IT-10	J1-10	J2-10	N/C	
AUXILIARY INPUTS				
J3-1	IN5 aux. input channel 5			
J3-2	IN6 aux. input channel 6			
J3-3	IN7 aux. input channel 7			
J3-4	IN8 aux. input channel 8			
AUXILIARY OUTPUTS				
J4-1	OT1 aux. output channel 1	For external siren		
J4-2	OT2 aux. output channel 2	For external lamp		
J4-3	OT3 aux. output channel 3	Future use		
SERIAL COMMUNICATION				
J5-1	RX	Pin 2	RS232	DB-9S
J5-2	TX	Pin 3	RS232	DB-9S
J5-3	ATN	Pin 4	RS232	DB-9S
J5-4	GND	Pin 5	RS232	DB-9S
J6-1	L1	Phone line connection		
J6-2	L2	Phone line connection		

receive capabilities, but without the controller of a conventional modem. The MX604 contains an FSK modulator and demodulator as well as filters, buffers, and timing and logic control functions. The diagram in Fig. 21-5 illustrates the block diagram of the of the MX604 modem chip. The MX604 pinout diagram is shown in Fig. 21-6, and Table 21-2 depicts the pinout list for the MX604.

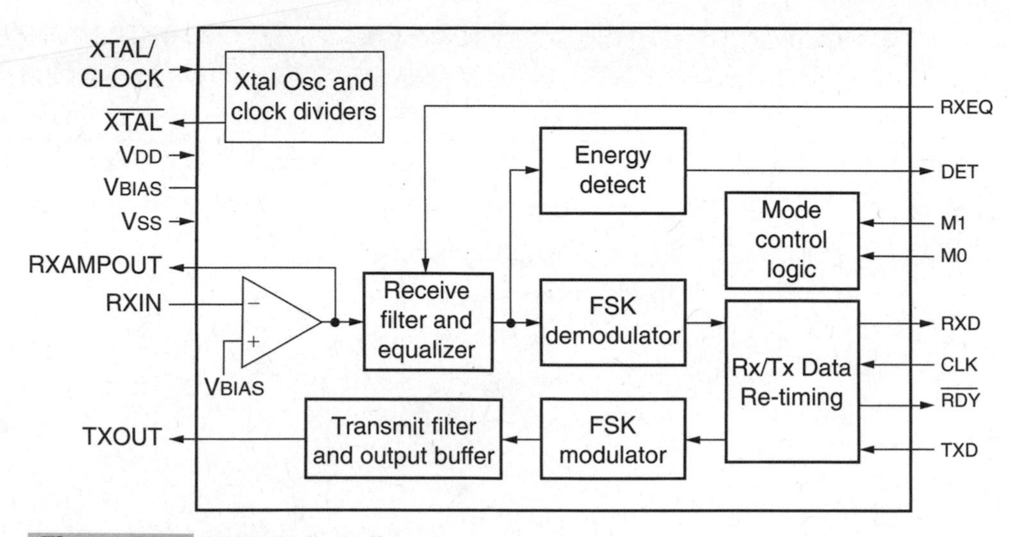

Figure 21-5 MX604 block diagram.

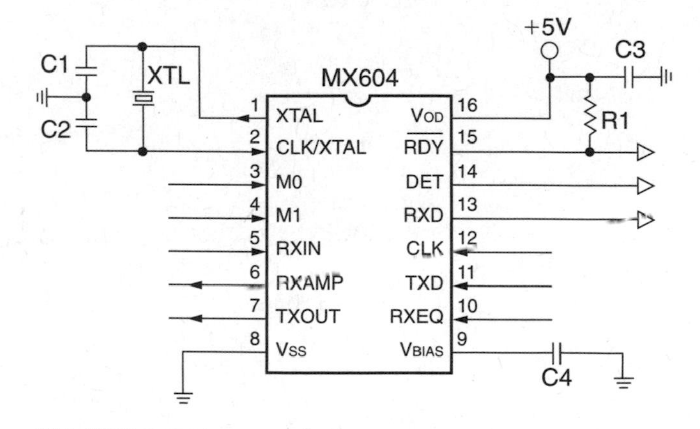

Figure 21-6 MX604 modem chip.

The ubiquitous 3.579-MHz color burst crystal at X2 is the brain of the modem chip that controls all the timing functions. The crystal is connected to pins 1 and 2 of the MX604, with two 18-pF capacitors connected across the crystal with a tap to ground. A bias capacitor is connected from pin 9 to ground, and a second 0.01-μF bypass capacitor is connected from $+V_{dd}$ to ground. Resistor R3 is connected from pin 15 to $+V_{dd}$ on pin 16. The MX604 chip as well as the real-time clock chip at U1 are powered from the 5-V regulator on board the main Data-Alert system board. Power to the modem is applied to pins 12 and 16, while ground is applied to pin 8.

The alarm data, i.e., ASCII alarm information, from each Data-Alert channel is fed to the modem on the TXD pin 11. The FSK-modulated output signal is sent out of the modem on the TXOUT pin 7. The FSK output is next fed to R4, which is coupled to U3, which acts as an output amplifier and signal conditioner. The output from U3 is fed first to R6

TABLE 21-2 MX-604 MODEM PINOUTS

PIN	NAME	DESCRIPTION
1	$\overline{\text{XTAL}}$	Output of the crystal oscillator inverter
2	CLK/XTAL	Input to the crystal oscillator inverter
3	M0	A logic input for mode setting
4	M1	A logic input for mode setting
5	RXIN	Input to the RX input amplifier
6	RXAMPOUT	Output of the RX input amplifier
7	TXOUT	Output of FSK generator
8	VSS	Negative supply (ground)
9	VBIAS	Internal bias voltage held at VDD/2
10	RXEQ	An input to enable/disable equalizer
11	TXD	An input for FSK modulator or retiming
12	CLK	An input, used to clock in/out bits
13	RXD	Output from FSK demod or retimed data
14	DET	An output of on-chip energy detector
15	$\overline{\text{RDY}}$	Ready for data, used for retiming
16	VDD	Positive voltage pin, decoupled to GND

and then to C7. The output from C7 presented at J1-8 is finally coupled to the output transformer on the Data-Alert board.

Pins 3 and 4 of the MX604 are the logic control pins, M0 and M1, respectively. When the modem is first energized, it must be set up in the "zero-power" mode before normal operation begins. The zero-power mode is first configured by setting both M0 and M1 to 1, by the microprocessor on the main Data-Alert board when the program is first run. The BS2 microprocessor handles all initial zero-power mode setup and, subsequently, the logic control of the MX604, to next place the MX604 in the TX mode. Once in the TX mode the modem chip is ready to receive alarm data information. When the modem is ready for regular operation, M1 must be set for 0 and M0 is set for 1. The STAMP 2 (BS2) is used to first set up the zero-power mode during initialization; the BS2 then switches to the TX mode in order to send out the serial information to the alarm central terminal. In this application only the TX mode is used.

Real-Time Clock Chip

Up to this point we have talked only about the modem chip and not the real-time clock chip. The NJ6355 real-time clock chip at U1 performs the important time reporting function on the modem module. The real-time clock is used to report the actual time of the triggered event to the remote Data-Term data terminal. The real-time clock requires only a few external components to operate. The 32,768-Hz crystal is fundamental to the operation of U1. Resistor R1 is used to protect the data output line, while resistor R2 is used hold U1 inactive while the microprocessor resets. Bypass capacitor C1 is connected from the pin 8 to ground. Both the MX604 modem and real-time clock chips are shown in Fig. 21-7, which shows the modem board.

Modem Module Operation

The operation of the modem module is quite straightforward. Once the Data-Alert has been triggered by an event, the Data-Alert board begins dialing the phone number programmed. The Data-Alert waits for a number of seconds to establish the phone connection. Once the connection has been made, the Data-Alert again waits a few more seconds and then begins sending a single FSK tone for 10 seconds followed by the alarm message of the particular channel that was activated. Once the alarm message has been sent, the microcontroller disconnects the telephone line and the system is reset. Now the Data-Alert is ready for the next event, and one complete alarm cycle has been completed.

The diagram in Fig. 21-8 depicts the modem module connection to the main Data-Alert circuit board. The modem module has a 10-pin male header, which is used for the main connections between the Data-Alert and the modem board. Note that the modem module plugs into the right or outside row of female header pins on the main Data-Alert board. The modem module also has some outboard connects. On the top left side of the modem board, above the 10-pin header, is a 4-pin header. This 4-pin header connects the clock chip at U1 to the auxiliary input header (AUX) on the main Data-Alert board. Pin J3-1 of the modem module goes to pin J3-1 on the main Data-Alert board and so on, as shown. The modem control lines M0 and M1 at the bottom of the modem module board are connected to the J4-3 and J4-2, respectively. Power for the modem module is obtained from the main Data-Alert board.

Your modem module is now almost ready. Once connected to the main Data-Alert board, you are now ready to test the Data-Alert with the modem module. Connect the programing cable to the Data-Alert and apply power to the board. Next, load the DATA.BS2 program (Listing 21-1) to the Data-Alert. Connect the Data-Alert to the phone line, pick up the phone, and trigger one of the Data-Alert input channels. You should be hearing the Data-Alert trying to dial the phone number programmed into the Data-Alert. After about 10 seconds you should hear a steady tone from the modem and then you should hear the serial alarm message coming from the modem. Once the message is sent, a hangup message is sent and the Data-Alert recycles for the next event. Your modem module is now complete and ready to use.

Setting Up the Data-Alert

Operation of the Data-Alert is simple, First you will need to connect a 9-V “wall wart” power supply to the circuit via the two power input pins at the top of the board. Next you will need to make up a serial communication or programming cable that connects the Data-Alert header J5 to the serial port of your programming computer.

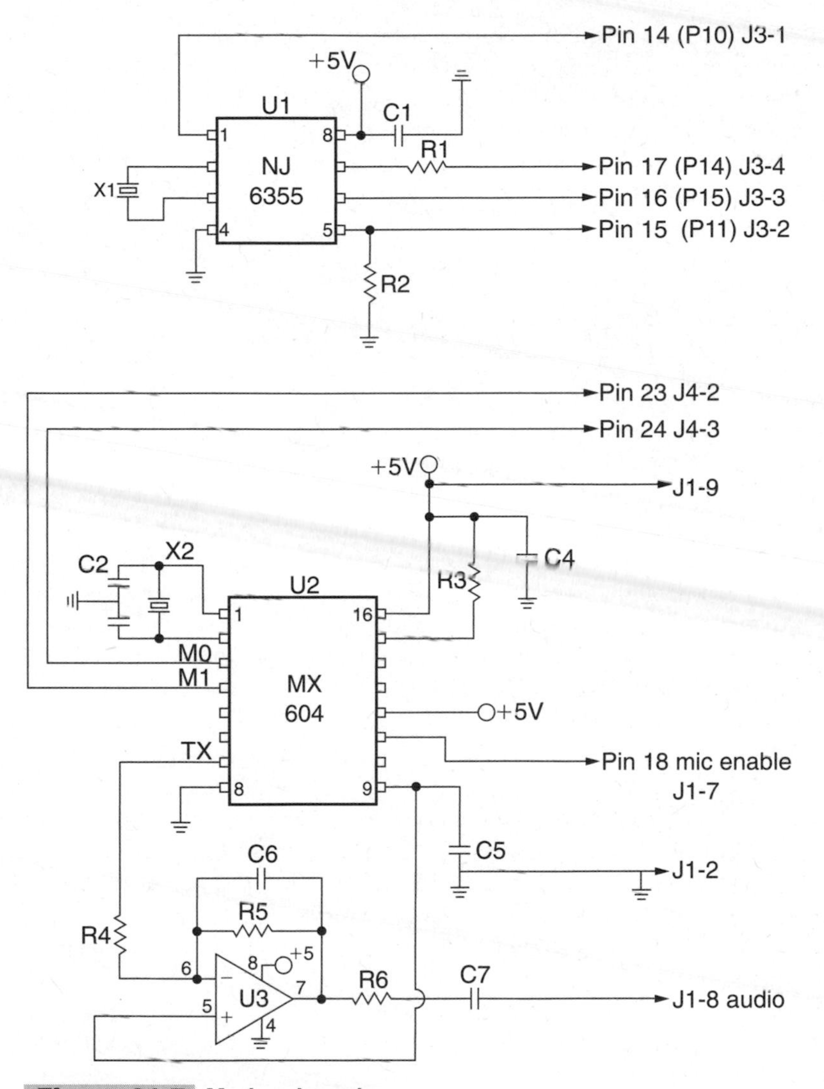

Figure 21-7 Modem board.

Now you will need to fire-up your PC and load the supplied disk into your computer. Once the communication cable has been attached, you will now need to load the STAMP2X or the STAMPW.EXE programs. These are the editor programs you will need to load the Data-Alert programs into the microprocessor. There are two Data-Alert programs included with this project: the DATA.BS2 program (Listing 21-1), which is the basic data reporting program, and MODEMTIME.BS2 (Listing 21-2), which reports data along with the time of day. Scroll down the displayed program, look for the simulated phone number, and replace it with *your own phone number.* Remember, a digit 10 is programmed by a zero. If you want to dial a long distance number, just add the full sequence of numbers. If you wish to dial out from a PBX telephone system, you will need to first put in the access number such as an 8 or 9 followed by a comma,

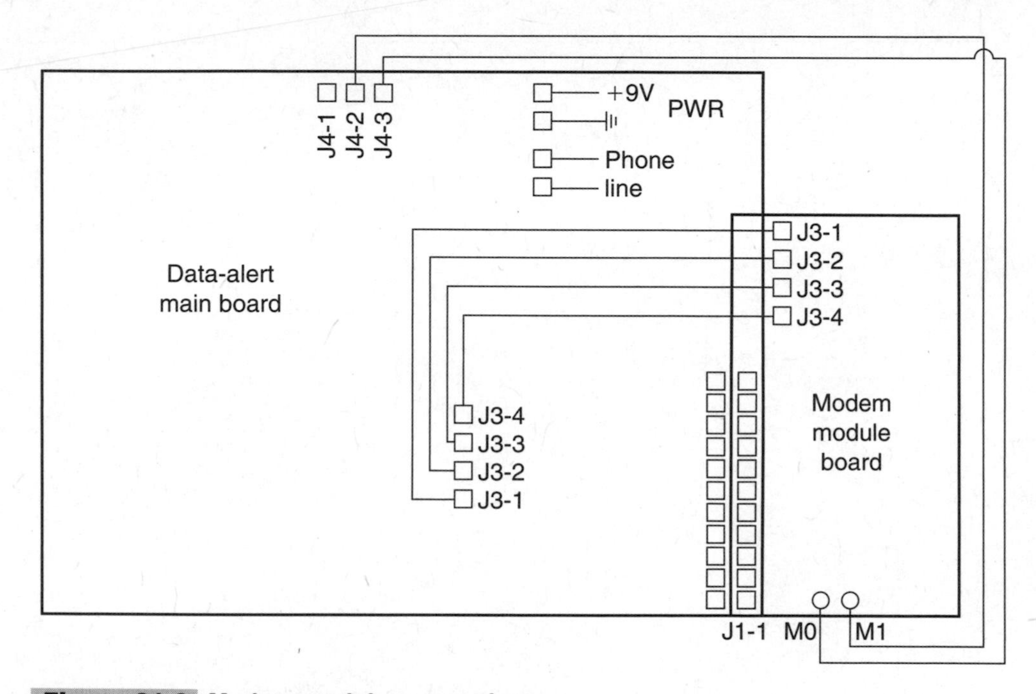

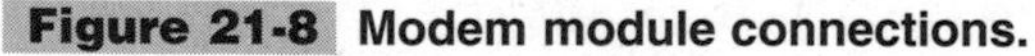
Figure 21-8 Modem module connections.

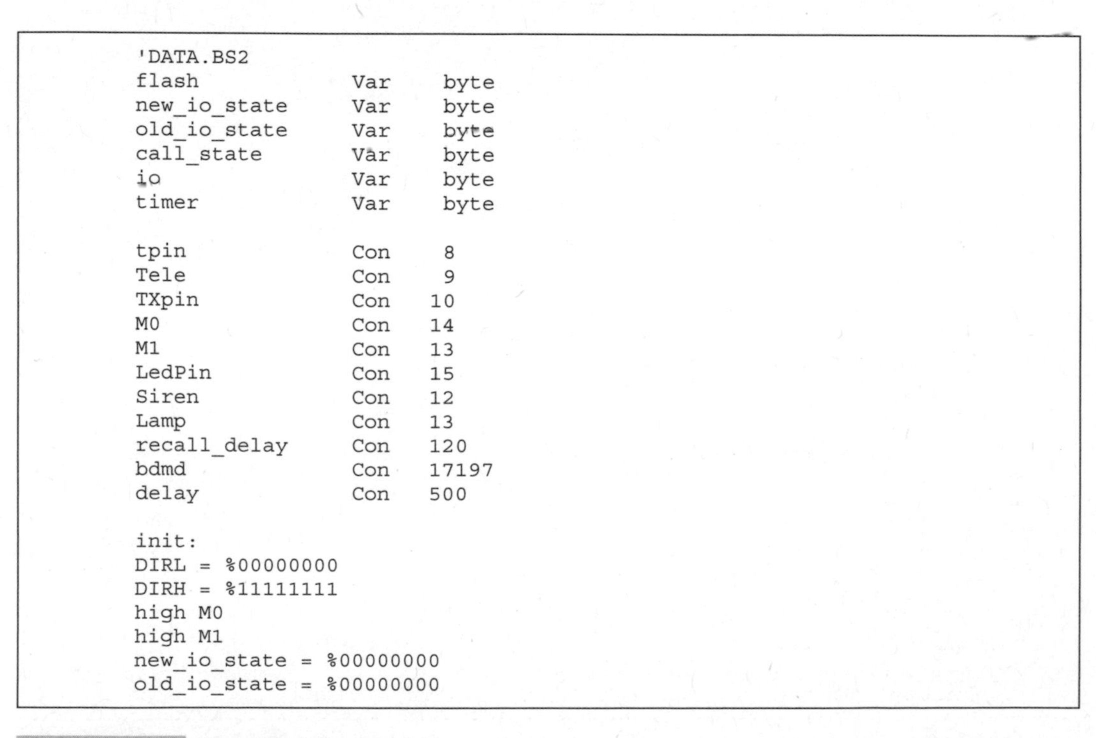

```
'DATA.BS2
flash            Var    byte
new_io_state     Var    byte
old_io_state     Var    byte
call_state       Var    byte
io               Var    byte
timer            Var    byte

tpin             Con     8
Tele             Con     9
TXpin            Con    10
M0               Con    14
M1               Con    13
LedPin           Con    15
Siren            Con    12
Lamp             Con    13
recall_delay     Con    120
bdmd             Con    17197
delay            Con    500

init:
DIRL = %00000000
DIRH = %11111111
high M0
high M1
new_io_state = %00000000
old_io_state = %00000000
```

Listing 21-1 Data-Alert Program 1

```
for flash = 1 to 3
 high LedPin
 pause 1500
 low LedPin
 pause 1500
next

main:
high LedPin
new_io_state=INL
pause 100
low LedPin
pause 100
if new_io_state <> %00000000 then dial_data
after_io1:
old_io_state = new_io_state
goto main
delay_and_scan
old_io_state = new_io_state
 for timer = 1 to recall_delay
 high LEDPin
 new_io_state=INL
 pause 100
 low LedPin
 pause 100
 if new_io_state <> old_io_state then dial_data
 old_io_state = new_io_state
 next
goto after_io1

dial_data:
if new_io_state = %00000000 then main
high LedPin
high Tele
'high Lamp
high Siren
pause 500
'dtmfout TXPin, 600, 600, [8]
dtmfout TXPin, 500, 100, [7,2,4,7,8,6,7]
sleep 10
gosub send_msg

goto delay_and_scan

send_msg:
io = new_io_state

if io.bit0=1 then gosub_chan_1
if io.bit1=1 then gosub_chan_2
if io.bit2=1 then gosub_chan_3
if io.bit3=1 then gosub_chan_4
pause 10

gosub_chan_1:
high M0
low M1
pause 6000
serout tpin,bdmd,50, [" CHAN1 BURGLAR ALARM",10,13]
```

Listing 21-1 Data-Alert Program 1 (*Continued*).

```
goto fin_1

gosub_chan_2:
high M0
low M1
pause 3500
serout tpin,bdmd,50, [" CHAN2 TEMPERATURE ALARM",10,13]
goto fin_2

gosub_chan_3:
high M0
low M1
pause 3500
serout tpin,bdmd,50, [" CHAN3 MOTION ALARM",10,13]
goto fin_3

gosub_chan_4:
high M0
low M1
pause 3500
serout tpin,bdmd,50, [" CHAN4 FIRE ALARM",10,13]
goto fin_4

fin_1:fin_2:fin_3:fin_4:
pause 2000
serout tpin,bdmd,10,["+++"]
pause 2000
serout tpin,bdmd,10,["ATH",10,13]
pause 100
high M0
high M1
pause 10000
low Tele
low LEDPin
low Siren
low Lamp
low M0
pause 100
return
end
```

Listing 21-1 **Data-Alert Program 1 (*Continued*).**

then the actual phone number. Next, press ALT-R to load the program into the Data-Alert circuit.

Your Data-Alert is now programmed and ready to operate. At last, you can connect your event or alarm inputs to the set of solder pads at the far edge of the circuit board, just below the phone line connection. Remember, you must use normally open circuit switches for the alarm inputs unless you are using the optional motion module, which provides both normally open and normally closed inputs. Connect your phone line via the phone jack and your Data-Alert is now ready to serve you.

Now you are ready to simulate an alarm condition to see if your Data-Alert functions. You can test the circuit in one of two ways. A simulated approach is to connect a crystal headphone across the secondary of T1 at J6, with the Data-Alert disconnected from the phone line. Apply power to the circuit, and connect a normally open switch across input

```
'modem time
flash           Var    byte
new_io_state    Var    byte
old_io_state    Var    byte
call_state      Var    byte
io              Var    byte
timer           Var    byte
temp            Var    byte

CLK             con     6
DATA_           con     7
NJU_CE          con     5
NJU_IO          con     4
tpin            con    8
Tele            con    9
TXpin           con   10
M0              con   11
M1              con   14
LcdPin          con   15
Lamp            con   13
Siren           con   12
recall_delay    con   120
bdmd            con   17197
Y10s   con     1      ' Array position of year 10s digit.
Y1s    con     0      ' "      "  "  year 1    "
Mo10s  con     3      ' "      "  "  month 10s  "
Mo1s   con     2      ' "      "  "  month 1s  "
D10s   con     5      ' "      "  "  day 10s   "
D1s    con     4      ' "      "  "  day 1s    "
H10s   con     8      ' "      "  "  hour 10s  "
H1s    con     7      ' "      "  "  hour 1s   "
M10s   con     10     ' "      "  "  minute 10s "
M1s    con     9      ' "      "  "  minute 1s  "
S10s   con     12     ' "      "  "  second 10s "
S1s    con     11     ' "      "  "  second 1s  "

day    con     6      ' "      "  "  day-of-week (1-7) digit.
digit  var     nib    ' Number of 4-bit BCD digits read/written.
DTG    var    nib(13) ' Array to hold "date/time group" BCD digits.

gosub read_clock               ' Get the DTG from the clock.

DTG(Y10s)=9: DTG(Y1s)=8              ' Year = 98.
DTG(Mo10s)=0: DTG(Mo1s)=8            ' Month = 08.
DTG(D10s)=0: DTG(D1s)=1              ' Day = 01.
DTG(day) = 5                         ' Day of week (1-7) = 1 (monday).
DTG(H10s)=0: DTG(H1s)=7              ' Hour = 08.
DTG(M10s)=1: DTG(M1s)=0              ' Minute = 30.
gosub write_clock               Write data to clock.
init:
DIRL = %00000000
DIRH = %11111111
high M0
high M1
new_io_state = %00000000
old_io_state = %00000000

for flash = 1 to 3
```

Listing 21-2 Data-Alert Program 2.

```
 high LedPin
 pause 1500
 low LedPin
 pause 1500
next

main:
high LedPin
new_io_state=INL
pause 100
low LedPin
pause 100
if new_io_state <> %00000000 then dial_data
after_io1:
old_io_state = new_io_state
goto main
delay_and_scan
old_io_state = new_io_state
 for timer = 1 to recall_delay
 high LEDPin
 new_io_state=INL
 pause 100
 low LedPin
 pause 100
 if new_io_state <> old_io_state then dial_data
 old_io_state = new_io_state
 next
goto after_io1

dial_data:
if new_io_state = %00000000 then main
high LedPin
high Tele
high Lamp
high Siren
pause 500
dtmfout TXPin, 600, 600, [8]
dtmfout TXPin, 500, 100, [7,5,4,7,7,6,7]
sleep 5
gosub send_msg

goto delay_and_scan
send_msg:
io = new_io_state

if io.bit0=1 then gosub chan_1
if io.bit1=1 then gosub chan_2
if io.bit2=1 then gosub chan_3
if io.bit3=1 then gosub chan_4
pause 10

read_clock:
 low NJU_IO                                         ' Set for read.
 high NJU_CE                                        ' Select the chip.
 for digit = 0 to 12                                ' Get 13 digits.
  shiftin DATA_,CLK,lsbpre,[DTG(digit)\4]           ' Shift in a digit.
 next                                               ' Next digit.
 low NJU_CE                                         ' Deselect the chip.
return                                              ' Return to program.
```

Listing 21-2 Data-Alert Program 2 (*Continued*).

```
write_clock:
 high NJU_IO                                        ' Set for write.
 high NJU_CE                                        ' Select the chip.
 for digit = 0 to 10                                ' Write 11 digits.
  shiftout DATA_,CLK,lsbfirst,[DTG(digit)\4]        ' Shift out a digit. '
 next                                               ' Next digit.
 low NJU_CE                                         ' Deselect the chip.
return                                              ' Return to program.
show_date:
 serout tpin,bdmd,[CR,HEX DTG(Mo10s), HEX DTG(Mo1s),"/",HEX DTG(D10s),
HEX
DTG(D1s),"/",HEX DTG(Y10s), HEX DTG(Y1s)]
return

show_time:
 serout tpin,bdmd,[CR,HEX DTG(H10s), HEX DTG(H1s),":",HEX DTG(M10s), HEX
DTG(M1s),":",HEX DTG(S10s), HEX DTG(S1s)]
return

gosub_chan_1:
high M0
low M1
pause 3000
gosub read_clock
gosub show_date
serout tpin,bdmd,[10,13]
gosub show_time
serout tpin,bdmd,[10,13]
serout tpin,bdmd,10, ["  ALERT CHANNEL 1 NOW ACTIVATED ",10,13]
serout tpin,bdmd,10, ["  BURGLAR ALARM ACTIVATION ",10,13]
serout tpin,bdmd,10, ["  TAKE ACTION IMMEDIATELY ",10,13]
goto fin_1

gosub_chan_2:
high M0
low M1
pause 3000
gosub read_clock
gosub show_date
serout tpin,bdmd,[10,13]
gosub show_time
serout tpin,bdmd,[10,13]
serout tpin,bdmd,10, ["ALERT CHANNEL 2 NOW ACTIVATED",10,13]
serout tpin,bdmd,10, [" TEMPERATURE ALARM  ",10,13]
serout tpin,bdmd,10, [" TAKE ACTION IMMEDIATELY ",10,13]
goto fin_2

gosub_chan_3:
high M0
low M1
pause 3000
gosub read_clock
gosub show_date
serout tpin,bdmd,[10,13]
gosub show_time
serout tpin,bdmd,[10,13]
serout tpin,bdmd,10, ["ALERT CHANNEL 3 NOW ACTIVATED",10,13]
serout tpin,bdmd,10, [" MOTION ALARM ",10,13]
serout tpin,bdmd,10, [" TAKE ACTION IMMEDIATELY ",10,13]
```

Listing 21-2 Data-Alert Program 2 (*Continued*).

```
goto fin_3

gosub_chan_4:
high M0
low M1
pause 3000
gosub read_clock
gosub show_date
serout tpin,bdmd,[10,13]
gosub show_time
serout tpin,bdmd,[10,13]
serout tpin,bdmd,10, ["ALERT CHANNEL 4 NOW ACTIVATED",10,13]
serout tpin,bdmd,10, [" FIRE ALARM MESSAGE ",10,13]
serout tpin,bdmd,10, [" TAKE ACTION IMMEDIATELY ",10,13]
goto fin_4

fin_1:fin_2:fin_3:fin_4:
pause 2000
serout tpin,bdmd,10,["+++"]
pause 2000
serout tpin,bdmd,10,["ATH",10,13]
pause 100
high M0
high M1
pause 10000
low Tele
low LEDPin
low Siren
low Lamp
low M0
pause 100
return
end
```

Listing 21-2 **Data-Alert Program 2 (*Continued*).**

IN1. Activate the switch at IN1 and you should begin hearing activity at the crystal headphone. First you should hear the touch-tone sequences followed by the alarm tone sequence. If this checks out, then you can move on to a "real" test by connecting the Data-Alert to the phone line, and repeat the same test over again.

The main Data-Alert circuit board has two 10-position female header jacks, i.e., J1 and J2, near the input terminal solder pads. These two 10-pin headers allow you to add the optional enhancement modules. The temperature/voltage module (TVL) or the motion module plugs into the second header row pin J2. The modem module is also installed at J1.

Uses for Data-Alert

The Data-Alert system is very flexible and can be used in a number of different alarm configurations. The Data-Alert could be used to monitor safes, doors, windows, floor mats, computers, movement, and temperature and voltage changes, as well as smoke and fire sensors. You can even monitor existing local alarms by using the Data-Alert as a multichannel dialer. With the Data-Alert and the motion module, you can easily protect an entire vacation house or cabin, using the existing phone line. You can even monitor an overheating computer or UPS failure. The Data-Alert can be utilized in almost any alarm configuration from multizoned alarm systems to simple multievent annunciator with a little imagination. The Data-Alert can be used day or night with your existing phone line; no additional phone lines or phone bills are incurred. The Data-Alert can be used to call a friend, a neighbor, a relative, or a coworker in the event of an alarm condition. If a second party is notified, the condition can be verified and then the police could be called depending on the severity of the alarm condition.

The Data-Alert is very flexible and can be tailored to many application, with a little imagination. The Data-Alert is ready to serve you!

Data-Alert (Main Board) Parts List

R1, R2, R3, R12	4.7-kΩ $^1/_4$-W resistor
R4, R5, R6, R7	10-kΩ $^1/_4$-W resistor
R8, R9, R10, R11	10-kΩ $^1/_4$-W resistor
R13	330-Ω $^1/_4$-W resistor
R14	620-Ω $^1/_4$-W resistor
R15	1-kΩ $^1/_4$-W resistor
R16	150-Ω $^1/_4$-W resistor
R17	130-V rms MOV
C1	10-μF 35-V electrolytic capacitor
C2, C7, C8, C9	0.1-μF 35-V capacitor
C3, C4, C5, C6	1-μF 35-V electrolytic capacitor
C10	0.001-μF 35-V capacitor
C11	0.1-μF 250-V Mylar capacitor
C12	47-μF 35-V electrolytic capacitor
D1	1N914 silicon diode
D2, D3	3.9-V zener diode
D4	LED
XTL	20-MHz ceramic resonator
T1	600- to 600-Ω audio transformer
J1, J2	10-position male header jacks
J3	3-position male header jack
J4, J5	4-position male header jack
U1	PIC 16C57 microprocessor

U2	24LC16B EEPROM memory
U3	MAX232 serial communication chip
U4	PVT412L MOS relay
U5	LM7805 5-Vdc regulator
Miscellaneous	PC board, wire, IC sockets

Modem Board (Daughter Board) Parts List

R1	1-kΩ $^1/_4$-W resistor
R2	10-kΩ $^1/_4$-W resistor
R3	100-kΩ $^1/_4$-W resistor
R4	12-kΩ $^1/_4$-W resistor
R5	120-kΩ $^1/_4$-W resistor
R6	500-Ω $^1/_4$-W resistor
C1, C4, C5, C7	0.1-μF 35-V ceramic capacitor
C2, C3	18-pF 35-V disk capacitor
C6	330-pF 35-V disk capacitor
X1	32.768-kHz crystal
X2	3.5795-MHz crystal
U1	NJ6355 real-time clock
U2	MX604 modem chip
U3	LM 358 dual op-amp
Miscellaneous	PC board, sockets, headers, wire

22

INPUT SENSOR MODULES

CONTENTS AT A GLANCE

Motion Module

The pyroelectric motion module is a great addition to the Tele-Alert or Page-Alert microprocessor alarm controller. The pyroelectric motion module centers around a pyroelectric, or body heat, infrared sensor and a controller chip, and it can be used to activate either the Tele-Alert or the Page-Alert alarm controller; it is shown in Fig. 22-1. The motion module can sense body heat from 10 to 50 ft away.

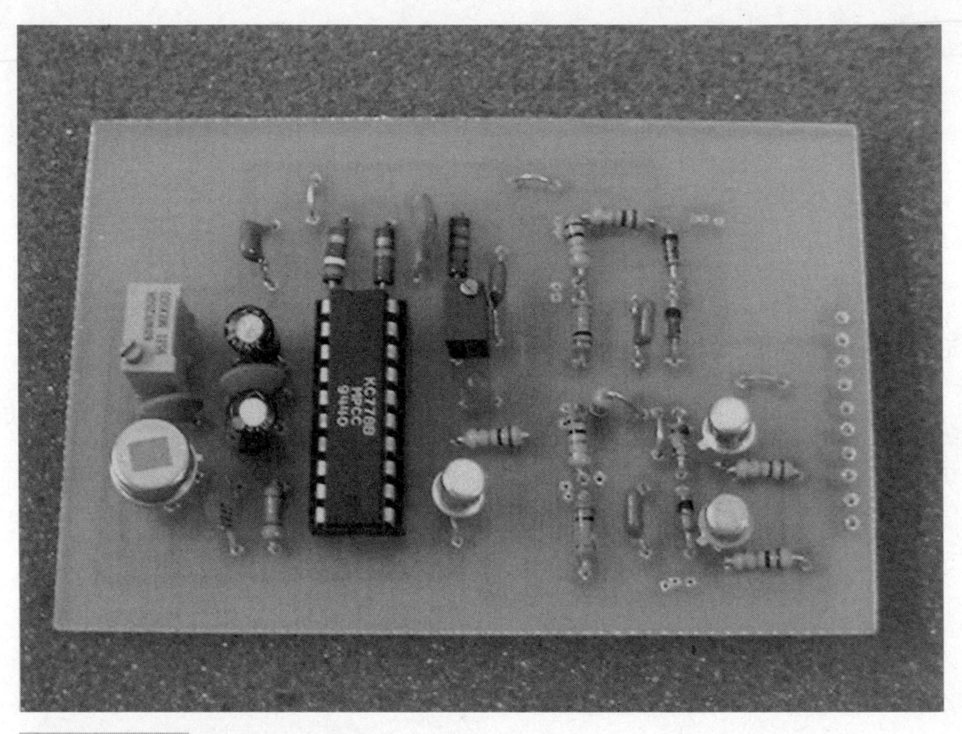

Figure 22-1 Motion module.

The pyroelectric infrared (PIR) controller chip block diagram is depicted in Fig. 22-2. The PIR controller is a complex chip that provides amplification, filtering, clock, comparators, a daylight detector, and a voltage regulator. Table 22-1 lists the pinouts of the 20-pin PIR controller chip.

The motion module circuit is illustrated in the system diagram in Fig. 22-3. The circuit begins with the sensitive PIR detector at U1. The pyroelectric infrared detector is a sensitive three-lead high-impedance sensor, which is shown with a 47-kΩ output resistor at R1. A sensitivity or range control for the PIR sensor is shown at R2. The PIR sensor is coupled to a specially designed PIR controller chip that is optimized for PIR alarm sensors. The motion module is powered from a 5-V source and can tap power from either the Tele-Alert or Page-Alert main controller boards. Once the PIR sensor picks up on the "scent" of a human, it determines if the signal is valid by using timing, comparators, and filters. If a valid signal is processed, it is output to pin 16 and is used to drive transistor Q1. Transistor Q1 is utilized to drive the input pin on the Tele-Alert or Page-Alert system boards.

Capacitors C5, C6, and C7 are offset filters, antialiasing filters, and a dc capacitor respectively. Resistors R3 and R4 are utilized as a frequency reference between pin 20 and V_{cc}. The PIR sensor is placed across pins 2, 7, and 8 of the controller chip. The actual alarm output pin of the controller is located at pin 16, which is coupled to the output driver transistor at Q1. The *R*/*C* time constant for the output "on time" of the controller is between pins 18 and 19, and is controlled by R5, R6, and C8. A movement indicator LED is located between pin 17 and pins 9 and 10, the analog and digital ground pins. The chip employs

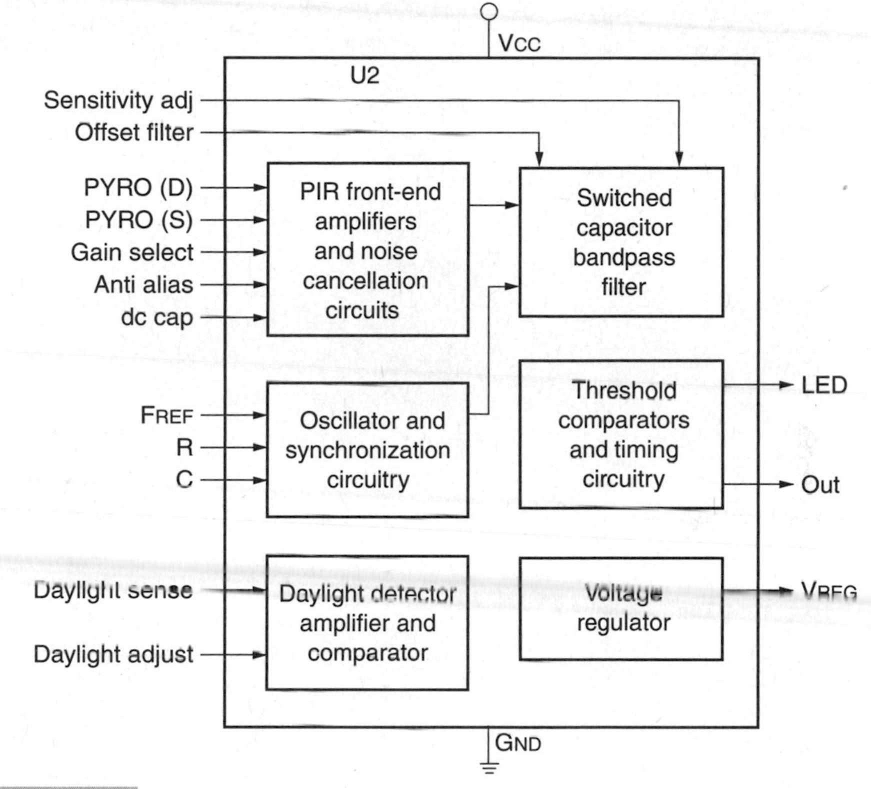

Figure 22-2 PIR controller chip.

a daylight detector, which is not implemented in our application. Power to the controller chip is brought to pin 1 of U1 via a 78l05 regulator on either the Tele-Alert or Page-Alert boards.

CONSTRUCTION OF THE MOTION MODULE

Construction of the PIR motion module is quite straightforward and can be completed in less than 1 hour. The motion module is fabricated on a $1^3/_4$- by $2^1/_2$-in circuit board. The PIR sensor is at the bottom, opposite the solder pad outputs. An IC socket is used for U1 to facilitate removal or exchange in the event of a malfunction of the controller chip. When constructing the motion module, be sure to observe the polarity of the capacitors, transistor, regulator, and the PIR sensor and the placement of U1 in its socket. The components, especially the controller chip and PIR sensor, can be damaged if they are placed incorrectly. Note that the PIR sensor is placed on the foil side of the circuit board. Pay particular attention to the layout diagram when installing the components. The LED indicator is located between the controller chip and the 10-pin header. Observe that pin 1 of the controller is located near the regulator at U1. At one edge of the PC board, you will find the 10-pin solder tabs used to secure the 10-pin male header pins, which couple the motion module to the Tele/Page-Alert. The top pin, 1, is reserved for the 9-V power input power

TABLE 22-1 PIR CONTROLLER PINOUTS

PIN	NAME	DESCRIPTION
1	V_{cc}	+5-V supply
2	Sensitivity adjust	PIR sensitivity input
3	Offset filter	PIR offset filter
4	Antialias	PIR antialias filter
5	DC capacitor	PIR gain stabilization filter
6	V_{reg}	Voltage regulator output
7	Pyrosensor (D)	Pyro drain reference pin
8	Pyrosensor (S)	Pyro source input
9	Ground (A)	Analog ground
10	Ground (D)	Digital ground
11	Daylight adjust	Daylight adjust and CDS
12	Daylight sense	Silicon photodiode input
13	Gain select	PIR gain
14	On/auto/off	Mode select state input
15	Toggle	Mode select toggle input
16	Out	Output
17	LED	Indicator LED
18	C	Off timer oscillator
19	R	Off timer oscillator
20	F_{ref}	Frequency reference oscillator

pin. Pin 2 is used as the system ground, while pin 3 is the output pin from the motion module to the Tele/Page-Alert main board. The rest of the header pins are not used. The 10-pin header is placed on the component side of the PC board and soldered on the foil side of the motion module PC board.

Once the motion module has been completed, you can easily test the board by applying a 9-V transistor battery to pins 1 and 2 of the header. Connect a voltmeter or oscilloscope to output pin 3 and ground pin 2 and you are ready to test the motion module board. Once power is applied, wait about 5 seconds, then wave your hand in front of the PIR sensor. The Fresnel lens doesn't have to be in front of the PIR for this test. The indicator LED should light, once your hand is waved in front of the PIR sensor. At this point in time, you should also see output pin 3 jump from 0 V to 5 V on your meter or scope. You may have to adjust the PIR sensitivity control at R2 or the time on control at R6 for optimum, once the Fresnel lens is in place and the enclosure is shut and you become familiar with the detector's operation. This completes the testing of your new motion module.

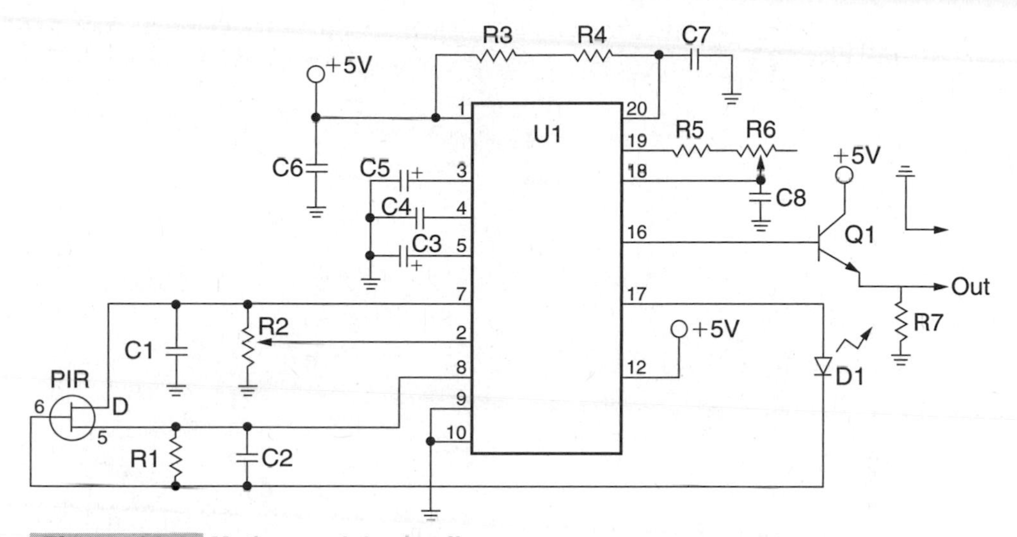

Figure 22-3 Motion module circuit.

ASSEMBLING THE UNIT

Next, you will need to connect the motion module to the Tele/Page-Alert controller. The motion module should be plugged into the 10-pin female header on the main Tele/Page-Alert board. On the Tele/Page-Alert board, the first header next to the screw terminals is the 15-pin header, which is used for the listen-in board, while the second header is a 10-pin one, which is used for the motion module board. The motion module board is plugged into the Tele/Page-Alert, with the components from both boards facing each other. Once the two boards are fastened together, you will need to place a hole in the Tele/Page-Alert enclosure to allow the PIR to see the room. You will at this point have to place the Fresnel lens ahead of the PIR sensor before securing the top of the Tele/Page-Alert enclosure. Table 22-2 is the motion module pinout chart.

The PIR sensor requires a Fresnel beam focusing lens. The Fresnel lens is placed ahcad of the PIR sensor to give the sensor a specific pattern of coverage. PIR Fresnel lenses come in two basic types. The most common beam pattern is the wide-angle lens which looks out to about 12 ft with a 30 to 50° beam pattern; the second Fresnel lens is the narrow-angle type which looks out to about 50 ft with a narrow beam pattern.

Your Tele-Alert or Page-Alert is now a complete motion detector alarm controller/dialer that can be used to protect a house, cottage, or workshop area. If you are clever in the placement of the PIR sensor, you can easily protect large areas. Note that you can also connect other alarm detectors or sensor switches to the three additional Tele/Page-Alert input channels (2 through 4) all at the same time.

MOTION MODULE PARTS LIST

R1	47-kΩ $^1/_4$-W resistor
R2	200-kΩ trim potentiometer

R3	3.9-kΩ $^1/_4$-W resistor
R4	56-kΩ $^1/_4$-W resistor
R5	10-kΩ $^1/_4$-W resistor
R6	1-MΩ trim potentiometer
R7	10-kΩ $^1/_4$-W resistor
C1, C4, C7	100-nF 35-V electrolytic capacitor
C2	220-pF 35-V Mylar capacitor
C3, C5	10-µF 35-V electrolytic capacitor
C6	0.1-µF 35-V disk capacitor
C8	4.7-nF 35-V disk capacitor
Q1	2N2222 transistor
LED	Indicator LED
U1	78L05 5-V regulator
U2	PIR controller chip KC778B
PIR	PIR sensor RE200B
F1, F2	PIR Fresnel lens, wide and narrow beams
Miscellaneous	PC board, header, IC socket

TABLE 22-2 MOTION MODULE PINOUTS

PIN	NAME	DESCRIPTION
1	+9 V	9-V power pin (not used)
2	GND	System ground pin
3	OUT1	Normally open/closed alarm loop
4	OUT2	Normally open/closed alarm loop
5	OUT3	PIR motion detector output
6	Not used	Not used
7	Not used	Not used
8	Not used	Not used
9	+5 V	Motion module system power

Alarm Sensor Module

NORMALLY OPEN/CLOSED INPUT LOOP SENSOR BOARD

The alarm loop sensor module can be used to connect any number of alarm sensors to the input of the Tele-Alert system. Using the alarm sensor module, you can easily turn your Tele-Alert or Page-Alert into a portable burglar alarm system that can call you and let you know someone has violated a particular space that you have protected.

The diagram in Fig. 22-4 depicts a clever method of utilizing both normally open and normally closed input sensors or alarm switches in a simple circuit. Normally open sensors or switch contacts are shown at S1, S2, and S3. Normally closed switches or alarm sensors are shown at S4, S5, S6, and S7. Any number of normally closed or normally open switches can be used in this input circuit. Not only can you use alarm switches such as window or door switches, but actual sensors can be substituted for switches—floor mats or driveway sensors, infrared heat detectors, etc. Two 1N914 diodes, an npn transistor, and a few resistors form the heart of this detector. The alarm loop sensor module is powered from a +5-V source and can obtain power directly from either the Tele-Alert or Page-Alert boards. You could elect to build one of these input circuits for each of the Tele-Alert or Page-Alert input pins if desired; this would permit very large alarm systems with many zones. The output of the alarm loop sensor module provides a 5-V signal to the input of the STAMP 2 on receiving an alarm signal. A few of these detectors circuits could be built on a single circuit board to create a multichannel alarm system using your Tele-Alert system.

ALARM SENSOR MODULE PARTS LIST

R1, R3	20-kΩ 1/4-W resistor
R2, R4	10-kΩ 1/4-W resistor
C1	0.01-μF 25-V disk capacitor
D1, D2	1N914 silicon diode
Q1	2N2222 npn transistor
Miscellaenous	Sensors or switches

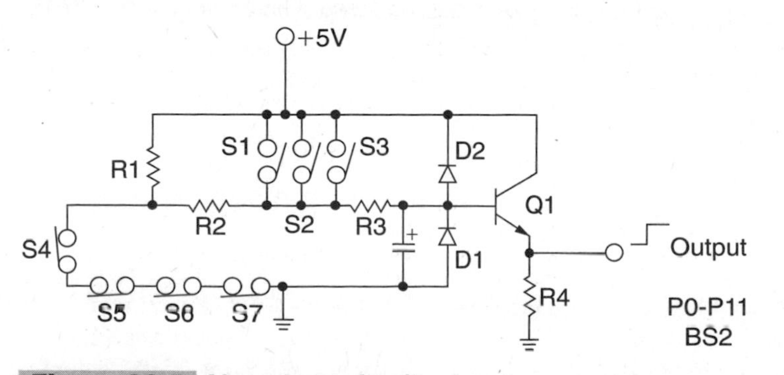

Figure 22-4 Alarm loop circuit.

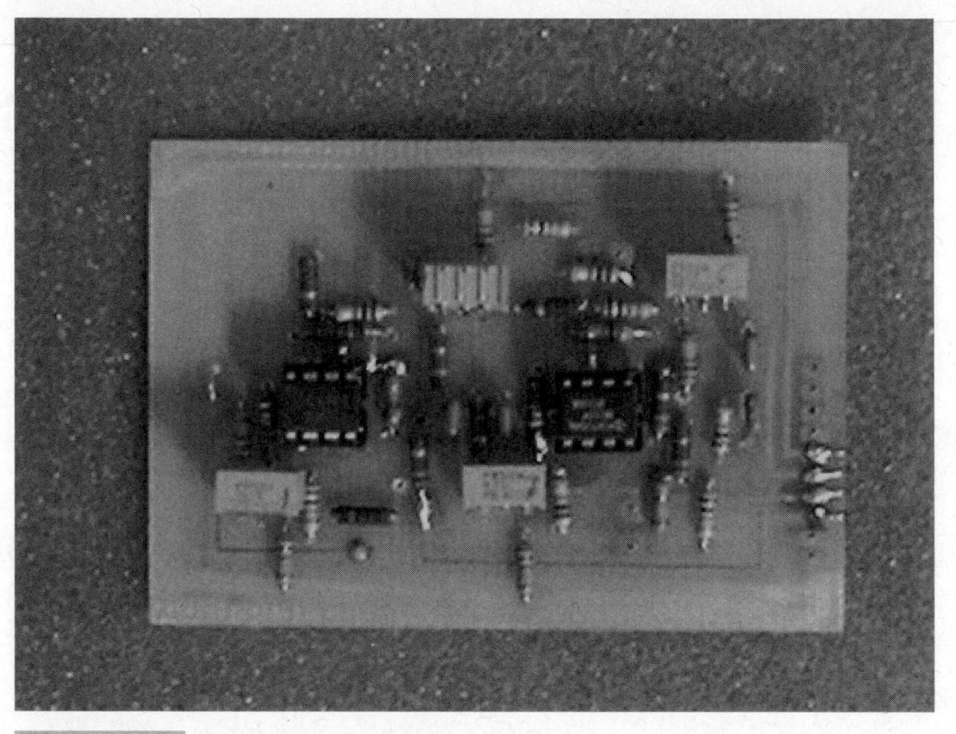

Figure 22-5 **Temperature/voltage level module.**

Temperature/Voltage Level Module Kit

The temperature/voltage level module board kit is a versatile addition to the Tele-Alert or Page-Alert microprocessor alarm controller, and is shown in Fig. 22-5. The temperature/voltage level module is a single-channel module that will permit you to monitor temperature level changes, either upward or downward, or it can allow you to measure voltage level changes by setting a user threshold control.

The temperature/voltage level (TVL) board consists of a single LM393 comparator integrated circuit and a handful of components (Fig. 22-6). The circuit begins with S1 (a resistive sensor, or thermistor) and R6. S1 and R6 form a voltage divider to ground. In this configuration, you are able to monitor upward temperature changes. If the positions of S1 and R6 are reversed, then you are able to monitor decreasing temperature changes. The thermistor output at pin 3 of U1 represents the plus (+) input to the comparator circuits. The resistor network at R12 acts as the threshold control on the minus (−) inputs of the comparator. The output of the comparator at pin 1 and pin 4 produces a (high) output when the threshold is sensed.

The TVL module can also be used to monitor voltage level changes by eliminating thermistor S1 and resistor R6 and replacing them with scaling resistors instead. Scaling resistors will now represent voltage dividers consisting of two resistors at each input channel. Ratios such as 100 kΩ to 1 kΩ for R1/R2 or 1 kΩ to 100 Ω for R3/R4 would be used as

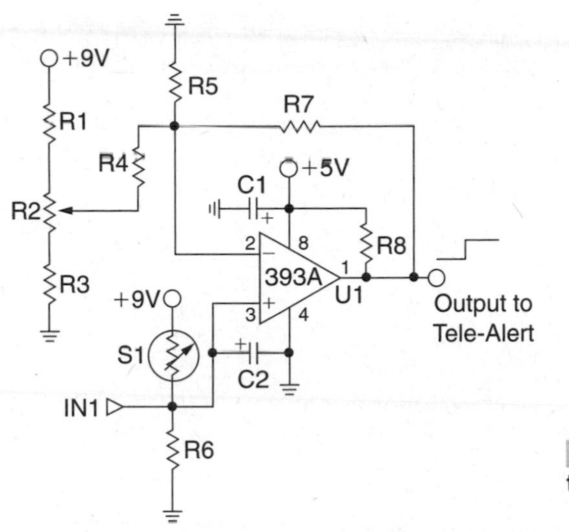

Figure 22-6 Threshold/comparator (TVL) module.

input scalers. The higher-value resistor would replace the thermistor, while the lower-value resistor would go to ground. Note, that capacitors C1 and C3 provide noise reduction when external inputs are used. To measure voltage level changes, you simply connect the input wire IN1 to the circuit or system being monitored, adjust the threshold control, and you are ready to go.

BUILDING THE TVL MODULE

Construction of the TVL board is quite straightforward and should take only about a half-hour or so. The TVL board consists of a single integrated circuit with an IC socket to facilitate changes if the board ever malfunctions. First, locate the layout diagram to help you orient the PC board and the component layout before starting to construct the board. Be sure to observe the orientation of the integrated circuit before inserting it into the socket to avoid damage at power-up. *Note:* for initial testing S1/R6 should not be inserted; see below. When later installing the thermistors or scaling resistors, be sure to carefully observe the placement of S1/R6, depending upon the temperature direction you desire to measure or if you want to sense voltage levels. Trim potentiometer R2 is used to set the threshold values for either temperature or voltage level sensing. The trim pots are located at the top of the circuit board. Remember, when voltage level detection is desired, scaling resistors are used in place of the thermistors, as mentioned earlier. The pinouts at the edge of the TVL board are described below and are shown in Table 22-3.

At the top of the board, the first solder pad used is pin 2, which is for the 9-V reference connection for the inputs at IN1 and trim potentiometer R2. Pin 3 is used for the system ground connection. Pins 4 and 5 are used for the two TVL outputs to the Tele-Alert board. Solder pads 6 and 7 are not used, while solder pads 8 and 9 are reserved for inputs IN1 and IN2 to the comparators. The last pin at position 10 is used to supply 5 V to the comparator.

Once the board has been completed, you will need to place the 10-pin header on the component side of the PC board and solder it on the foil side of the PC board. The 10-pin header allows connection with the main Tele/Page-Alert board. The 10-pin header on the

TABLE 22-3 TEMPERATURE/VOLTAGE LEVEL MODULE PINOUTS

PIN NUMBER	PIN DESIGNATION
1	+9-V reference voltage
2	System ground
3	Channel 1 output to main board
4	Channel 2 output to main board
5	Channel 3 output to main board
6	Channel 4 output to main board
7	No connection
8	No connection
9	+5-Vdc system power
10	+5-Vdc system power
IN1	Channel 1 input
IN2	Channel 2 input
IN3	Channel 3 input
IN4	Channel 4 input

TVL board then can be inserted into the 10-pin female header (second inside header) on the main Tele/Page-Alert board once you are ready. Note that pins 8 and 9 are inputs to the comparator, and these pins should be either clipped or bent so as not to be inserted into the Tele/Page-Alert board. The voltage level inputs from the circuit being remotely monitored are connected to pins 8 and 9, i.e., inputs IN1 and IN2, respectively.

Your temperature/voltage level module is now complete and ready to go. It is advised that you test your new TVL board before you connect it to the main Tele/Page-Alert board.

TESTING THE TVL BOARD

Connect a scope or multimeter to the channel 1 output on the TVL board. Next, connect the plus (+) lead from a 5-V power supply to pins 2 and 10, and then connect the minus (−) lead from the power supply to the system ground at pin 3. For this initial testing, resistors R1/R2 and R3/R4 are not inserted. Now, locate an adjustable low-voltage power supply. Connect the plus lead (+) from the adjustable supply to the input at IN1 or pin 8, and connect the minus (−) lead to the system ground at pin 3. Be sure the adjustable power supply is initially adjusted for zero volts before applying to terminals at the edge of the TVL board. Now, turn on the first 5-V power supply and slowly advance the adjustable power supply voltage from 0 to 1 or 2 V. You may have to set the threshold to set the trip point at which you will begin to see an output on the scope. Once the comparator is tripped, the scope reading should change from 0 V to 5 V. Once channel 1 has been tested, you can move on to testing channel 2 in the same manner. Once the TVL board has been tested and you have decided whether you wish temperature or voltage level sensing, you can insert the input resistors S1/R2 or the scalers.

The TVL board can now be inserted into the 10-pin female header on the main Tele/Page-Alert board. Note that the components on both circuit boards should face each other when connected together.

TEMPERATURE/VOLTAGE LEVEL MODULE PARTS LIST

U1	LM393 comparator IC
R1, R3	2-kΩ $^1/_4$-W resistor
R2	50-kΩ potentiometer
R5	1-MΩ $^1/_4$-W resistor
R4, R6	10-kΩ $^1/_4$-W resistor
R7	10-MΩ $^1/_4$-W resistor
R8	3.3-kΩ $^1/_4$-W resistor
C1	0.1-µF 25-V disk capacitor
C2	0.01-µF 25-V disk capacitor
S1	Resistive sensor, thermistor, etc.
Miscellaneous	Circuit board, header, wire, etc.

Listen-In Module Kit

The listen-in module kit allows you to listen in to the room or area being monitored by the Tele-Alert alarm controller (Fig. 22-7). The listen-in board can be used only with the Tele-Alert system board. With the listen-in board you can listen in for up to 2 minutes after you have been called by the Tele-Alert unit. The circuit for the listen-in board is shown in Fig. 22-8.

BASIC COMPONENTS

The listen-in board, begins with the sensitive electret microphone at MIC. The sensitive electret microphone is first biased via R1, which is enabled via the microprocessor. The sound output from the electret microphone is coupled through C1 and R2 to the dual op-amp at U1:A, which acts as an audio preamplifier. The LM358 op-amp at U1 is a single-supply device, thus eliminating the need for the usual two power supplies used with most op-amps. The gain of U1 is essentially controlled by resistors R3 and R6. The output of the first preamp stage is next coupled to the second amplifier stage at U1:B, via C2 and R5. The output of the second audio amplifier stage at U1:B is coupled directly to the telephone coupling transformer T1 via C3. The listen-in board kit is powered by the 5-V system power source through the main Tele-Alert board.

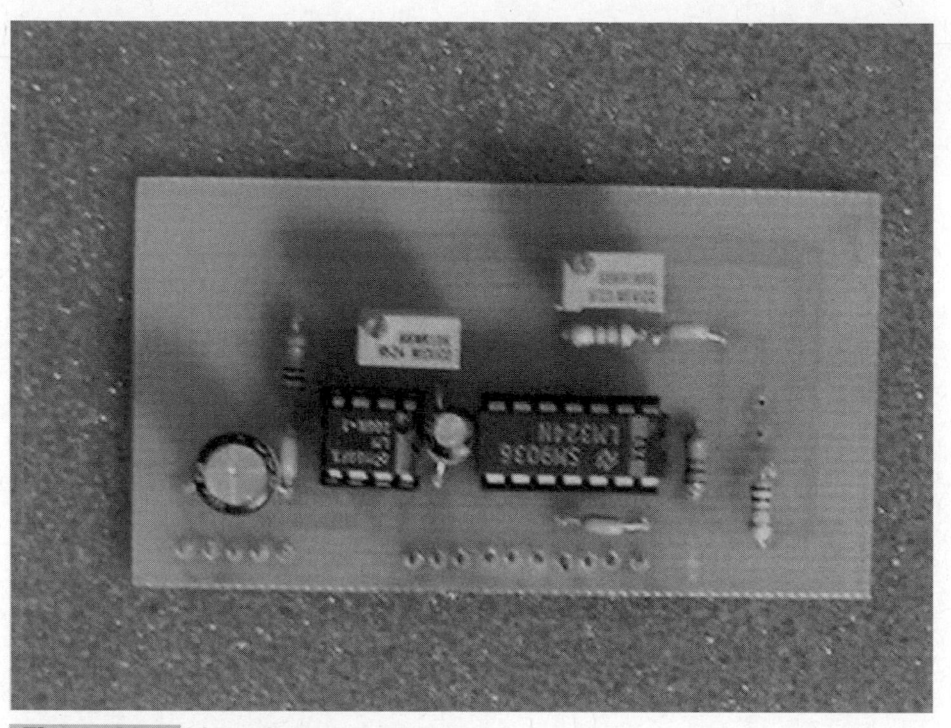

Figure 22-7 Listen-in module kit.

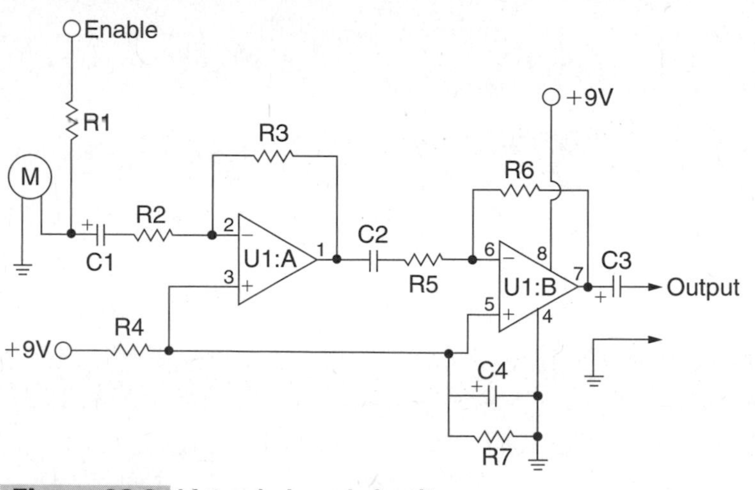

Figure 22-8 Listen-in board circuit.

CONSTRUCTING THE MODULE

The listen-in board kit measures 2 by $1^1/_4$ in. It was designed to easily plug into the Tele-Alert. The listen-in board is simple to construct, and can be completed in about a half-hour or so. Locate all the components and display them in front of you. Begin first by installing the resistors, followed by the capacitors. When constructing the kit, be sure to carefully observe the polarity of the capacitors and the integrated circuits before placing them on the circuit board. Integrated circuits are generally marked by an indented recessed circle or a square notch at the top of the chip. Usually the top left pin closest to the circle or notch is pin 1. Integrated circuit sockets are used to facilitate component changes in the event of a later problem. The electret microphone has two leads, which first must be identified before placing it on the circuit board. One lead is fastened to the case or body of the microphone, and this is the ground lead.

The listen-in module contains a single in-line row of 10 male header pins at one edge of the circuit board, which allows the board to connect to the main Tele-Alert board. The first of the 10 pins begins with the 9-V supply at pin 1 followed by the system ground connection at pin 2. Many of the solder pads are not used except for the power pins, the audio output at pin 8, and the audio enable connection at pin 7. The 10-pin male header is inserted and soldered on the foil side of the circuit board, once the board has been completed. Both circuit boards should have the components facing upwards when connected together and finished. Table 22-4 lists the pinouts.

Once completed, the listen-in board can be tested, by applying a 5-volt source to the two power pins, system ground at P1-2 and 5 volts at P1-9. You can then connect a headphone to the output pin P1-8 and ground, and apply 5 V to the enable pin at P1-7, in order to activate the electret microphone. You should now be able to hear room sounds in the headphone, and the listen-in board is now complete and ready to use.

ASSEMBLING TO THE TELE-ALERT BOARD

You are now ready to attach the listen-in board to the main Tele-Alert board. Take a look at the Tele-Alert board; at one edge you will notice the input terminal connections at the outside edge of the board, followed by two rows of header sockets. The first row of female 10-pin header sockets at the outside edge is used to hold the listen-in board, while the second set of inside header pins is used to power the TVL board and the motion module board. The TVL or motion module daughter boards can be utilized at the same time as the listen-in board if desired. Grasp the Tele-Alert board in one hand and plug in the listen-in module.

Now you are now ready to utilize the listen-in board with your Tele-Alert. Connect a 9-V "wall wart" power supply, phone line, and alarm switch or sensor contacts to the Tele-Alert

TABLE 22-4 LISTEN-IN MODULE PINOUTS

PIN	NAME	DESCRIPTION
1	+9 V	+9 V (not used)
2	GND	System ground
7	Enable	Microphone enable
8	Audio	Audio ouput pin
9	+5 V	+5-V system power

main board and you are ready to go! The listen-in board can be used in conjunction with motion module to detect movement and then let you listen in to room sounds once motion is detected, so your Tele-Alert can be used as a complete self-contained alarm system.

LISTEN-IN MODULE PARTS LIST

R1	2.2-kΩ $^1/_4$-W resistor
R2, R7	10-kΩ $^1/_4$-W resistor
R3, R5	100-kΩ $^1/_4$-W resistor
R4	5.6-kΩ $^1/_4$-W resistor
R6	5-MΩ $^1/_4$-W resistor
C1	2.2-μF 35-V electrolytic capacitor
C2	0.1-μF 35-V disk capacitor
C3, C4	10-μF 35-V electrolytic capacitor
U1	LM358 dual op-amp
MIC	Electret microphone
Miscellaneous	PC board,10-pin header, IC sockets

23

DATA-TERM UNIT

CONTENTS AT A GLANCE

The Data-Term unit is a versatile companion to the Data-Alert System. The Data-Term unit, when called by the Data-Alert system, will provide remote control of home appliances or remote alarm data display. The Data-Term is placed across your phone line at home or away and waits for a call from the Data-Term unit. The Data-Term "picks up" the phone when called and can be used to display remote data information. The Data-Term is shown in Fig. 23-1.

The heart of the Data-Term is the PIC 16C57 BASIC STAMP 2 alternative microcontroller. The complete Data-Term circuit is shown in Fig. 23-2. The PIC 16C57 chip emulates the BASIC STAMP 2 computer. The PIC 16C57 is preloaded with a BASIC interpreter much like the Parallax BASIC STAMP 2 (BS2). The PIC 16C57, however, is a much less expensive approach to solving a problem, but requires a few more external parts than the BASIC STAMP 2. The PIC 16C57 requires very little in the way of support chips to make it function—only two support chips, a 24LC16B EEPROM for storages, and a MAX232 for communication. Only a few extra components are needed to form a functional microprocessor with serial input/outputs. A 20-MHz ceramic resonator and a few resistors and a diode are all that are needed to use the processor in its most simple form.

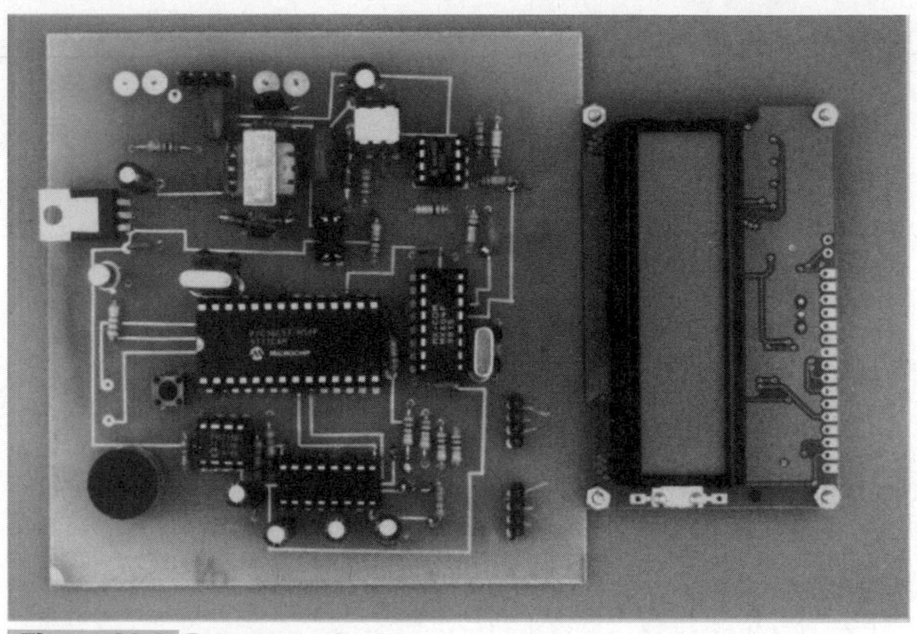

Figure 23-1 Data term unit.

The 28-pin PIC 16C57 is a very capable and versatile little microprocessor that can perform many tasks.

Circuit Description

The circuit begins with the telephone input at L1 And L2. A metal oxide varistor (MOV) at R1 is used to protect the circuit from phone line spikes. Ahead of the audio coupling transformer T1 are two control circuits, which are optically coupled to the microprocessor. The optocoupler at U3 is used to sense phone line ringing voltage, and it sends a 5-V signal to the STAMP 2 to awaken the BS2 controller to a phone call. Optocoupler U2 is used to answer, or "pick up the phone," upon ringing. The STAMP 2 senses ringing voltage, and then uses the optorelay at U2 to answer the phone. Capacitor C3 and resistor R2 are used coupled to the transformer T1, but the phone line signals are essentially blocked from reaching the transformer by C3. When the phone rings and the microprocessor senses the ringing, the program is instructed to apply a 5-V signal to U2, which is used to answer the phone by shorting out capacitor C3. This effectively couples the phone line to the Data-Term circuit.

The incoming phone data signals are clipped and filtered by D3, D4, and C5. The op-amp at U4 is used to amplify the incoming audio signal from the phone. The op-amp U4 also performs coupling duty between T1 and the modem at U5. Next the audio signals are directed at the MXCOM MX604 data modem. The modem chip is a full FSK modem with transmit and receive capabilities just like the controller of a conventional modem. The

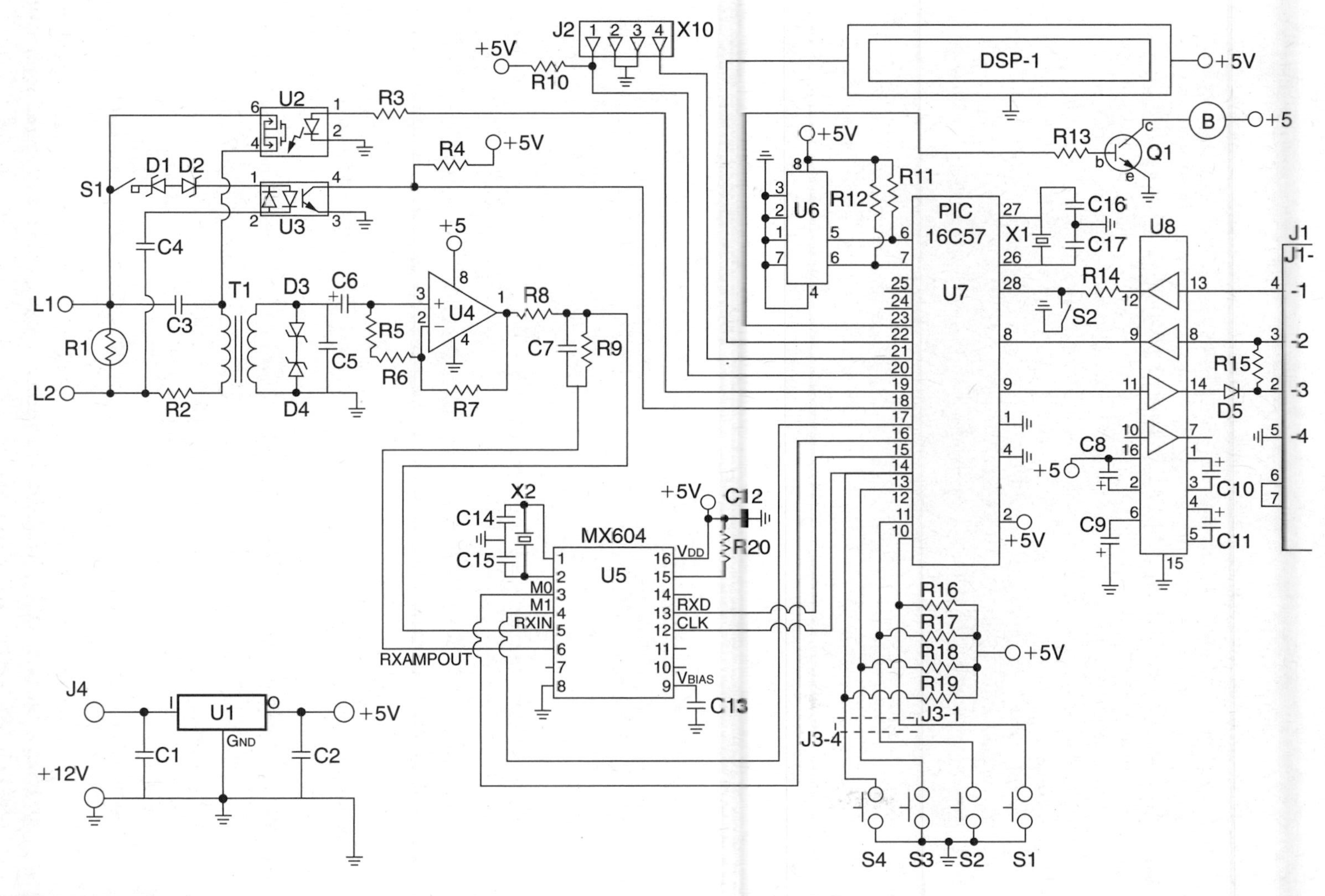

Figure 23-2 Data-Term circuit.

MX604 contains an FSK modulator and demodulator as well as filters, buffers, timing, and logic control functions. Figure 23-3 is a block diagram of the of the MX604 modem chip. Table 23-1 is the pinout list for the MX604 modem chip. Crystal X2 is used as a frequency reference for the modem. Incoming audio signals are coupled via pins 5 and 6, while outgoing data signals leave the modem on pin 13. M0 and M1, i.e., pins 3 and 4, are used to program the modem states; see Table 23-2. Initially the BASIC STAMP 2 applies a high, or 5-V, signal to pins 3 and 4 to initialize the modem. Then the mode is set to receive at 1200 baud. The STAMP 2 applies the clock control signals to the modem at pin 12.

In order to turn the BS2 look-alike into a Data-Term project, a few more parts must be added to the basic configuration. Table 23-3 lists the BS2 pinouts First, you need to configure pins 10 through 13 as user inputs. Momentary pushbutton switches S1 to S4 are used for future activation of Data-Term functions, such as display scrolling and manual control. A resistor is placed across each input line to 5 V. Additionally, pins 18, 8, and 28 are configured as inputs. Pin 18 is used for ringer detect, while pins 8 and 28 are used for serial communication inputs. The Data-Term project has many pins that are used for outputs. Pins 16 and 17 are outputs configured to animate the modem, while pin 19 is used to connect the phone line. Pins 20 and 21 are used to send X10 information to the power line controller. Pin 22 drives the LCD display, while pin 23 controls the buzzer. Pin 24 is used to light the status LED, and pin 25 is free for future use. Pin 9 of U7 is used for serial communication output. The 2-kbyte external memory chip at U6 is coupled to the microprocessor via pins 6 and 7. A 20-MHz reference crystal is connected across pins 27 and 28 of the processor. Power for the processor is a regulated 5-V supply provided by U1. A 9- to 12-V "wall wart" power supply can be used to power the regulator input. Five-volt power is supplied to pin 2 of U7, while ground is applied to pins 1 and 4. A microprocessor reset is furnished between pin 28 and ground. The microprocessor communicates serially via U8, a MAX-232 chip. The MAX-232 requires only four 1-μF capacitors to operate. The output of the MAX-232 requires a diode D5 and resistor R15 to interface to your personal computer in order to download the software.

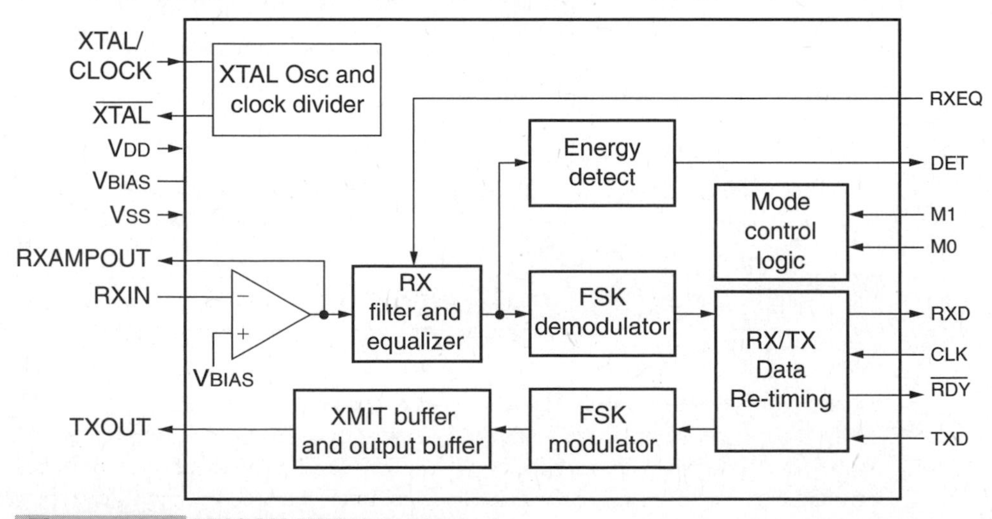

Figure 23-3 MXCOM MX604 modem.

TABLE 23-1 MXCOM MX604 MODEM PINOUTS

PIN NUMBER	NAME	DESCRIPTION
1	XTAL output	Output of the crystal oscillator inverter
2	XTAL/CLOCK	Input to the crystal oscillator inverter
3	M0	A logic input for mode setting
4	M1	A logic input for mode setting
5	RXIN	Input to the RX input amplifier
6	RXAMPOUT	Output of the RX input amplifier
7	TXOUT	Output of FSK generator
8	VSS	Negative supply voltage
9	VBIAS	Internal bias voltage held at $V_{dd}/2$
10	RXEQ	An input to enable/disable equalizer
11	TXD	An input for FSK modulator /retiming
12	CLK	An input used to clock in/out bits
13	TXD	Output from FSK demod or retiming data
14	DET	An output of on-chip energy detector
15	RDY	Ready for data, used for retiming
16	VDD	Positive voltage pin

TABLE 23-2 MODEM LOGIC CONTROL

M1	M0	RX/TX MODE CONTROL
0	0	1200 bps TX, 75 bps RX
0	1	1200 bps TX off
1	1	1200 bps RX off

Building the Data-Term Unit

Construction of the Data-Term unit is straightforward. A single 4- by 6-in glass-epoxy circuit board is used for this project. Single-row male header pins are used for J1 and J2 as well as for the input switches at J3; see Table 23-4. A two-screw-position terminal is used for the telephone line input. Male header pins could also be used for the display and buzzer if desired. As mentioned, a 5-V regulator powers the Data-Term unit, and the regulator is powered by a 12-V power supply. The LCD display, LED, and buzzer, as well as the push button switches,

TABLE 23-3 STAMP 2 PINOUTS (U7)

BS2 PIN IN	OUTPUT	FUNCTION	DIRECTION
10	P0	S1	IN
11	P1	S2	IN
12	P2	S3	IN
13	P3	S4	IN
14	P4	CLK	OUT
15	P5	RXD	IN
16	P6	M0	OUT
17	P7	M1	OUT
18	P8	RING	OUT
19	P9	LINE	OUT
20	P10	Zpin	OUT
21	P11	Mpin	OUT
22	P12	Display	OUT
23	P13	Buzzer	OUT
24	P14	Lamp	OUT
25	P15	Future	IN/OUT

are all mounted at the front panel of a plastic enclosure. Two telephone jacks are placed on the rear panel, one for the telephone line and the other for the X10 controller. A coaxial power jack is also placed on the rear panel to accept connections from the 12-Vdc wall wart power supply. The entire circuit is placed in a 6- by 6- by 2-in plastic enclosure.

Operating the Unit

Operation of the Data-Term is really quite simple. Connect a serial cable between your personal computer and the Data-Term; see Fig. 23-4. First apply power to the Data-Term circuit. Next, load the Data-Term software into the Data-Term circuit. Connect the phone line and the X10 controller and you are ready. You can now use your Data-Alert to call the Data-Term unit to report an alarm condition, which can be used to turn on appliances in your home or office. Install the DATA.BS2 software into the Data-Alert, and connect the Data-Alert to a phone line. Next connect your Data-Term to a second phone or neighbor's phone line and you can test the communication between your Data-Alert and your Data-Term units.

Locate the STAMP 2 Windows editor program titled STAMPW.EXE. Next connect the programming cable between your PC and the Data-Term. Apply power to the circuit and

TABLE 23-4 J-SERIES CONNECTOR/HEADER PINOUTS

SERIAL I/O—PROGRAMMING CABLE	
J1-1	ATN
J1-2	TX
J1-3	RX
J1-4	Ground
X10 OUTPUTS	
J2-1	Zpin
J2-2	GND
J2-3	GND
J2-4	Mpin
USER SWITCHES	
J3-1	S1
J3-2	S2
J3-3	S3
J3-4	S4

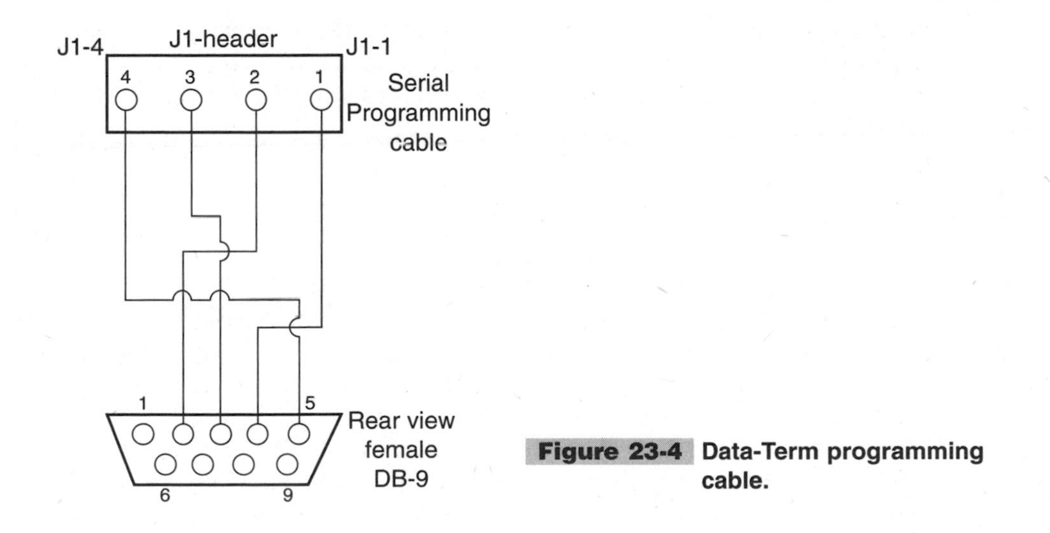

Figure 23-4 Data-Term programming cable.

download one of the RXMODEM.BS2 programs (Listing 23-1, 23-2, or 23-3) to the Data-Term unit. You are now ready to test communications between the Data-Alert and the Data-Term units.

Figure 23-5 shows the X10 control cable between the Data-Term and the X10 controller unit. The software provided with the Data-Term is evolving and can be expanded by you for more functions as well as future applications.

The Data-Alert system and Data-Term together form a powerful alarm reporting system that can be used to monitor many different types of sensors and/or alarm conditions and report to a remote Data-Term station located at your next door neighbor or across the country. Now you are free to go about your business or leisure and not worry about monitoring important devices.

```
'RXMODEM4.BS2
chan      var      byte
msg       var      byte(20)
houseA    con       0
unit1     con       0
houseB    con       1
houseC    con       2
houseD    con       3
clk       con       4
ipin      con       5
M0        con       6
M1        con       7
RI        var      in8
Tele      con       9
clrLCD    con       12
opin      con       12
buzz      con       13
lite      con       14
bdmd      con       17197
zpin      con       10
mpin      con       11
init:
DIRL = %00000000
DIRH = %00000000
low  clk
High M0      'modem init
High M1      'modem init
pause 1000
serout opin,17197,[clrLCD]

hold: if IN8 = 1 then hold
'hold: IF in0 = 1 then hold
main:
pause 1000
high Tele     'telephone line
pause 2000
high clk      'modem clock
pause 19000
low M0        'modem setup
pause 1000
```

Listing 23-1 Data-Term program 1.

```
SERIN ipin,17197,[WAIT(" CHAN"), dec1 chan, STR msg\19\10]
debug STR msg
pause 200
serout opin,17197,[STR msg\19]
branch chan-1,[st1,st2,st3,st4]

st1:
xout mpin,zpin,[houseA\unit1]
xout mpin,zpin,[houseA\uniton]
pause 15000
xout mpin,zpin,[houseA\unitoff]
pause 20000
low tele
goto init

st2:
xout mpin,zpin,[houseB\unit1]
xout mpin,zpin,[houseB\uniton]

pause 15000
xout mpin,zpin,[houseB\unitoff]
pause 20000
low tele
goto init

st3:
xout mpin,zpin,[houseC\unit1]
xout mpin,zpin,[houseC\uniton]
pause 15000
xout mpin,zpin,[houseC\unitoff]
pause 20000
low tele
goto init

st4:
xout mpin,zpin,[houseD\unit1]
xout mpin,zpin,[houseD\uniton]
pause 15000
xout mpin,zpin,[houseD\unitoff]
pause 20000
low tele
goto init
```

Listing 23-1 Data-Term program 1 (*Continued*).

```
'rxmodem5
chan      var     byte
msg       var     byte(20)
houseA    con      0
unit1     con      0
houseB    con      1
houseC    con      2
houseD    con      3
clk       con      4
ipin      con      5
```

Listing 23-2 Data-Term program 2.

```
M0          con       6
M1          con       7
RI          var      in8
Tele        con      9
clrLCD      con      12
opin        con      12
buzz        con      13
lite        con      14
bdmd        con      17197
zpin        con      10
mpin        con      11
serDat      var      word
EEaddr       var      word
samples      con      35
log          data     (samples)
endLog       con      log+samples-1
init:
DIRL = %00000000
DIRH = %00000000
low  clk
High M0           'modem init
High M1           'modem init
pause 1000
serout opin,17197,[clrLCD]
hold: if IN8 = 1 then hold
main:
pause 1000
high Tele         'telephone line
pause 2000
high clk          'modem clock
pause 19000
low M0            'modem setup
pause 1000
SERIN ipin,17197,[WAIT(" CHAN"), dec1 chan, STR msg\19\10]
'debug STR msg
pause 200
serout opin,17197,[STR msg\19]
branch chan-1,[st1,st2,st3,st4]
st1:
xout mpin,zpin,[houseA\unit1]
xout mpin,zpin,[houseA\uniton]
pause 15000
xout mpin,zpin,[houseA\unitoff]
pause 20000
low tele
goto init
st2:
xout mpin,zpin,[houseB\unit1]
xout mpin,zpin,[houseB\uniton]
pause 15000
xout mpin,zpin,[houseB\unitoff]
pause 20000
low tele
goto init

st3:
xout mpin,zpin,[houseC\unit1]
xout mpin,zpin,[houseC\uniton]
```

Listing 23-2 Data-Term program 2 (Continued).

```
pause 15000
xout mpin,zpin,[houseC\unitoff]
pause 20000
low tele
goto init
st4:
xout mpin,zpin,[houseD\unit1]
xout mpin,zpin,[houseD\uniton]
pause 15000
xout mpin,zpin,[houseD\unitoff]
pause 20000
low tele
goto init
for EEaddr=log to endLog    'Store sample
  serin ipin,17197,[serDat]
  debug "input ",serDat,tab,EEaddr,cr
  Write  EEaddr,serDat
  pause 500
next
pause 500
 serout opin,bdmd2,[12]
for EEaddr=log to endLog  'get sample
 read EEaddr,serDat
 debug "output ",SerDat,tab,EEaddr,cr
 pause 100
serout opin,19107,[serDat]
 pause 500
 next
```

Listing 23-2 Data-Term program 2 (*Continued*).

```
'RXMODEM6.BS2

chan    var  byte
msg     var  byte(20)
houseA   con    0
unit1   con    0
houseB   con    1
houseC   con    2
houseD   con    3
clk      con    4
ipin     con    5
M0        con    6
M1        con    7
S1       var    IN2
RI       var    IN8
Tele     con     9
clrLCD     con    12
opin     con    12
buzz     con    13
lite    con    14
bdmd     con     17197
zpin    con     10
mpin     con     11
```

Listing 23-3 Data-Term program 3.

```
serDat  var    word
EEaddr   var    word
samples  con    2
log   data    (samples)
endLog  con     log+samples-1

init:
DIRL = %00000000
DIRH = %00000000
low  clk
High M0    'modem init
High M1    'modem init
pause 1000
if IN2 = 0 then lamp
serout opin,17197,[clrLCD]
hold: if IN8 = 1 then hold
'hold: if IN0 = 1 then hold
main:
pause 1000
high Tele   'telephone line
pause 2000
high clk    'modem clock
pause 19000
low M0      'modem setup
pause 1000
SERIN ipin,17197,[WAIT(" CHAN"), dec1 chan, STR msg\19\10]
'debug STR msg
for EEaddr=log to endLog    'Store sample
   serin ipin,17197,[msg]
   debug "input ",STR msg,tab,EEaddr,cr
   Write  EEaddr, msg
 'pause 500
next
pause 200
serout opin,17197,[STR msg\19]
branch chan-1,[st1,st2,st3,st4]
st1:
xout mpin,zpin,[houseA\unit1]
xout mpin,zpin,[houseA\uniton]
pause 15000
xout mpin,zpin,[houseA\unitoff]
pause 20000
low tele
goto init

st2:
xout mpin,zpin,[houseB\unit1]
xout mpin,zpin,[houseB\uniton]
pause 15000
xout mpin,zpin,[houseB\unitoff]
pause 20000
low tele
goto init
st3:
xout mpin,zpin,[houseC\unit1]
xout mpin,zpin,[houseC\uniton]
pause 15000
xout mpin,zpin,[houseC\unitoff]
```

Listing 23-3 Data-Term program 3 (*Continued*).

```
pause 20000
low tele
goto init
st4:
xout mpin,zpin,[houseD\unit1]
xout mpin,zpin,[houseD\uniton]
pause 15000
xout mpin,zpin,[houseD\unitoff]
pause 20000
low tele
goto init
lamp:
'high lite
'pause 3000
'low lite
pause 500
 serout opin,17197,[12]
for EEaddr=log to endLog  'get sample
 read EEaddr,msg
 debug "output ",STR msg,tab,EEaddr,cr
 pause 100
serout opin,19197,[ msg]
 pause 500
 next
```

Listing 23-3 Data-Term program 3 (*Continued*).

Data-Term Parts List

R1	MOV130V
R2	150-Ω $^1/_4$-W resistor
R3	620-Ω $^1/_4$-W resistor
R4, R10, R16	10-kΩ $^1/_4$-W resistor
R17, R18, R19	10-kΩ $^1/_4$-W resistor
R5	600-Ω $^1/_4$-W resistor
R6, R7, R9	100-kΩ $^1/_4$-W resistor
R8	130-kΩ $^1/_4$-W resistor
R11, R12, R14, R15	4.7-kΩ $^1/_4$-W resistor
R13	1-kΩ $^1/_4$-W resistor
C1, C8, C9, C10, C11	1-μF 50-V electrolytic capacitor
C2, C12, C13	0.1-μF 35-V disk capacitor
C3, C4	0.1-μF 200-V capacitor
C5	0.001-μF 35-V disk capacitor
C7	100-pF 35-V disk capacitor

C14, C15, C16, C17	18-pF 35-V disk capacitor
X1	20-MHz crystal
X2	3.5-MHz crystal
T1	600-Ω transformer
D1, D2, D3, D4	3.9-V zener diode
D5	1N914 silicon diode
D6	LED
S5	PC pushbutton
S2, S3, S4, S5	Chassis pushbutton
J1, J2, J3	4-pin male headers
J4	Coaxial power jack
U1	LM3805 5-V regulator
U2	PVT312 optorelay
U3	PS2505-1 optoisolator
U4	LM358 op-amp
U5	MX604 modem chip
U6	24LC16B memory
U7	PIC 16C57 with BASIC interpreter, Parallax Inc.
U8	MAX-232 RS-232 converter chip
X10	TW523 or PL513 X10 interface (X10 sender unit)
Miscellaneous	Printed circuit board, IC sockets, wire, enclosure

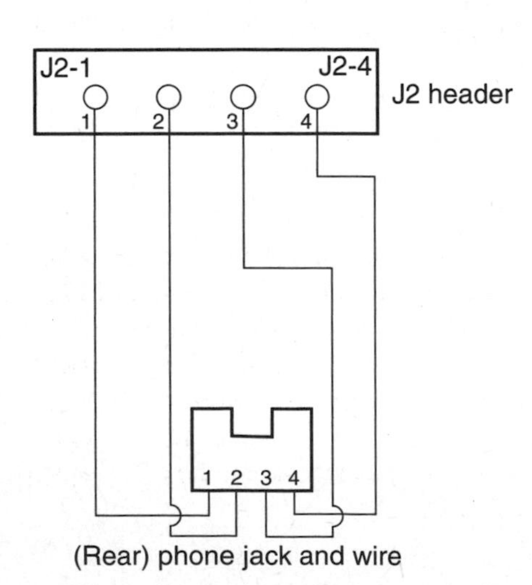

Figure 23-5 X10 header cable.

WEBLINK MONITORIN

CONTENTS AT A GLANCE

Imagine remotely monitoring up to 12 channels of information and displaying it anywhere in the world, via the Internet with the Weblink multichannel monitoring system; see Fig. 24-1. The Weblink system consists of a BASIC STAMP 2 microprocessor and a mini web-server module. The Weblink system allows enormous flexibility in creating a remote monitoring system. The Weblink monitoring system can monitor normally open and normally closed alarm switches for perimeter detection and/or monitor a wide range of sensors such as light, humidity, temperature, pH, and magnetic sensors. Once activated, the alarm system will notify you via the Internet anywhere in the world.

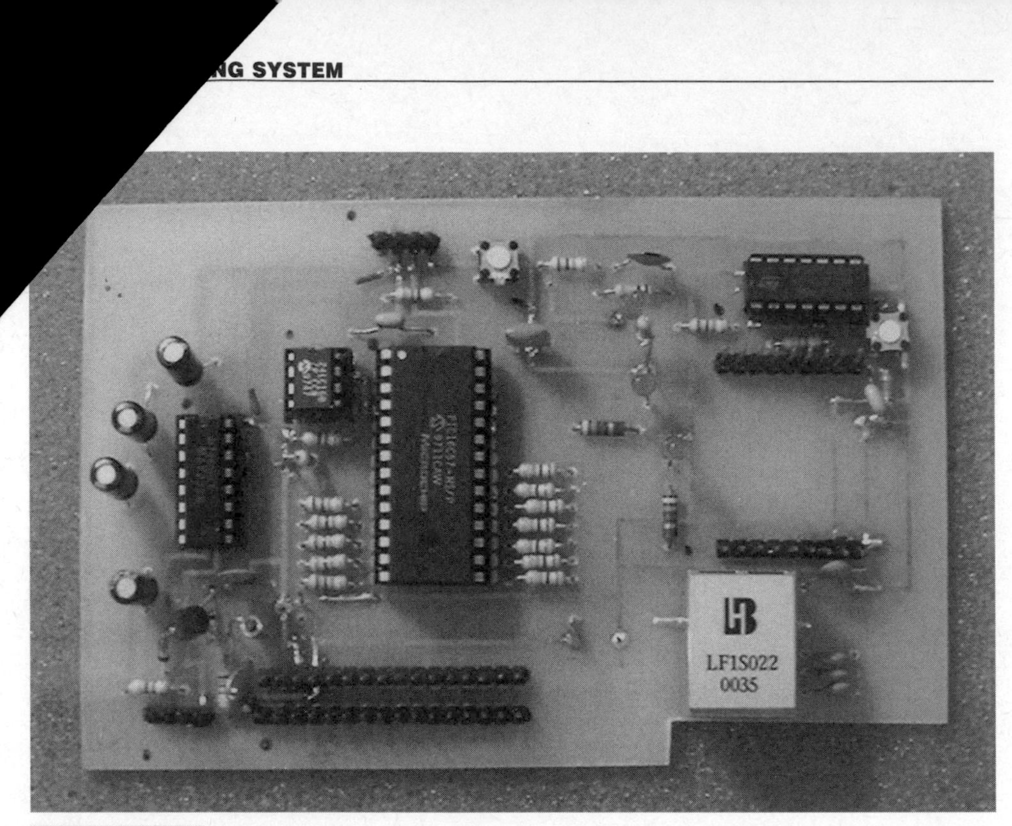

Figure 24-1 **Weblink monitoring system.**

The Weblink monitoring system provides a dedicated web page, incorporated in the on-board mini web-server module (WSM), which can be used to remotely display the 12 alarm conditions on any web browser. The multichannel Weblink monitoring system can also send an e-mail message or a file to a remote computer, informing you of an alarm condition.

Basic Components

The Weblink monitoring system is centered around the BASIC STAMP 2 controller section, shown in Fig. 24-2, and a mini web server, shown in Fig. 24-3. The project revolves around the lower-cost BASIC STAMP 2 "alternative" (the 16C57 PIC BASIC interpreter chip, a 28-pin DIP style STAMP 2), rather than the stock BASIC STAMP 2 module. The lower-cost 28-pin PIC BASIC interpreter chip requires a few external support chips and occupies a bit more real estate, but the cost is significantly less. The 28-pin BASIC STAMP 2 "alternative" requires a few external support components, a 24LC6B memory chip, a MAX232 serial interface chip, and a ceramic resonator. The MAX232 requires four 1-μF capacitors, a diode, and a resistor and provides a conventional serial I/O, which is used to program the Weblink microprocessor. The "alternative" STAMP 2 was chosen instead of a non-BASIC PIC chip, since this is a work in progress and features can be added at any time in the future. The Weblink system provides 12 normally open/closed inputs which can be used for sensor detection. The system has four output pins, a status

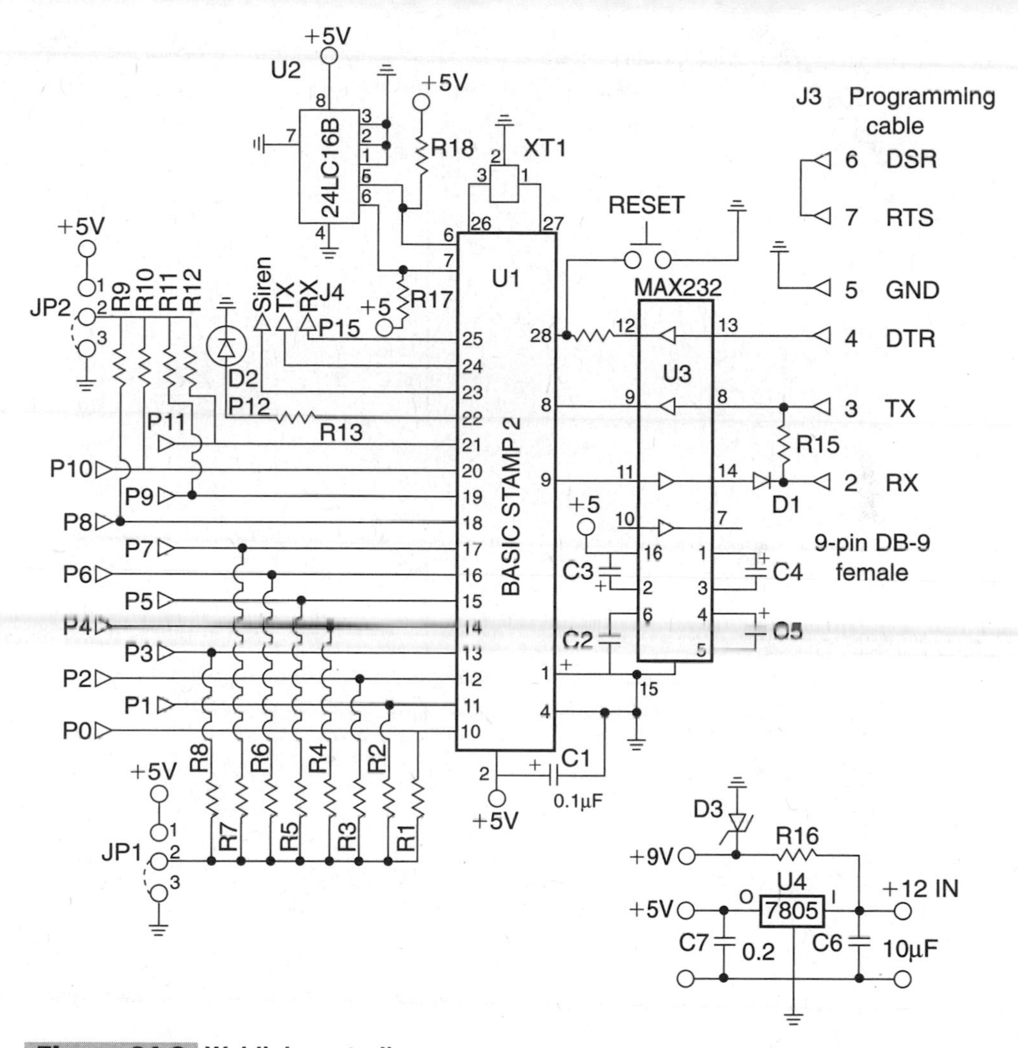

Figure 24-2 Weblink controller.

indicator, an output for alarm siren, annunciator, strobe lamps, etc. and two additional pins for serial communication to an always-on PC which can act as a server, if you elect to send e-mail notification to a remote computer, cell phone, or alphanumeric pager.

Figure 24-3 illustrates the mini web-server module (WSM), which plugs into the main Weblink monitoring board. The mini web-server module is a real web server in a 1-in-square package. The module itself consists of two main chips—a microprocessor for control and an Ethernet protocol server—and a crystal. Additional support components consist of two exclusive-OR gates and a pulse transformer/filter with an integrated RJ-45 jack, a few capacitors and resistors, and a "link" indicator LED. This powerful little web-server module is a true web server that will hold a custom 48-kbyte web page. The Weblink monitor web page consists of a 12-channel status chart of alarm stations with green/red indicator lights. Anyone going to the web page address of the Weblink mini-server module

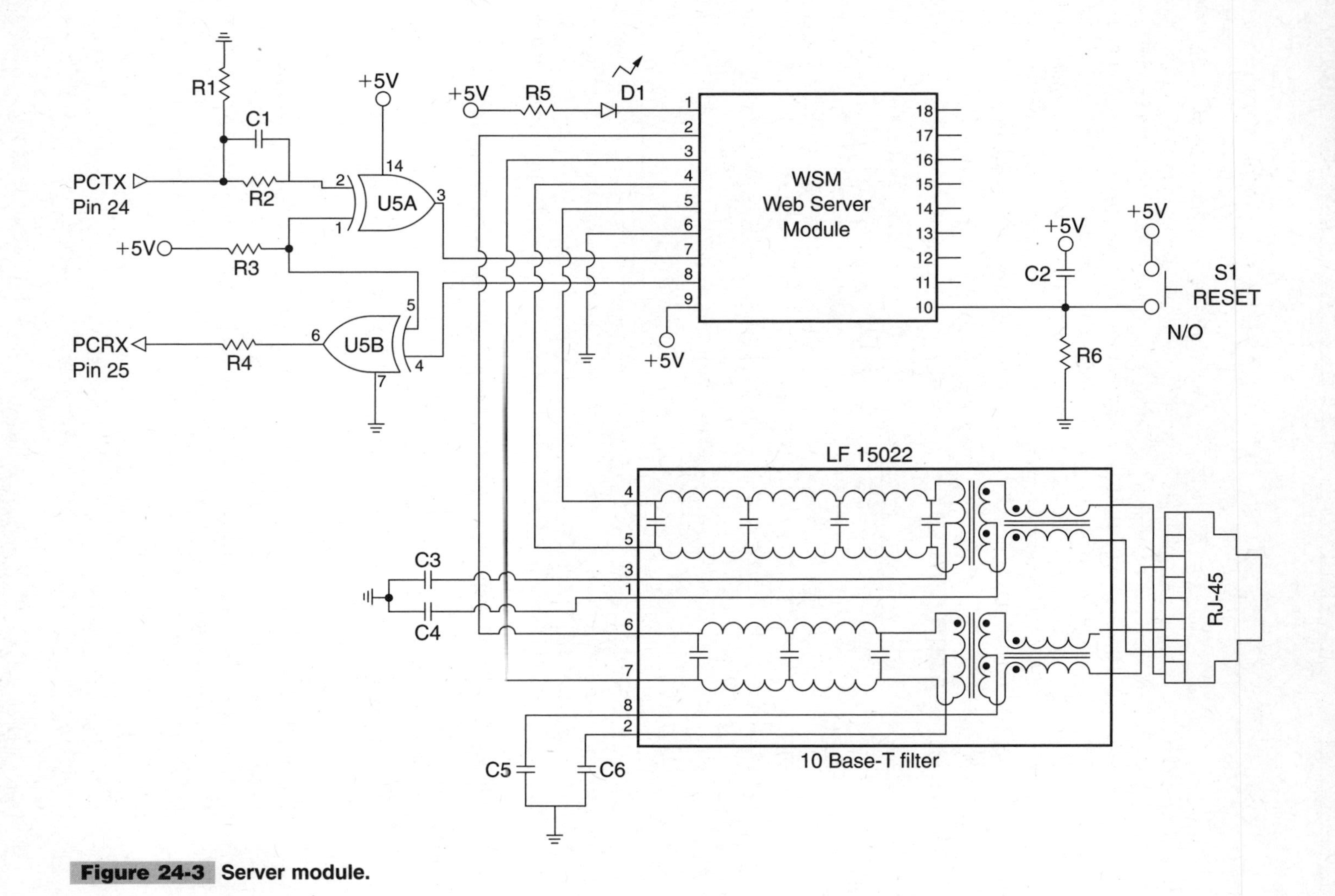

Figure 24-3 Server module.

would be able to view the status of the Weblink multichannel alarm from anywhere in the world.

The diagram in Fig. 24-4 depicts a clever method of providing both normally open and normally closed input sensors or alarm switches. Any number of normally closed or normally open switches can be used in this input circuit. Two 1N914 diodes, an npn transistor, and a 10-kΩ resistor are all that are required for the input device. You could elect to build one of these input circuits for each of the BASIC STAMP's input pins P0 through P11 if you desire. The output of this input converter circuit provides a 5-V signal to the input of the STAMP 2 upon an alarm signal. A daughter input board could be designed to plug into the header pins at J2 at the bottom of the main Weblink PC board.

COMPARATOR CIRCUIT

The sensor threshold/comparator circuit, shown in Fig. 24-5, can be used to monitor a wide array of input sensor devices. The threshold/comparator circuit consists of a dual LM393 comparator chip; only a single op-amp is shown. Two threshold comparator circuits can be constructed using this one chip and few support components. The negative input of the comparator chip at pin 2 is used as the feedback path as well as the threshold adjustment point based on a 50-kΩ potentiometer control. The threshold control can be biased with either 5 or 9 V. A 9-V reference source, on the main board, provides a wider-range input. A sensor device is connected to the positive input of the comparator at pin 3. Almost any resistive-type sensor could be utilized in this circuit. A temperature sensor, a resistive light sensor, a humidity sensor, a pH sensor, or a Hall effect switch could be used at S1. You could also elect to use the input at In1 to monitor a 5-V signal source and control the threshold for voltage change detection at this input. The output of the threshold detection circuit is a 5-V output, which could be connected to any number of BASIC STAMP 2 inputs. For example, you could monitor four temperature sensors, light sensors, and voltage sources all at the same time with the multichannel Weblink system. A comparator daughter board could be fabricated and plugged into the header pins at J2.

CONNECTING DEVICES

The diagram in Fig. 24-6 highlights how to connect various output devices to the multichannel Weblink system. As mentioned, the alarm monitoring system provides an output

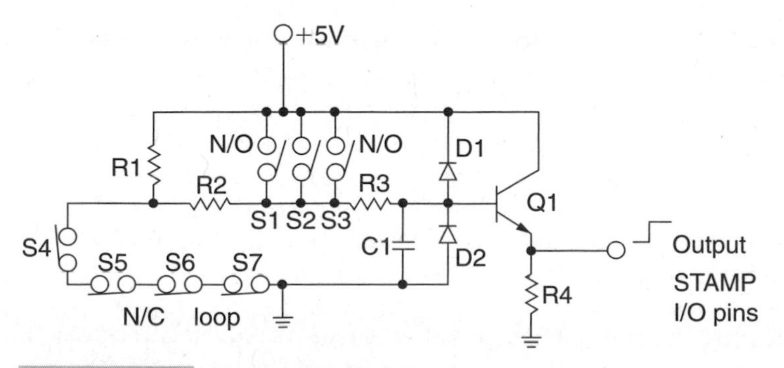

Figure 24-4 Alarm loop circuit.

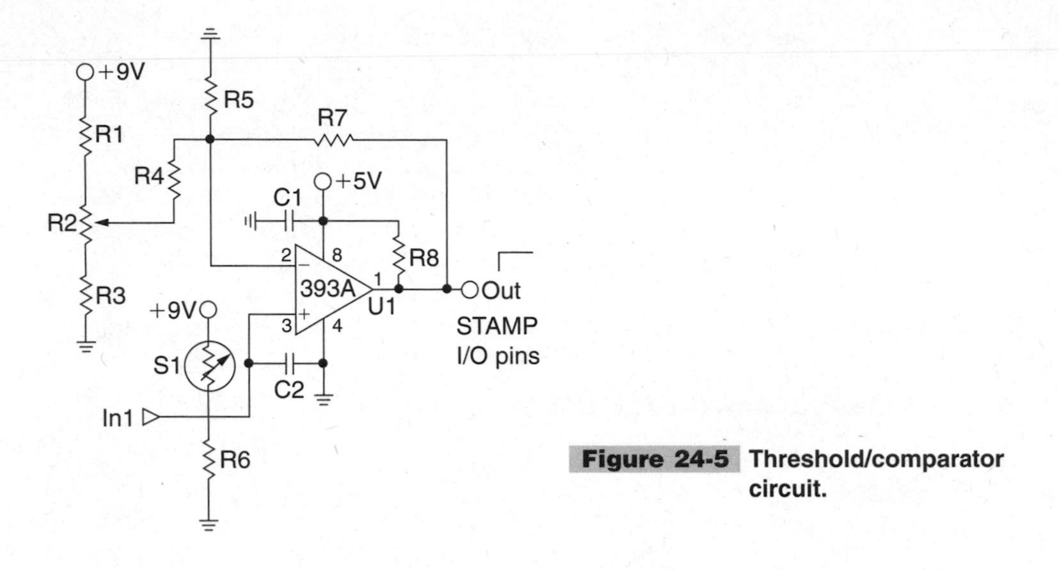

Figure 24-5 Threshold/comparator circuit.

for external alarm devices. The output at pin 23 (P13) from the controller is connected through a 1-kΩ resistor to the base of the 2N3904 relay driver transistor. Relay RL1 is a 5-V driver relay that can be used to turn on Sonalert or other small sounders or lamps. You can also use relay 1 to drive a second, higher-current relay, which can be used to drive a large motor siren or strobe lamp, etc. The smaller 5-V relay at RL1 is used to drive the larger-current 110-V relay RL2. So, for low current loads, only one relay is needed, while higher current loads must have two relays.

Construction of the System

The Weblink monitoring system is fabricated on a 3- by 5-in double-sided circuit board, which houses the STAMP 2 microprocessor, optional Weblink server, and the "glue," or support, components. A 5-V regulator at U4 powers the STAMP 2, the mini web server, and associated support components. The 9-V zener is the reference source for the sensor comparator module, if one is utilized.

Construction of the Weblink PC board is quite straightforward. First, install all the IC sockets. Next you can install the resistors; be sure to observe the color codes on the resistors in order to correctly identify and install them in their proper locations. Next install the capacitors; be sure to observe the polarity marking on the capacitors. Locate the diodes and the indicator LED. Be careful to identify the zener versus signal diodes and observe the correct polarity. Install the voltage regulator; once again be sure to observe the input versus output pins when installing this component on the board. Finally, install the header pins for the web-server module, the input pins, and power as well as programming connector J3 and server connection J4. Last, you will need to install the input programming jumpers JP1 and JP2, which tie the STAMP 2 input pins either high or low. The software provided (Listing 24-1) is configured for inputs to be tied low or to ground and go high upon activation.

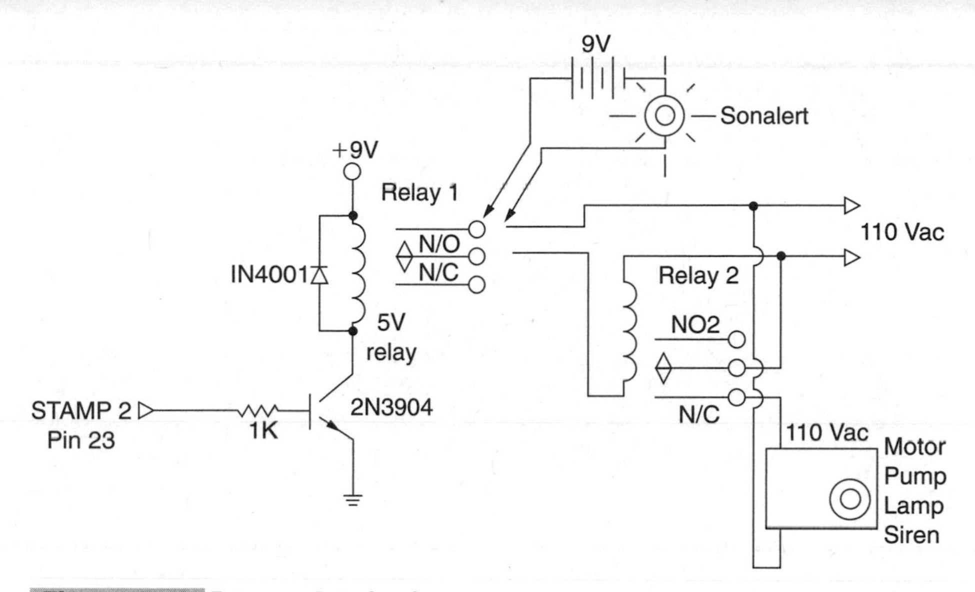

Figure 24-6 Power relay circuit.

```
'WLINK.BS2
'Weblink Monitoring System
'all inputs tied to ground and go high +5v upon activation
flash             Var   byte
io                Var   word
timer             Var   byte
tpin              Con   14
LedPi             Con   12
Siren             Con   13
recall_delay      Con   120
bdmd              Con   16468     '9600 baud
delay             Con   500
init:
INA=0000
INB=0000
INC=0000
DIR13=1
DIR14=1
DIR15=1
for flash = 1 to 3
 high LedPin
 pause 1500
 low LedPin
 pause 1500
next
main:
high LedPin
pause 100
low LedPin
if IN0 = 1 then gosub_chan_1
```

Listing 24-1 Weblink monitoring software.

```
if IN1 = 1 then gosub_chan_2
if IN2 = 1 then gosub_chan_3
if IN3 = 1 then gosub_chan_4
if IN4 = 1 then gosub_chan_5
if IN5 = 1 then gosub_chan_6
if IN6 = 1 then gosub_chan_7
if IN7 = 1 then gosub_chan_8
if IN8 = 1 then gosub_chan_9
if IN9 = 1 then gosub_chan_10
if IN10 = 1 then gosub_chan_11
if IN11 = 1 then gosub_chan_12
pause 100
goto main
gosub_chan_1:
high Siren
pause 2000
serout tpin,bdmd,50,["$80,$00,$01"]
pause 6000
low Siren
low LedPin
goto fin_1
gosub_chan_2:
high Siren
pause 2000
serout tpin,bdmd,50,[$80,$01,$01]
pause 6000
low Siren
low LedPin
goto fin_2
gosub_chan_3:
high Siren
pause 2000
serout tpin,bdmd,50,[$80,$02,$01]
pause 6000
low Siren
low LedPin
goto fin_3
gosub_chan_4:
high Siren
pause 2000
serout tpin,bdmd,50,[$80,$03,$01]
pause 6000
low Siren
low LedPin
goto fin_4
gosub_chan_5:
high Siren
pause 2000
serout tpin,bdmd,50,[$80,$04,$01]
pause 6000
low Siren
low LedPin
goto fin_5
gosub_chan_6:
high Siren
pause 2000
serout tpin,bdmd,50,[$80,$05,$01]
```

Listing 24-1 **Weblink monitoring software (*Continued*).**

```
pause 6000
low Siren
low LedPin
goto fin_6
gosub_chan_7:
high Siren
pause 2000
serout tpin,bdmd,50,[$80,$06,$01]
pause 6000
low Siren
low LedPin
goto fin_7
gosub_chan_8:
high Siren
pause 2000
serout tpin,bdmd,50,[$80,$07,$01]
pause 6000
low Siren
low LedPin
goto fin_8
gosub_chan_9:
high Siren
pause 2000
serout tpin,bdmd,50,[$80,$08,$01]
pause 6000
low Siren
low LedPin
goto fin_9
gosub_chan_10:
high Siren
pause 2000
serout tpin,bdmd,50,[$80,$09,$01]
pause 6000
low Siren
low LedPin
goto fin_10
gosub_chan_11:
high Siren
pause 2000
serout tpin,bdmd,50,[$80,$a,$01]
pause 6000
low Siren
low LedPin
goto fin_11
gosub_chan_12:
high Siren
pause 2000
serout tpin,bdmd,50,["$80,$b,$01"]
pause 6000
low Siren
low LedPin
goto fin_12
'system clear & reset
fin_1
pause 20000
serout tpin,bdmd,50,["$80,00,$00"]
```

Listing 24-1 **Weblink monitoring software (*Continued*).**

```
return
end
fin_2
pause 20000
serout tpin,bdmd,50,["$80,01,$00"]
return
end
fin_3
pause 20000
serout tpin,bdmd,50,["$80,02,$00"]
return
end
fin_4pause 20000
serout tpin,bdmd,50,["$80,03,$00"]
return
end
fin_5
pause 20000
serout tpin,bdmd,50,["$80,04,$00"]
return
end
fin_6
pause 20000
serout tpin,bdmd,50,["$80,05,$00"]
return
end
fin_7
pause 20000
serout tpin,bdmd,50,["$80,06,$00"]
return
end
fin_8
pause 20000
serout tpin,bdmd,50,["$80,07,$00"]
return
end
fin_9
pause 20000
serout tpin,bdmd,50,["$80,08,$00"]
return
end
fin_10
pause 20000
serout tpin,bdmd,50,["$80,09,$00"]
return
end
fin_11
pause 20000
serout tpin,bdmd,50,["$80,$a,$00"]
return
end
fin_12
pause 20000
serout tpin,bdmd,50,["$80,$b,$00"]
return
end
```

Listing 24-1 Weblink monitoring software (*Continued*).

Modes of Operation

The Weblink monitoring system can be configured for three different modes of operation. The first and most elegant configuration of the multichannel Weblink consists of the mini web-server module (WSM) plugged into the Weblink controller board. With the mini web-server module installed, your Weblink monitoring system will present your sensor/alarm conditions to a dedicated web page, which is hosted in your mini web-server module. The Weblink monitor's input sensor conditions can be viewed by you or anyone with a web page browser such as Internet Explorer or Netscape from any location around the world. The mini web-server module eliminates the use of a dedicated computer and is very compact. Your always-on Ethernet connection plugs into the multichannel Weblink board and you're ready to go!

The second Weblink configuration assumes you already have a PC that is always running. Installing NetPorter software, from *http://www.al-williams.com/netporter.htm,* will allow the Weblink monitor system to send alarm conditions via e-mail to a remotely specified computer. In this configuration you would not need the mini web-server module installed on the Weblink system board; see Fig. 24-7. This configuration also assumes an always-on Ethernet connection.

The third configuration of the Weblink monitoring controller allows the Weblink board to be utilized as a stand-alone alarm system, with a siren, for local alarm indication. The Weblink output could also be used to activate a telephone dialer to summon a friend, neighbor, relative, or police if desired.

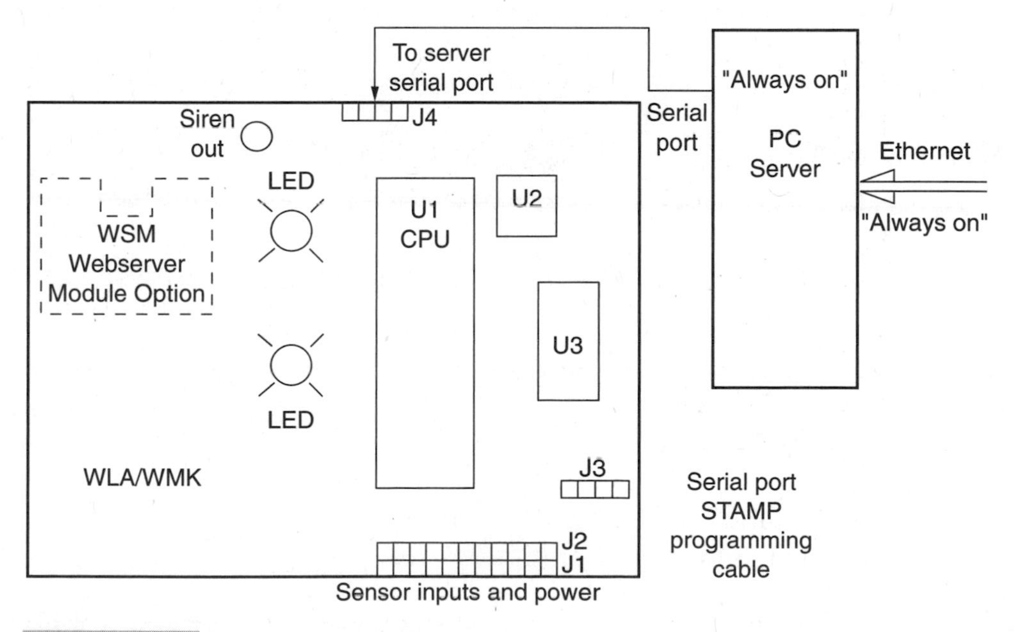

Figure 24-7 Weblink PC server.

Setting up and Operating the System

Setup and operation of the multichannel network alarm system are quite straightforward. First you will need to decide which configuration you will be using. Do you wish to view the alarm conditions via a web page with a self-contained system or do you have an always-on computer running that can be used to send alarm conditions to a remote computer by e-mail or File Transfer Protocol (FTP)? If you elect to use the self-contained mini web-server module option, you must first have an always-on Ethernet connection such as Road Runner Cable modem service or DSL.

The diagram in Fig. 24-8 illustrates the mini web-server module installed on the main Weblink monitor system board. Plug the WSM into the main Weblink board. Note the proper orientation of the WSM: the serial number in the top right corner faces the STAMP 2. The Weblink mini web server is preloaded with a 12-channel status monitor web page. You will first need to figure out what types of sensors or switches you wish to use as input devices. Sensors inputs and power are connected at the bottom of the circuit board at J1 and J2. Remember, J2 is provided in the event that you wish to build a sensor input daughter board. Next, you will need to connect your sensors and a 9- to 12-V power supply or "wall wart." Connect your programming cable between a PC or laptop and your Weblink board at J3; see Fig. 24-9.

First you will need to locate the Windows STAMP 2 editor program called STAMPW.EXE. Next you will need to connect up the programming cable between your programming PC and the Weblink board. Now locate and install the program titled WLINK1.BS2 (Listing 24-1), and load the program into the STAMP 2. Your always-on Ethernet connection is now plugged into the Weblink board, at the RJ-45 jack. Once the controller is powered up and the Ethernet connection established, your Weblink monitor

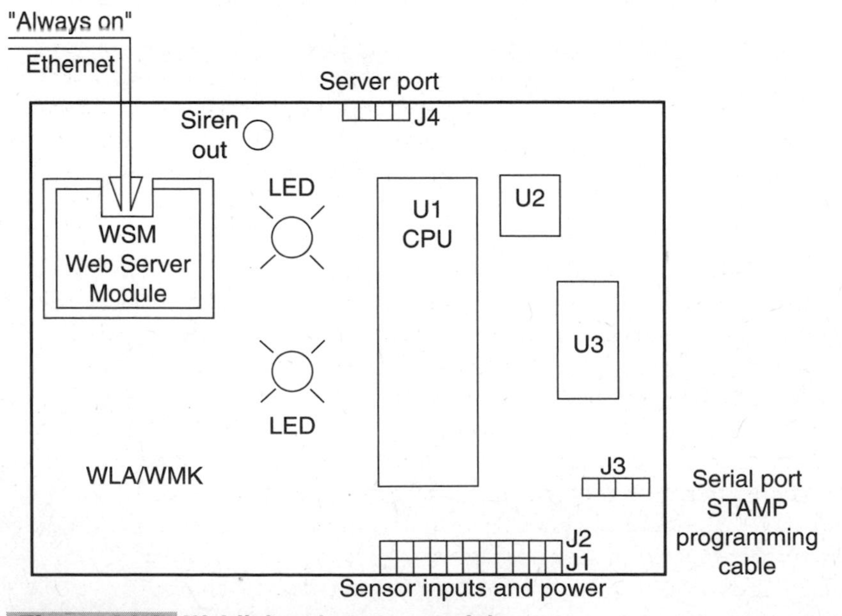

Figure 24-8 **Weblink web-server module.**

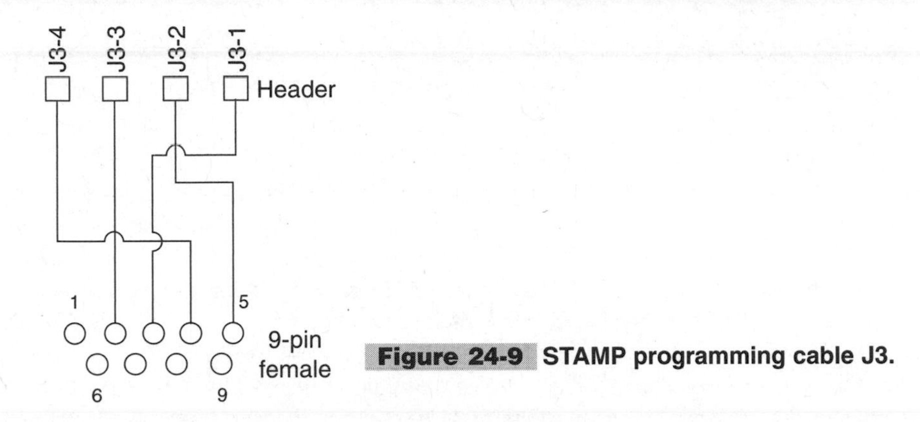

Figure 24-9 STAMP programming cable J3.

alarm is ready for operation. Note there is a reset switch for both STAMP 2 and the mini server module, if you encounter a system start-up problem.

If you elect to use the Weblink monitoring system to remotely view alarm conditions, you will need to load the software into the Siteplayer or WSM. Note the SWM Weblink software is on the supplied CD-ROM. In order to install the Weblink software into the WSM, you will need to obtain a Siteplayer development board available at $99.95 and the WSM module at $29.95 from http:www.siteplayer.com. Two other options are also available: You can obtain the WSM preprogrammed for $45.00 from josh@petruzz.net or have your own WSM programmed for $15.95 plus $6.00 shipping.

IDENTIFYING NETWORK LOCATION

If you have elected to use the Weblink monitor system board with the WSM server, you will need to identify its location on the network, and this can be tricky, depending on your network configuration. If you anticipate using the Weblink monitoring system behind a Linksys 4-port router/switch, then you will need go to the Linksys web page—on board the router, via your browser, go to the main page, then advanced, then forwarding. In the service port range column, you type the port number that will be translated to a Weblink (for this example use 2001); then type the FIXED IP address of the Weblink on the internal network. You must fix the IP address of the Weblink or else it could wander around with Dynamic Host Configuration Protocol (DHCP) and get lost. Now you have attached the outside port 2001 to the inside IP address of the Weblink.

From the outside world you then surf the Weblink by typing http://xxx.xxx.xxx.xxx:2001, where xxx.xxx.xxx.xxx is your outside world IP address. The Linksys will translate this port request to the following: http://inside.Weblink.ip.address:2001.

Since the Weblink responds to all ports, the 2001 request is just like any other normal HTTP port 80 request. You can do this with up to 10 entries in the Linksys table. (I thought you could do more, but it seems that they have limited it.) The only wrinkle that I see is that if you are using PPoe for your DSL connection, this may not work, since the PPoe connection is not always direct. You may be going through some routers in your ISP.

Try forwarding port 80 in the Linksys to your PC's internal IP address and run WeblinkPC. If someone can surf WeblinkPC from the outside world, then you should be able to surf the hardware Weblinks.

If instead you elected to use your own PC server instead of the on-board mini web-server module on the Weblink monitor board to send e-mail notification, you will need to supply your own always-running computer, which will act as your server. This configuration also requires an always-on Ethernet connection in order to send e-mail or FTP notifications to your remote monitoring computer, cell phone, or alpha pager. You will need to make up serial connection cable from the output of the Weblink controller board to your always-on server PC via header J4. Figure 24-8 depicts the Weblink-to-PC server connections. Sensors as well as power are connected on the bottom of the circuit board at J1. Connector J2 is provided for a sensor daughter board. Once your serial port cable is connected between the Weblink monitor board and your PC server, you will need to power up the Weblink board and load the EMDAT1.BS2 into your STAMP 2 controller, for alarm e-mail notification. Note that the program contains a sample e-mail address, which will have to be changed to your own e-mail address. Next you will need to install the mini server program WEBSERV.EXE into your always-running PC, and configure it for e-mail files. Your Weblink monitoring system is now ready to serve you.

ADDITIONAL SOFTWARE

Also included with the software files on the CD-ROM are three additional programs: SITEPLAYERPC, which is used to create/change web pages in the WSM module, SITELINKER, which is used to download the binary image to the WSM module, and a program titled Siteplayer Serial Port Tester. The last program can be used to GET or SET the WSM's address.

If you have the need to remotely monitor alarm or sensor conditions across your home, office, or the Internet, then you will definitely want to build the Weblink multichannel system. This low-cost system offers great flexibility in monitoring remote alarm conditions.

Weblink Monitoring System Board Parts List

R1–R12	10-kΩ $^1/_4$-W resistor
R13	330-Ω $^1/_4$-W resistor
R14, R15, R17, R18	4.7-kΩ $^1/_4$-W resistor
R16	47-Ω $^1/_4$-W resistor
C1	0.1-μF 35-V disk capacitor
C2, C3, C4, C5	1-μF 35-V electrolytic capacitor
C6	10-μF 35-V electrolytic capacitor
C7	0.22-μF 35-V disk capacitor
D1	1N914 diode
D2	LED

D3	9-V zener diode
U1	PIC 16C57 BASIC STAMP 2
U2	24LC16B memory chip
U3	MAX232 serial interface chip
U4	LM7805 5-V regulator
S1	PC momentary pushbutton
J1	Input/output header I (female)
J2	Input header II—input modules (female)
J3	Serial header (programming) (male)
J4	Serial header (PC server) (male)
JP1, JP2	Input programming pins default (pin 1 to ground)
Miscellaneous	Header pins, circuit board, IC sockets

Web-Server Module Assembly Parts List

WSM	Web-server module
R19, R23	330-Ω $^1/_4$-W resistor
R20	10-kΩ $^1/_4$-W resistor
R21	100-kΩ $^1/_4$-W resistor
R24	47-kΩ $^1/_4$-W resistor
C8, C9, C10, C11	0.01-μF 35-V disk capacitor
C12	100-pF 35-V disk capacitor
C13	0.1-μF 25-V disk capacitor
D4	LED
U5	74HC86 quad exclusive-OR gate
T1	LF1S022 pulse transformer and RJ-45 jack
S1	Reset switch—momentary pushbutton

Threshold/Comparator Board Parts List

U1	LM393 comparator IC
R1, R3	2-kΩ $^1/_4$-W resistor

R2	50-kΩ potentiometer
R4	10-kΩ 1/4-W resistor
R5, R6	1-MΩ 1/4-W resistor
R7	10-MΩ 1/4-W resistor
R8	3.3-kΩ 1/4-W resistor
C1	0.1-μF 35-V disk capacitor
C2	0.01-μF 35-V disk capacitor
S1	Thermistor or resistive sensor
Miscellaneous	Header, wire, circuit board

Alarm Loop Input Circuit Parts List

R1, R3	20-kΩ 1/4-W resistor
R2, R4	10-kΩ 1/4-W resistor
C1	0.01-μF 25-V disk capacitor
D1, D2	1N914 silicon diode
Q1	2N2222 npn transistor
S1, S2, S3	Normally open switches or sensors
S4, S5, S6, S7	Normally closed switches
Miscellaneous	Header, wire, circuit board

25

X-LINK INTERNET CONTROL SYSTEM

CONTENTS AT A GLANCE

Imagine remotely controlling up to 30 X10 power line devices around your home, shop, or office. In addition, you can control six additional devices via local relay control from anywhere in the world, via the Internet with the X-link control system, shown in Fig. 25-1. The X-link system consists of a BASIC STAMP 2 microprocessor and a mini web-server module. From your laptop in California you can control pumps, lamps, motors, cameras, and heating and cooling systems back at your home or office. The X-link could also be used to help an aging parent remember to turn things off. The X-link system allows enormous flexibility in creating a remote control system.

The X-link control system provides a dedicated web page, incorporated in the on-board mini web-server module (WSM), which can be used display 30 X10 devices plus an additional six local devices from any remote web browser. The remote web display consists of a 6-by-6 matrix.

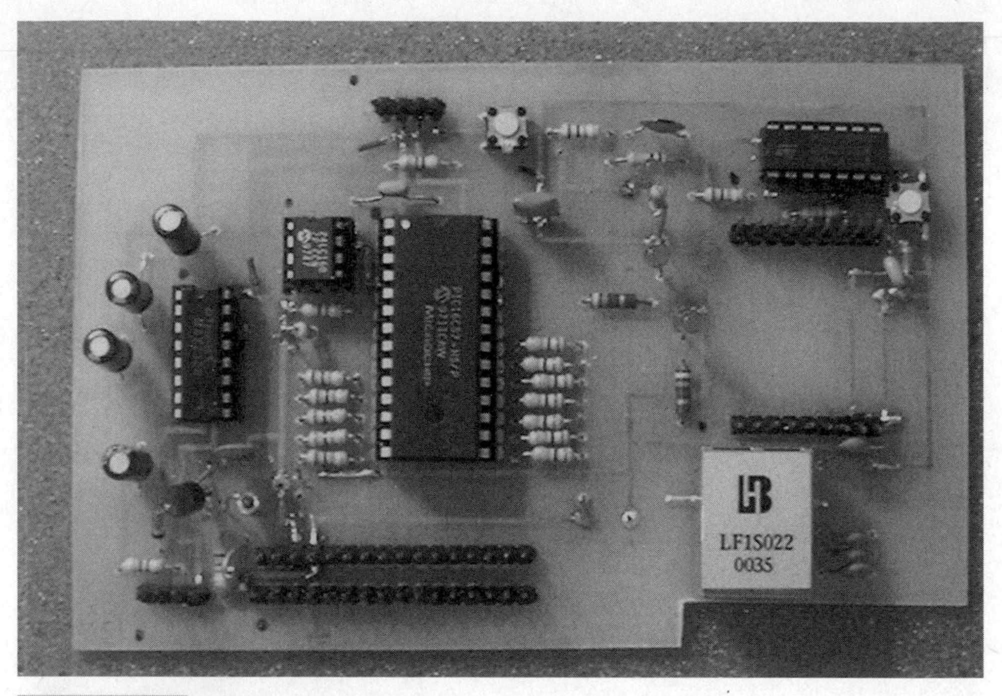

Figure 25-1 X-link controller.

Basic Components

The X-link control system is centered around the BASIC STAMP 2 "alternative," shown in Fig. 25-2 and the mini web server shown in Fig. 25-3. The project revolves around the lower-cost 16C57 PIC BASIC interpreter chip, a 28-pin DIP style STAMP 2, rather than the stock BASIC STAMP 2 module. The lower-cost 28-pin PIC BASIC interpreter chip requires a few external support chips, a 24LC6B memory chip, a MAX232 serial interface chip, and a ceramic resonator and occupies a bit more real estate, but the cost is significantly less. The MAX232 requires four 1-μF capacitors, a diode, and a resistor and provides a conventional serial I/O, which is used to program the X-link microprocessor. The STAMP 2 was chosen instead of a non-BASIC PIC, since this is a work in progress and features can be added at any time in the future. The X-link system utilizes two STAMP output pins, 10 (P0) and 11 (P1), to control the X10 interface. Up to six local devices can be controlled by STAMP output pins 18 (P8) through 23 (P13). Pumps, lamps, sirens, and remote cameras can be controlled by these output pins. STAMP 2 pins 12 (P2) through 17 (P7) are spare and not used in this application. STAMP 2 pins 24 (P14) and 25 (P25) are used for serial communication between the STAMP 2 and the web-server module; P14 is the transmit pin, while P15 is the receive pin. In this application P15 is utilized to send serial data from the web-server module to the STAMP 2 controller.

The diagram in Fig. 25-3 illustrates the mini web-server module (WSM), which plugs into the main X-link monitoring board. The mini web-server module is a real web server in a 1-in-square package. The module itself consists of two main chips, a micro-

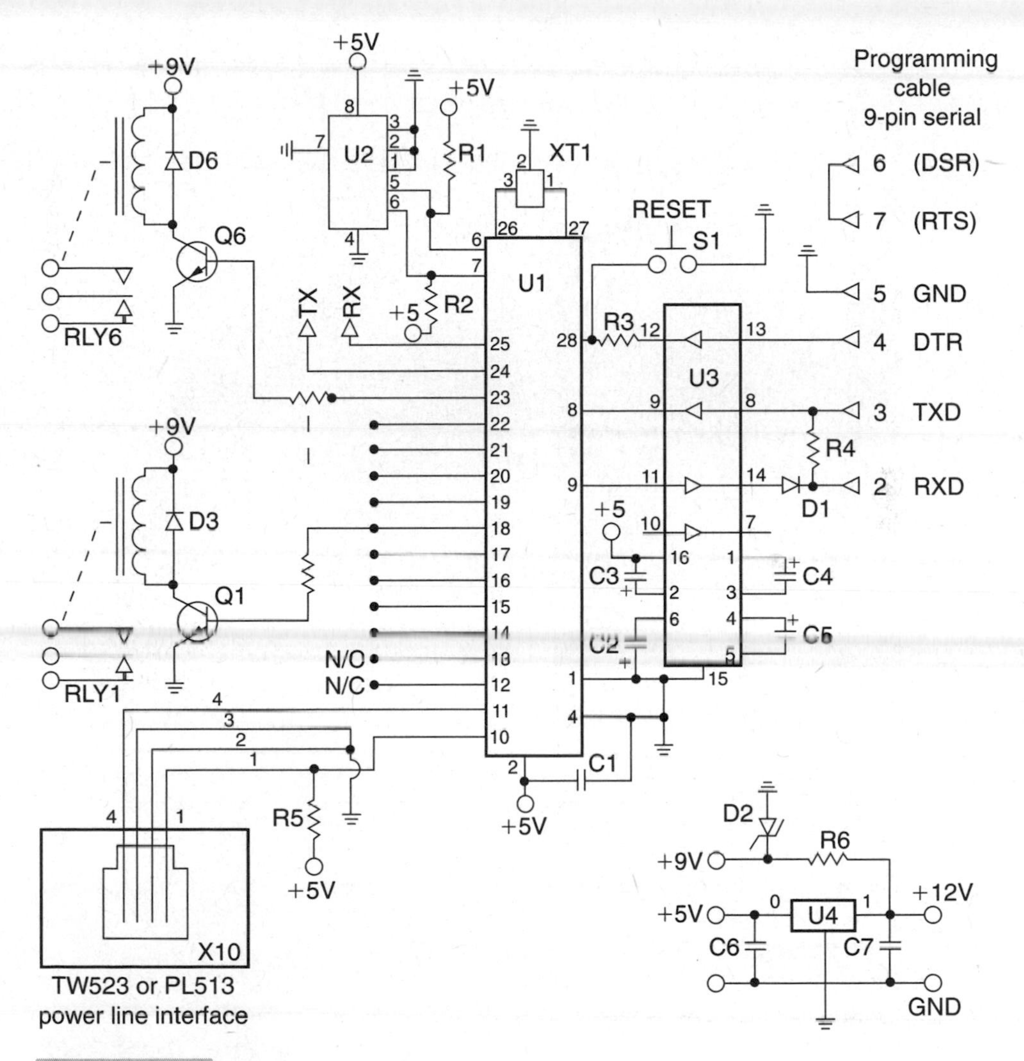

Figure 25-2 X-link main board.

processor for control and an Ethernet protocol server, and a crystal. Additional support components consist of two exclusive-OR gates and a pulse transformer/filter with an integrated RJ-45 jack, a few capacitors and resistors, and a link indicator LED. This powerful little web-server module is a true web server that will hold a custom 48-kbyte web page.

The X-link control system can be configured for two different modes of operation, remote and local. The first and most elegant configuration of the X-link control system consists of the mini web-server module plugged into the X-link controller board. With the mini web-server module installed, your X-link system will present control status conditions to a dedicated web page, which is hosted in your mini web-server module. The X-link control conditions can be viewed by you with any web page browser such as Internet Explorer or Netscape from any location around the world. The mini web-server module

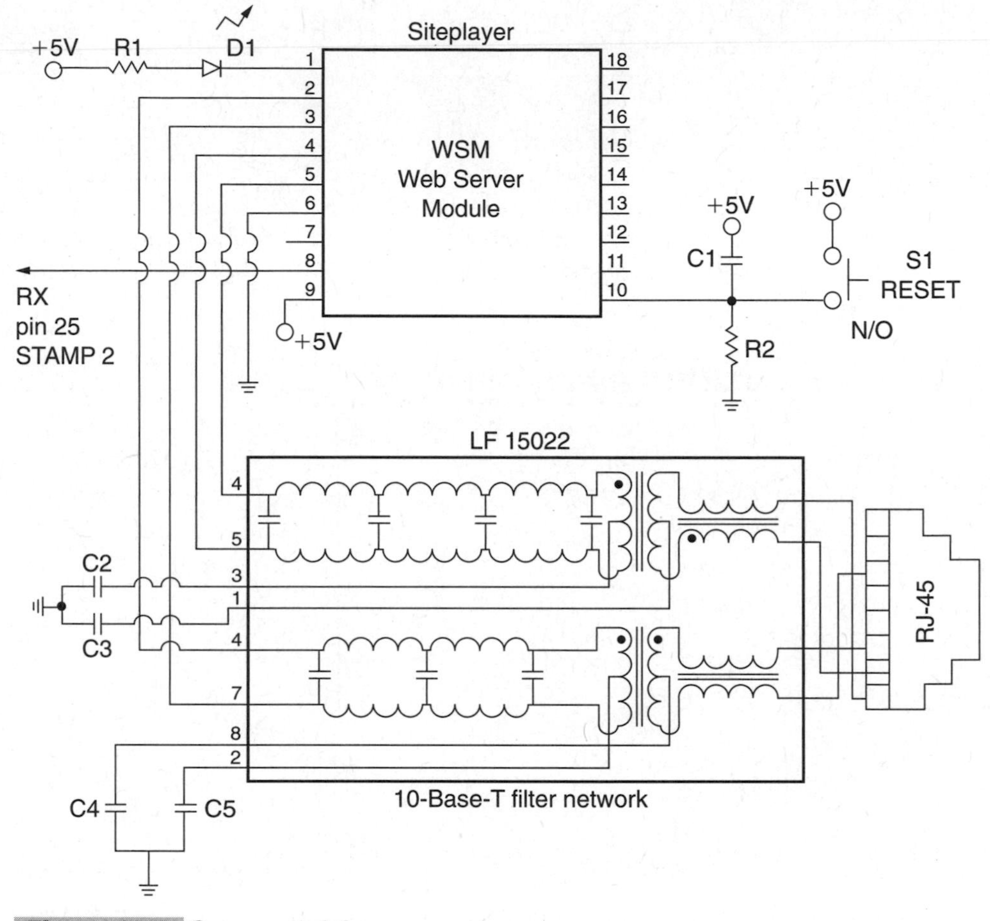

Figure 25-3 Server module.

eliminates the need for a dedicated computer and is very compact. Your always-on Ethernet connection plugs into the multichannel X-link board and you're ready to go!

In a second configuration of the X-link the STAMP 2 controller allows the X-link board to be utilized as a stand-alone, or local, control system. The X-link outputs can also be used to activate a any number of local X10 devices such as such as lamps, sirens, pumps, heating and cooling devices, and local relay control devices such as buzzers, cameras, and recording devices.

KEYBOARD CONTROLLER

Figure 25-4 illustrates a preprogrammed keyboard chip from ham radio operator K1EL. (Type K1EL in you web browser.) This chip will allow you to use a standard IBM PS/2 keyboard to input controls to the STAMP 2 controller in order to locally control the X-link system, if you do not wish to use the remote Internet feature. The 8-pin encoder chip is powered by 5 V. The keyboard is connected to ground and a 5-V source, as is the chip. Pin 2 is a reset pin and is not used. Pin 3 is the output to the STAMP 2 controller. Pin 4 is the baud rate selector and is connected to 5 V for 9600-baud operation. Pin 6 is the mark/sense input and is programmed to ground. Pin 5 is connected to the keyboard CLK pin, and pin 7 of the Katkbd chip is connected to the data pin of the keyboard. Also note the output of the Katkbd could be connected to a mini RF link transmitter such as one of the Abacom models. A compatible RF receiver could then be connected to the input of the STAMP 2 controller serial input pin, for a wireless X10 system. To utilize local keyboard control of X10 devices, you will need to connect the output from pin 3 of the Katkbd chip to pin 25 (P15) of the STAMP 2 controller for local operation.

CONNECTING POWER DEVICES

The diagram in Fig. 25-5 highlights how to connect high-power output devices to the X-link control system. As mentioned, the X-link system provides an output for external devices. Output pins 18 (P8) through 23 (P13) from the controller are connected through 1-kΩ resistors to the base of the 2N3904 relay driver transistors. Relay 1 is a 5-V driver relay that can be used to turn on Sonalert or other small sounders or lamps. You can also use relay 1 to drive a second higher-current relay, which can be used to drive a large motor siren or strobe lamp, etc. The smaller 5-V relay at relay 1 is used to drive the larger-current 110-V relay 2. So, for low-current loads only one relay is needed, while higher-current loads must have two relays, to control motors, pumps, and heating and cooling devices.

Construction of the Board

The X-link control system was fabricated on a 3- by 5-in double-sided circuit board, which houses the STAMP 2 microprocessor, the optional X-link server, and the "glue," or support components (Fig. 25-6). A 5-V regulator, at U4, powers the STAMP 2, the mini web server, and associated support components. The 9-V zener is used for the a reference source for relay control.

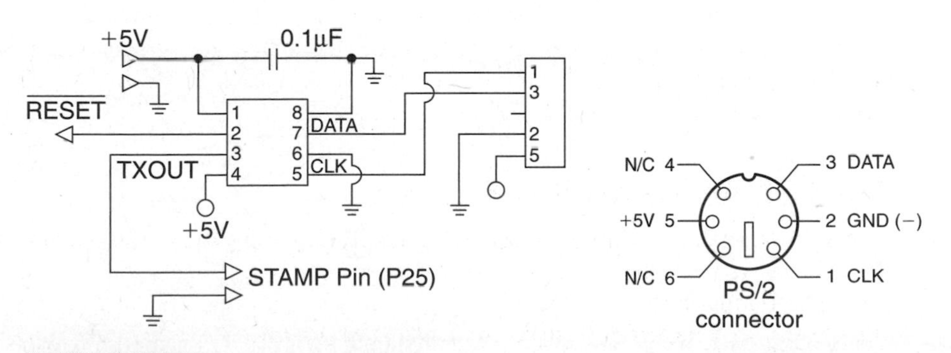

Figure 25-4 Keyboard controller.

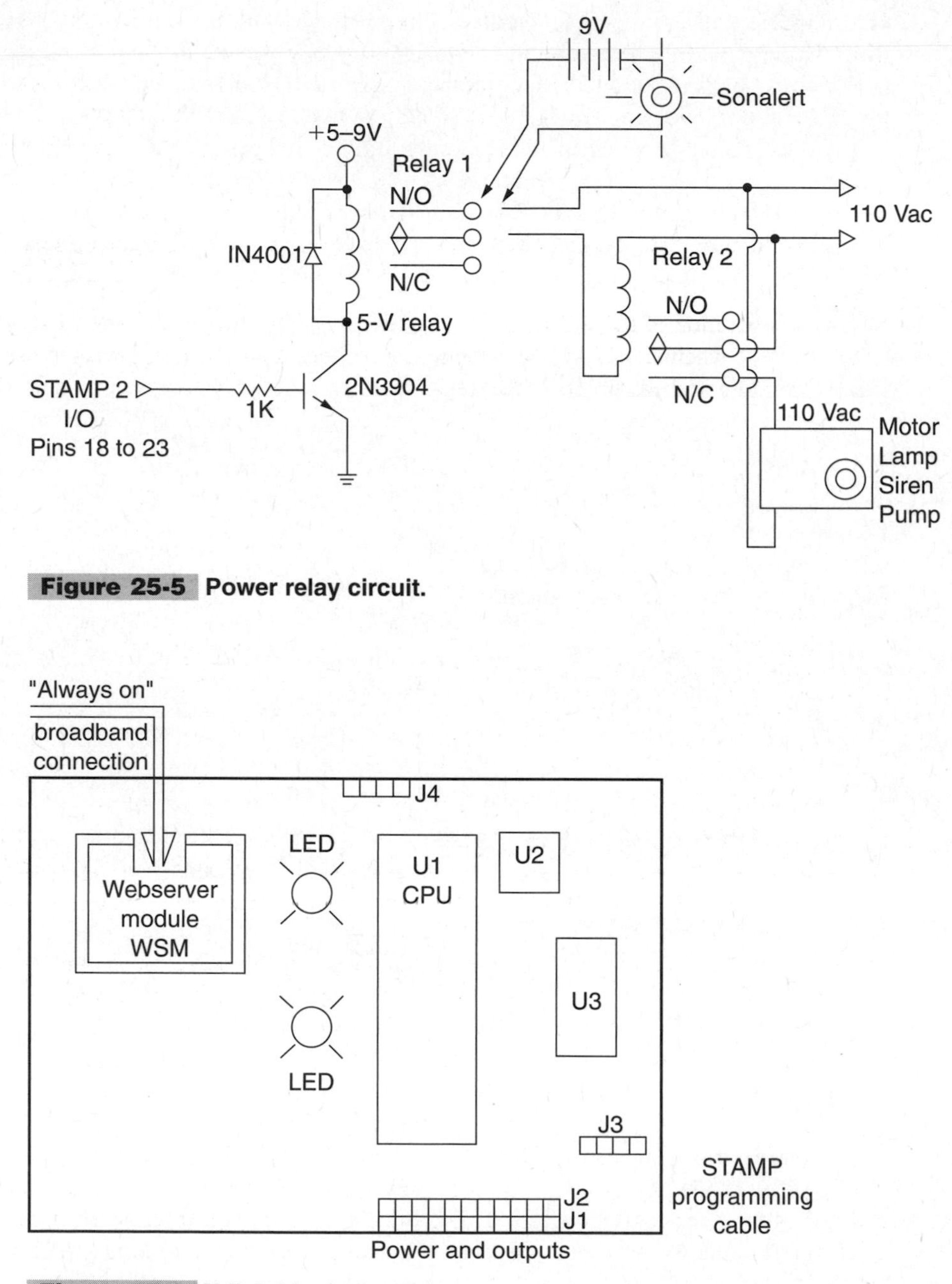

Figure 25-5 Power relay circuit.

Figure 25-6 X-link block diagram.

Construction of the X-link PC board is quite straightforward. First, install all the IC sockets. Next you can install the resistors; be sure to observe the color codes on the resistors in order to correctly identify and install them in their proper locations. Next install the capacitors; be sure to observe the polarity marking on the capacitors. Locate the diodes and the indicator LED. Be careful to identify the zener versus signal diodes and observe the correct polarity. Install the voltage regulator; once again be sure to observe the input versus output pins when installing this component on the board. Finally, install the header pins for the web-server module, the input pins, and power as well as programming connector J3 and server connection J4.

Setting up and Operating the System

Setup and operation of the X-link control system are quite straightforward. First, you will need to decide whether you will be using remote or local control. If you wish to use the X-link for remote web control via the self-contained system web-server module (WSM) option, you must have an always-on Ethernet connection such as Road Runner Cable modem service or DSL, as a host for the X-link system board, at the control site that you wish to control.

Next, you will need to connect your X-link board to a 9- to 12-V power supply or "wall wart." Connect your programming cable between a PC or laptop and your X-link board; see Fig. 25-7. You will need to locate the STAMP 2 program titled XLink.BS2 (Listing 25-1), and load the program into the STAMP 2. Your always-on Ethernet connection is now plugged into the X-link board, at the RJ-45 jack. Once the controller is powered up and the Ethernet connection established, your X-link control system is ready for operation. To activate a particular device, simply press the number/letter for that device; to turn off a particular device, just press the number/letter for the device once again and the device will turn off. Note there is a reset switch for both STAMP 2 and the mini server module, in the event you encounter a system start-up problem.

The diagram in Fig. 25-6 illustrates the mini web-server module installed on the main X-link system board. Plug the WSM module into the main X-link board. Note the proper orientation of the (WSM) module: the serial number in the top right corner faces the STAMP 2.

In order for the X-link system to operate, the WSM or Siteplayer mini web server must be programmed with a control status monitor web page. Software for the web-server module is supplied in the X-link directory on the supplied CD-ROM. In order to install the Siteplayer or web-server module software, you will need to obtain the Siteplayer development board priced at $99.95. The Siteplayer development board and Siteplayer mini web-server module are available from http://www.siteplayer.com.

Two other options are also available. You can purchase a WSM preprogrammed with the X-link software already installed for $45 from josh@petruzz.net, or you could have your own WSM module programmed for $15.45 plus $6.00 shipping. If you wish to purchase a Siteplayer module or have your Siteplayer module programmed without purchasing the development board, contact josh@petruzz.net.

Next, you will need to figure out what types of outputs you wish to control. The status matrix is laid out from A0 to A5 followed by B0 to B5, then C0 to C5, D0 to D5, and E0 to E5, while the local (non-X10) controls are labeled F0 to F5. Letters A through E with numbers 0 through 5 were designed to represent X10 control devices, so you can control up to 30 X-10 devices. Letter F along with numbers 0 through 5 allow for local control of up to six devices or relay-controlled devices. The X-link monitor web page is shown in Table 25-1. The

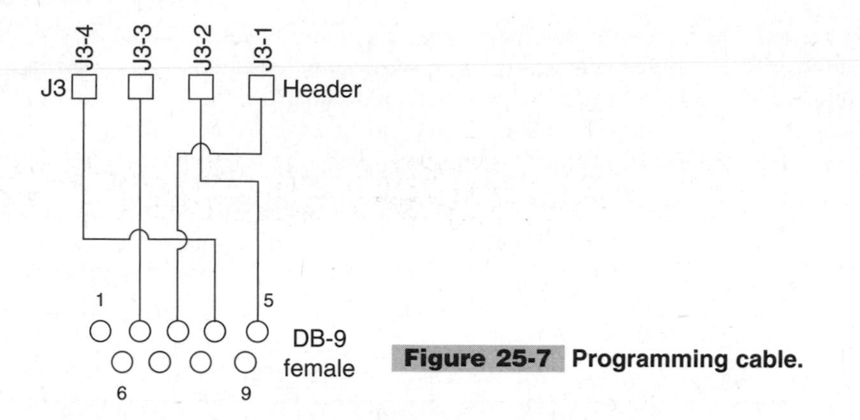

Figure 25-7 Programming cable.

control matrix display on the web page provides green button/indicator lights. In order to activate a device, you would first press a letter light and then a number light. For example, if you wished to activate the first X-10 device, designated A1, you would press button A, then button 0. Remember, in computer programming terms, zero is often used to start or as a one.

The X10 receiver modules come in many variations from lamp to appliance modules; they are available from RadioShack as well as many other suppliers. If you take a close look at the X10 receiver modules, you will see that there are both letter and number designations. The letter designations represent the X10 house control, while the number designations are associated with the X10 unit number. Each of the X10 control/device modules contain both a letter and a number switch. The house control designation can be used for say, first floor, second floor, basement, or garage, while the number designation can be used for devices within a particular floor or area.

IDENTIFYING NETWORK LOCATION

If you have elected to use the X-link monitor system board with the WSM server, you will need to identify its location on the network, and this can be tricky, depending on your network configuration. If you anticipate using the X-link monitoring system behind a Linksys 4-port router/switch, then you will need to go to the Linksys web page—on board the router, via your browser, go to the main page, then advanced, then forwarding. In the service port range column, you type the port number that will be translated to a X-link; for this example use 2001. Then type the FIXED IP address of the X-link on the internal network. You must fix the IP address of X-link or else it could wander around with DHCP and get lost. Now you have attached the outside port 2001 to the inside IP address of the X-link.

From the outside world you then surf the X-link by typing http://xxx.xxx.xxx.xxx:2001, where xxx.xxx.xxx.xxx is your outside world IP address. The Linksys will translate this port request to the following: http://inside.Weblink.ip.address:2001.

Since the X-link responds to all ports, the 2001 request is just like any other normal HTTP port 80 request. You can do this with up to 10 entries in the Linksys table. (I thought you could do more, but it seems that they have limited it.) The only wrinkle that I see is that if you are using PPoe for your DSL connection, this may not work, since the PPoe connection is not always direct. You may be going through some routers in your ISP.

Try forwarding port 80 in the Linksys to your PC's internal IP address and run X-linkPC. If someone can surf X-linkPC from the outside world, then you should be able to surf the hardware X-links.

TABLE 25-1 X-LINK CONTROL SYSTEM STATUS PAGE

•A	•0
•B	•1
•C	•2
•D	•3
•E	•4
•F	•5

```
'X-Link Internet Control System
'Xlink.bs2
stateAN0  var  bit
stateAN1  var  bit
stateAN2  var  bit
stateAN3  var  bit
stateAN4  var  bit
stateAN5  var  bit
stateBN0  var  bit
stateBN1  var  bit
stateBN2  var  bit
stateBN3  var  bit
stateBN4  var  bit
stateBN5  var  bit
stateCN0  var  bit
stateCN1  var  bit
stateCN2  var  bit
stateCN3  var  bit
stateCN4  var  bit
stateCN5  var  bit
stateDN0  var  bit
stateDN1  var  bit
stateDN2  var  bit
stateDN3  var  bit
stateDN4  var  bit
stateDN5  var  bit
stateEN0  var  bit
stateEN1  var  bit
stateEN2  var  bit
stateEN3  var  bit
stateEN4  var  bit
stateEN5  var  bit
stateFN0  var  bit
stateFN1  var  bit
stateFN2  var  bit
stateFN3  var  bit
stateFN4  var  bit
stateFN5  var  bit
letter    var  byte
number    var  byte
```

Listing 25-1 X-link Internet control system software.

```
baud      Con  813  '1200 baud
pin       Con  15
zPin      Con  0    '3
mPin      Con  1    '4
houseA    Con  0
houseB    Con  1
houseC    Con  2
houseD    Con  3
houseE    Con  4
houseF    Con  5
Unit1     Con  0
Unit2     Con  1
Unit3     Con  2
Unit4     Con  3
Unit5     Con  4
Unit6     Con  5
lamp1     Con  13
lamp2     Con  12
lamp3     Con  11
lamp4     Con  10
lamp5     Con  9
lamp6     Con  8
 letterin:
   serin pin,baud,[letter]                    ' expect ascii 65,..,75
   'branch letter-65,[LA]
   debug letter
   branch letter-97,[La,Lb,Lc,Ld,Le,Lf]       ' letter-65 is 0,1,2,..,10
   goto letterin                              ' bad letter
   getnumber:                                 ' subroutine
   serin pin,baud,[number]
   return
 LA
   gosub getnumber                            ' expect ascii 48,50,..,57
   branch number-48  ,[AN0,AN1,AN2,AN3,AN4,AN5]              ' routine to
execute 0,1,,9
   goto letterin                               ' bad number
  LB:
   gosub getnumber                             ' expect ascii 48,50,..,57
   branch number-48  ,[BN0,BN1,BN2,BN3,BN4,BN5]              ' B routine to
execute
   goto letterin                               ' bad number
 LC:
   gosub getnumber                             ' expect ascii 48,50,..,57
   branch number-48  ,[CN0,CN1,CN2,CN3,CN4,CN5]              ' C routine to
execute
   goto letterin                                ' bad number
  LD:
   gosub getnumber                              ' expect ascii 48,50,..,57
   branch number-48  ,[DN0,DN1,DN2,DN3,DN4,DN5]              ' D routine to
execute
   goto letterin                                ' bad number


  LE:
   gosub getnumber
   branch number-48  ,[EN0,EN1,EN2,EN3,EN4,EN5]
   goto letterin
  LF:
   gosub getnumber
```

Listing 25-1 X-link Internet control system software (*Continued*).

```
    branch number-48  ,[FN0,FN1,FN2,FN3,FN4,FN5]
    goto letterin
   AN0:
    xout mPin,zPin,[houseA\unit1]
    branch StateAN0,[AN0on,AN0off]
   AN0on:
    StateAN0=1
    xout mPin,zPin,[houseA\uniton]
    goto letterin
   AN0off:
    StateAN0=0
    xout mPin,zPin,[houseA\unitoff]
    goto letterin
   AN1:
    xout mPin,zPin,[houseA\unit2]
    branch StateAN1,[AN1on,AN1off]
AN1on:
    StateAN1=1
    xout mPin,zPin,[houseA\uniton]
    goto letterin
   AN1off:
    StateAN1=0
    xout mPin,zPin,[houseA\unitoff]
    goto letterin
   AN2:
    xout mPin,zPin,[houseA\unit3]
    branch StateAN2,[AN2on,AN2off]
   AN2on:
    StateAN2=1
    xout mPin,zPin,[houseA\uniton]
    goto letterin
   AN2off:
    StateAN2=0
    xout mPin,zPin,[houseA\unitoff]
    goto letterin
   AN3:
    xout mPin,zPin,[houseA\unit4]
    branch StateAN3,[AN3on,AN3off]
   AN3on:
    StateAN3=1
    xout mPin,zPin,[houseA\uniton]
    goto letterin
   AN3off:
    StateAN3=0
    xout mPin,zPin,[houseA\unitoff]
    goto letterin
   AN4:
    xout mPin,zPin,[houseA\unit5]
    branch StateAN4,[AN4on,AN4off]
   AN4on:
    StateAN4=1
    xout mPin,zPin,[houseA\uniton]
    goto letterin
   AN4off:
    StateAN4=0
    xout mPin,zPin,[houseA\unitoff]
    goto letterin
   AN5:
```

Listing 25-1 **X-link Internet control system software (*Continued*).**

```
 xout mPin,zPin,[houseA\unit6]
 branch StateAN5,[AN5on,AN5off]
AN5on:
 StateAN5=1
 xout mPin,zPin,[houseA\uniton]
 goto letterin
AN5off:
 StateAN5=0
 xout mPin,zPin,[houseA\unitoff]
 goto letterin
BN0:
 xout mPin,zPin,[houseB\unit1]
 branch StateBN0,[BN0on,BN0off]
BN0on:
 StateBN0=1
 xout mPin,zPin,[houseB\uniton]
 goto letterin
BN0off:
 StateBN0=0
 xout mPin,zPin,[houseB\unitoff]
 goto letterin
BN1:
 xout mPin,zPin,[houseB\unit2]
 branch StateBN1,[BN1on,BN1off]
BN1on:
 StateBN1=1
 xout mPin,zPin,[houseB\uniton]
 goto letterin
BN1off:
 stateBN1=0
 xout mPin,zPin,[houseB\unitoff]
 goto letterin
BN2:
 xout mPin,zPin,[houseB\unit3]
 branch StateBN2,[BN2on,BN2off]
BN2on:
 StateBN2=1
 xout mPin,zPin,[houseB\uniton]
 goto letterin
BN2off:
 stateBN2=0
 xout mPin,zPin,[houseB\unitoff]
 goto letterin
BN3:
 xout mPin,zPin,[houseB\unit4]
 branch StateBN3,[BN3on,BN3off]
BN3on:
 StateBN3=1
 xout mPin,zPin,[houseB\uniton]
 goto letterin
BN3off:
 stateBN3=0
 xout mPin,zPin,[houseB\unitoff]
 goto letterin
BN4:
 xout mPin,zPin,[houseB\unit5]
 branch StateBN4,[BN4on,BN4off]
BN4on:
```

Listing 25-1 X-link Internet control system software (*Continued*).

```
 StateBN4-1
 xout mPin,zPin,[houseB\uniton]
 goto letterin
BN4off:
 stateBN4=0
 xout mPin,zPin,[houseB\unitoff]
 goto letterin
BN5:
 xout mPin,zPin,[houseB\unit6]
 branch StateAN5,[BN5on,BN5off]
BN5on:
 StateBN5=1
 xout mPin,zPin,[houseB\uniton]
 goto letterin
BN5off:
 StateAN5=0
 xout mPin,zPin,[houseB\unitoff]
 goto letterin
CN0:
 xout mPin,zPin,[houseC\unit1]
 branch StateCN0,[CN0on,CN0off]
CN0on:
 StateCN0=1
 xout mPin,zPin,[houseC\uniton]
 goto letterin
CN0off:
 StateCN0=0
 xout mPin,zPin,[houseC\unitoff]
 goto letterin
CN1:
 xout mPin,zPin,[houseC\unit2]
 branch StateCN1,[CN1on,CN1off]
CN1on:
 StateCN1=1
 xout mPin,zPin,[houseC\uniton]
 goto letterin
CN1off:
 StateCN1=0
 xout mPin,zPin,[houseC\unitoff]
 goto letterin
CN2:
 xout mPin,zPin,[houseC\unit3]
 branch StateCN2,[CN2on,CN2off]
CN2on:
 StateCN2=1
 xout mPin,zPin,[houseC\uniton]
 goto letterin
CN2off:
 StateCN2=0
 xout mPin,zPin,[houseC\unitoff]
 goto letterin
CN3:
 xout mPin,zPin,[houseC\unit4]
 branch StateCN3,[CN3on,CN3off]
CN3on:
 StateCN3=1
 xout mPin,zPin,[houseC\uniton]
 goto letterin
```

Listing 25-1 **X-link Internet control system software (*Continued*).**

```
CN3off:
 StateCN3=0
 xout mPin,zPin,[houseC\unitoff]
 goto letterin
CN4:
 xout mPin,zPin,[houseC\unit5]
 branch StateCN4,[CN4on,CN4off]
CN4on:
 StateCN4=1
 xout mPin,zPin,[houseC\uniton]
 goto letterin
CN4off:
 StateCN4=0
 xout mPin,zPin,[houseC\unitoff]
 goto letterin
CN5:
 xout mPin,zPin,[houseC\unit6]
 branch StateAN5,[CN5on,CN5off]
CN5on:
 StateCN5=1
 xout mPin,zPin,[houseC\uniton]
 goto letterin
CN5off:
 StateCN5=0
 xout mPin,zPin,[houseC\unitoff]
 goto letterin
DN0:
 xout mPin,zPin,[houseD\unit1]
 branch StateDN0,[DN0on,DN0off]
DN0on:
 StateDN0=1
 xout mPin,zPin,[houseD\uniton]
 goto letterin
DN0off:
 StateDN0=0
 xout mPin,zPin,[houseD\unitoff]
 goto letterin
DN1:
 xout mPin,zPin,[houseD\unit2]
 branch StateDN1,[DN1on,DN1off]
DN1on:
 StateDN1=1
 xout mPin,zPin,[houseD\uniton]
 goto letterin
DN1off:
 StateDN1=0
 xout mPin,zPin,[houseD\unitoff]
 goto letterin
DN2:
 xout mPin,zPin,[houseD\unit3]
 branch StateDN2,[DN2on,DN2off]
DN2on:
 StateDN2=1
 xout mPin,zPin,[houseD\uniton]
 goto letterin
DN2off:
 StateDN2=0
 xout mPin,zPin,[houseD\unitoff]
```

Listing 25-1 **X-link Internet control system software (*Continued*).**

```
 goto letterin
DN3:
 xout mPin,zPin,[houseD\unit4]
 branch StateDN3,[DN3on,DN3off]
DN3on:
 StateDN3=1
 xout mPin,zPin,[houseD\uniton]
 goto letterin
DN3off:
 StateDN3=0
 xout mPin,zPin,[houseD\unitoff]
 goto letterin
DN4:
 xout mPin,zPin,[houseD\unit5]
 branch StateDN4,[DN4on,DN4off]
DN4on:
 StateDN4=1
 xout mPin,zPin,[houseD\uniton]
 goto letterin
DN4off:
 StateDN4=0
 xout mPin,zPin,[houseD\unitoff]
 goto letterin
DN5:
 xout mPin,zPin,[houseD\unit6]
 branch StateDN5,[DN5on,DN5off]
DN5on:
 StateDN5=1
 xout mPin,zPin,[houseD\uniton]
 goto letterin
DN5off:
 StateDN5=0
 xout mPin,zPin,[houseD\unitoff]
 goto letterin
EN0:
 xout mPin,zPin,[houseE\unit1]
 branch StateEN0,[EN0on,EN0off]
EN0on:
 StateEN0=1
 xout mPin,zPin,[houseE\uniton]
 goto letterin
EN0off:
 StateEN0=0
 xout mPin,zPin,[houseE\unitoff]
 goto letterin
EN1:
 xout mPin,zPin,[houseE\unit2]
 branch StateEN1,[EN1on,EN1off]
EN1on:
 StateEN1=1
 xout mPin,zPin,[houseE\uniton]
 goto letterin
EN1off:
 StateEN1=0
 xout mPin,zPin,[houseE\unitoff]
 goto letterin
EN2:
```

Listing 25-1 X-link Internet control system software (*Continued*).

```
 xout mPin,zPin,[houseE\unit3]
 branch StateEN2,[EN2on,EN2off]
EN2on:
 StateEN2=1
 xout mPin,zPin,[houseE\uniton]
 goto letterin
EN2off:
 StateEN2=0
 xout mPin,zPin,[houseE\unitoff]
 goto letterin
EN3:
 xout mPin,zPin,[houseE\unit4]
 branch StateEN3,[EN3on,EN3off]
EN3on:
 StateEN3=1
 xout mPin,zPin,[houseE\uniton]
 goto letterin
EN3off:
 StateEN3=0
 xout mPin,zPin,[houseE\unitoff]
 goto letterin
EN4:
 xout mPin,zPin,[houseE\unit5]
 branch StateEN4,[EN4on,EN4off]
EN4on:
 StateEN4=1
 xout mPin,zPin,[houseE\uniton]
 goto letterin
EN4off:
 StateEN4=0
 xout mPin,zPin,[houseE\unitoff]
 goto letterin
EN5:
 xout mPin,zPin,[houseE\unit6]
 branch StateAN5,[EN5on,EN5off]
EN5on:
 StateEN5=1
 xout mPin,zPin,[houseE\uniton]
 goto letterin
EN5off:
 StateEN5=0
 xout mPin,zPin,[houseE\unitoff]
 goto letterin
FN0:
 high lamp1
 branch StateFN0,[FN0on,FN0off]
FN0on:
 StateFN0=1
 high lamp1
 goto letterin
FN0off:
 StateFN0=0
 low lamp1
 goto letterin
FN1:
 high lamp2
 branch StateFN1,[FN1on,FN1off]
FN1on:
```

Listing 25-1 X-link Internet control system software (*Continued*).

```
  StateFN1=1
  high lamp2
  goto letterin
 FN1off:
  StateFN1=0
  low lamp2
  goto letterin
 FN2:high lamp3
  branch StateFN2,[FN2on,FN2off]
 FN2on:
  StateFN2=1
  high lamp3
  goto letterin
 FN2off:
  StateFN2=0
  low lamp3
  goto letterin
 FN3:high lamp4
  branch StateFN3,[FN3on,FN3off]
 FN3on:
  StateFN3=1
  high lamp4
  goto letterin
 FN3off:
  StateFN3=0
  low lamp4
  goto letterin
 FN4:high lamp5
  branch StateFN4,[FN4on,FN4off]
 FN4on:
  StateFN4=1
  high lamp5
  goto letterin
 FN4off:
  StateFN4=0
  low lamp5
  goto letterin
 FN5:high lamp6
  branch StateFN5,[FN5on,FN5off]
 FN5on:
  StateFN5=1
  high lamp6
  goto letterin
 FN5off:
  StateFN5=0
  low lamp6
  goto letterin
```

Listing 25-1 **X-link Internet control system software (*Continued*).**

ADDITIONAL PROGRAMS

Also included with the software files are three additional programs: SITEPLAYERPC, which is used to create/change web pages in the WSM module, the SITELINKER program, which is used to download the binary image to the WSM module, and a program titled Siteplayer Serial Port Tester, which can be used to GET or SET the WSM's address.

The X-link system is a great tool for the "control freak" who needs to be able to remotely control many devices with a single self-contained system. The X-link is an invaluable control system that offers enormous flexibility in monitoring and controlling remote devices.

X-Link Control System Board Parts List

R1, R2, R3, R4	4.7-kΩ 1/4-W resistor
R5	10-kΩ 1/4-W resistor
R6	100-Ω 1/2-W resistor
C1, C7	0.1-μF 25-V disk capacitor
C2, C3, C4, C5	1-μF 35-V electrolytic capacitor
C6	2-μF 35-V electrolytic capacitor
D1	1N914 silicon diode
D2	9-V 1/2-W zener diode
D3, D4, D5, D6, D7, D8	1N4002 silicon diode
Q1, Q2, Q3, Q4, Q5, Q6	2N3904 transistor
U1	PIC16C57 STAMP 2
U2	24LC16B memory
U3	MAX232 serial interface
U4	78LO5 5-V regulator
RL1, RL2, RL3, Rl4, RL5, RL6	5–6-V mini relay
S1	Momentary pushbutton switch, normally open
XTL	20-MHz ceramic resonator
X10	TW523 or PL513 X10 interface

X-Link Web-Server Module Assembly Parts List

R1	330-Ω 1/4-W resistor
R2	47-kΩ 1/4-W resistor
C1	0.1-μF 50-V disk capacitor
C2, C3, C4, C5	0.01-μF 35-V disk capacitor
T1	LFIS022 pulse transformer/RJ-45 jack

S1	Reset pushbutton, normally open
D1	LED
WSM	Siteplayer web-server module

INDEX

About the Author

Tom Petruzzellis is an engineer whose assignments currently include working with geophysical field equipment at the University of Binghamton (NY). A resident of Vestal, NY, Tom has more than 30 years of experience in the field of electronics and has authored many articles and books on the subject.

CD-ROM WARRANTY

This software is protected by both United States copyright law and international copyright treaty provision. You must treat this software just like a book. By saying "just like a book," McGraw-Hill means, for example, that this software may be used by any number of people and may be freely moved from one computer location to another, so long as there is no possibility of its being used at one location or on one computer while it also is being used at another. Just as a book cannot be read by two different people in two different places at the same time, neither can the software be used by two different people in two different places at the same time (unless, of course, McGraw-Hill's copyright is being violated).

LIMITED WARRANTY

Customers who have problems installing or running a McGraw-Hill CD should consult our online technical support site at http://books.mcgraw-hill.com/techsupport. McGraw-Hill takes great care to provide you with top-quality software, thoroughly checked to prevent virus infections. McGraw-Hill warrants the physical CD-ROM contained herein to be free of defects in materials and workmanship for a period of sixty days from the purchase date. If McGraw-Hill receives written notification within the warranty period of defects in materials or workmanship, and such notification is determined by McGraw-Hill to be correct, McGraw-Hill will replace the defective CD-ROM. Send requests to:

McGraw-Hill
Customer Services
P.O. Box 545
Blacklick, OH 43004-0545

The entire and exclusive liability and remedy for breach of this Limited Warranty shall be limited to replacement of a defective CD-ROM and shall not include or extend to any claim for or right to cover any other damages, including, but not limited to, loss of profit, data, or use of the software, or special, incidental, or consequential damages or other similar claims, even if McGraw-Hill has been specifically advised of the possibility of such damages. In no event will McGraw-Hill's liability for any damages to you or any other person ever exceed the lower of suggested list price or actual price paid for the license to use the software, regardless of any form of the claim.

McGRAW-HILL SPECIFICALLY DISCLAIMS ALL OTHER WARRANTIES, EXPRESS OR IMPLIED, INCLUDING, BUT NOT LIMITED TO, ANY IMPLIED WARRANTY OF MERCHANTABILITY OR FITNESS FOR A PARTICULAR PURPOSE.

Specifically, McGraw-Hill makes no representation or warranty that the software is fit for any particular purpose and any implied warranty of merchantability is limited to the sixty-day duration of the Limited Warranty covering the physical CD-ROM only (and not the software) and is otherwise expressly and specifically disclaimed.

This limited warranty gives you specific legal rights; you may have others which may vary from state to state. Some states do not allow the exclusion of incidental or consequential damages, or the limitation on how long an implied warranty lasts, so some of the above may not apply to you.